In this treatise, the authors present the general theory of orthogonal polynomials on the complex plane and several of its applications. The assumptions on the measure of orthogonality are general, the only restriction is that it has compact support on the complex plane. In the development of the theory the main emphasis is on asymptotic behavior and the distribution of zeros.

In the first two chapters exact upper and lower bounds are given for the orthonormal polynomials and for the location of their zeros. The next three chapters deal with regular nth-root asymptotic behavior, which plays a key role both in the theory and in its applications. Orthogonal polynomials with this behavior correspond to classical orthogonal polynomials in the general case, and many extremal properties of measures in mathematical analysis and approximation theory turn out to be equivalent to this type of regularity. Several easy-to-use criteria are presented for regular behavior.

The last chapter contains applications of the theory, including exact rates for convergence of rational interpolants, best rational approximants, and nondiagonal Padé approximants to Markov functions (Cauchy transforms of measures). The results are based on potential-theoretic methods, so both the methods and the results can be extended to extremal polynomials in norms other than L^2 norms. The Appendix contains a brief introduction to potential theory.

ENCYCLOPEDIA OF MATHEMATICS AND ITS APPLICATIONS

EDITED BY G.-C. ROTA

Volume 43

General Orthogonal Polynomials

ENCYCLOPEDIA OF MATHEMATICS AND ITS APPLICATIONS

1 Luis A. Santalo *Integral geometry and geometric probability*
2 George E. Andrews *The theory of partitions*
3 Robert J. McEliece *The theory of information and coding: A mathematical framework for communication*
4 Willard Miller, Jr. *Symmetry and separation of variables*
5 David Ruelle *Thermodynamic formalism: The mathematical structures of classical equilibrium statistical mechanics*
6 Henryk Minc *Permanents*
7 Fred S. Roberts *Measurement theory with applications to decisionmaking, utility, and the social services*
8 L. C. Biedenham and J. D. Louck *Angular momentum in quantum physics: Theory and application*
9 L. C. Biedenham and J. D. Louck *The Racah–Wigner algebra in quantum theory*
10 W. Dollard and Charles N. Friedman *Product integration with application to differential equations*
11 William B. Jones and W. J. Thron *Continued fractions: Analytic theory and applications*
12 Nathaniel F. G. Martin and James W. England *Mathematical theory of entropy*
13 George A. Baker, Jr and Peter R. Graves-Morris *Padé approximants, Part I: Basic theory*
14 George A. Baker, Jr and Peter R. Graves-Morris *Padé approximants, Part II: Extensions and applications*
15 E. C. Beltrametti and G. Cassinelli *The logic of quantum mechanics*
16 G. D. James and A. Kerber *The representation theory of the symmetric group*
17 M. Lothaire *Combinatorics on words*
18 H. O. Fattorini *The Cauchy problem*
19 G. G. Lorentz, K. Jetter, and S. D. Riemenschneider *Birkhoff interpolation*
20 Rudolf Lidl and Harald Niederreiter *Finite fields*
21 William T. Tutte *Graph theory*
22 Julio R. Bastida *Field extensions and Galois theory*
23 John R. Cannon *The one-dimensional heat equation*
24 Stan Wagon *The Banach–Tarski paradox*
25 Arto Salomaa *Computation and automata*
26 Neil White (ed) *Theory of matroids*
27 N. H. Bingham, C. M. Goldie, and J. L. Teugels *Regular variation*
28 P. P. Petrushev and V. A. Popov *Rational approximation of real functions*
29 Neil White (ed) *Combinatorial geometries*
30 M. Pohst and H. Zassenhaus *Algorithmic algebraic number theory*
31 J. Aczel and J. Dhombres *Functional equations containing several variables*
32 Marek Kuczma, Bogden Chozewski, and Roman Ger *Iterative functional equations*
33 R. V. Ambartzumian *Factorization calculus and geometric probability*
34 G. Gripenberg, S.-O. Londen, and O. Staffans *Volterra integral and functional equations*
35 George Gasper and Mizan Rahman *Basic hypergeometric series*
36 Erik Torgersen *Comparison of statistical experiments*
37 Arnold Neumaier *Interval methods for systems of equations*
38 N. Korneichuk *Exact constants in approximation theory*
39 Richard A. Brualdi and Herbert J. Ryser *Combinatorial matrix theory*
40 N. White (ed) *Matroid applications*
41 Shoichiro Sakai *Operator algebras in dynamical systems: The theory of unbounded derivations in C^*-algebras*
42 W. Hodges *Model theory*

ENCYCLOPEDIA OF MATHEMATICS AND ITS APPLICATIONS

General Orthogonal Polynomials

HERBERT STAHL
Technische Fachhochschule Berlin

VILMOS TOTIK
Bolyai Institute, József Attila University and
Department of Mathematics, University of South Florida

CAMBRIDGE UNIVERSITY PRESS
Cambridge, New York, Melbourne, Madrid, Cape Town, Singapore,
São Paulo, Delhi, Dubai, Tokyo

Cambridge University Press
The Edinburgh Building, Cambridge CB2 8RU, UK

Published in the United States of America by Cambridge University Press, New York

www.cambridge.org
Information on this title: www.cambridge.org/9780521135047

First published 1992
This digitally printed version 2010

A catalogue record for this publication is available from the British Library

Library of Congress Cataloguing in Publication data
Stahl, Herbert.
General orthogonal polynomials / Herbert Stahl, Vilmos Totik.
p. cm. – (Encyclopedia of mathematics and its applications ; v. 43)
Includes bibliographical references and index.
ISBN 0-521-41534-9 (hardback)
1. Orthogonal polynomials. I. Totik, V. II. Title.
III. Series.
QA404.5.S73 1992
512′.55 – dc20 91-27733
CIP

ISBN 978-0-521-41534-7 Hardback
ISBN 978-0-521-13504-7 Paperback

Contents

Preface

The theory of orthogonal polynomials can be divided into two loosely related parts. One of them is the formal, algebraic aspect of the theory, which has close connections with special functions, combinatorics, and algebra, and it is mainly devoted to concrete orthogonal systems or hierarchies of systems such as the Jacobi, Hahn, Askey–Wilson, ... polynomials.

The investigation of more general orthogonal polynomials with methods of mathematical analysis belongs to the other part of the theory. Here the central questions are the asymptotic behavior of the polynomials and their zeros, recovering the measure of orthogonality, and so forth. This part has applications to approximation processes such as polynomial and rational interpolation, Padé approximation, and best rational approximation, to Fourier expansions, quadrature processes, eigenvalue problems, and so forth.

Textbooks on orthogonal polynomials usually cover material from both parts of the theory but give different emphasis in accordance with individual preference. Only the classical book [Sz3] by Gábor Szegő aims at a treatment of the subject in an encyclopedic manner. The present book is exclusively devoted to the second part of the theory. The main emphasis is on the investigation of the asymptotic behavior of *general* orthogonal polynomials, but related questions as, for instance, the distribution of zeros are also taken into consideration. A whole chapter is devoted to applications of the results in other areas.

Until now most of the asymptotic theory of orthogonal polynomials has concentrated on orthogonal systems for which the measure of orthogonality is supported on the real line or on the unit circle. Even then it has usually been assumed that the measure of orthogonality is sufficiently thick on its support. The present work is devoted to orthogonal polynomials with respect to *general measures* μ. The only requirement on μ is that it has compact support in $\mathbb{C}$, that is, both the support $S(\mu)$ of μ and the "thickness" of μ can be arbitrary. Thus, we allow both the measure and its support to be "wild," each of which has its own reflection in the general theory.

For orthonormal polynomials $p_n(\mu; z)$, as for sequences of polynomials in general, there exists a hierarchy of types of asymptotic behavior. We mention here the most common ones, which are called *power* (or Szegő), *ratio*, and *nth-root* asymptotic behavior. Roughly speaking, these mean that the sequences

$$\left\{ \frac{p_n(\mu; z)}{\varphi(z)^n} \,\middle|\, n \in \mathbb{N} \right\}, \tag{P.1}$$

$$\left\{ \frac{p_{n+1}(\mu; z)}{p_n(\mu; z)} \,\middle|\, n \in \mathbb{N} \right\}, \tag{P.2}$$

and

$$\{ \sqrt[n]{p_n(\mu; z)} \mid n \in \mathbb{N} \}, \tag{P.3}$$

respectively, tend to a limit on a certain set of values $z \in \mathbb{C}$ as $n \to \infty$ (in (P.1) the function φ has to be appropriately chosen). It is easy to see that each type of asymptotics in the hierarchy (P.1) to (P.3) implies the next one. Consequently, the nth-root asymptotic behavior is the most general of the three types and it requires the weakest assumptions. At the same time it is sufficient for many applications, as, for instance, the convergence of polynomial (Chebyshev–Fourier) expansions based on the system $\{p_n(\mu; z) \mid n \in \mathbb{N}\}$, or the convergence of continued fractions or Padé approximants to Markov functions (for other applications see Chapter 6).

In this context the present work can be classified as a monograph on nth-root asymptotic behavior. Earlier research in this direction has been mainly due to P. P. Korovkin, J. Ullman, P. Erdős, G. Freud, P. Turán, and H. Widom, although the case of general support has barely been touched upon. It was especially J. Ullman who systematically studied different bounds and asymptotics on orthogonal polynomials with respect to arbitrary measures μ on $[-1, 1]$, and we owe a lot to his research and personally to him for initiating and keeping alive the subject.

The present monograph synthesizes and considerably extends earlier research concerning general orthogonal polynomials. A large part of it (cf. Chapters 3–6) contains new results very often without any precedence (cf. Chapter 5, Sections 3.2–3.4, 4.2–4.6). We have put special emphasis on examples illustrating that our results are sharp. It has also been important for us to illustrate the possible connection with and the applicability of our theory to other fields of mathematical analysis (see Chapter 6).

Our proofs use potential-theoretic considerations. The usefulness of logarithmic potentials in the general theory can be easily understood if we recall that the modulus of a polynomial is basically nothing else than the exponential of a discrete potential. We very rarely use explicitly the orthogonality property – most of our proofs are based on the L^2 minimality of the monic orthogonal polynomials. Hence, our method works for L^p-extremal polynomials as well.

The content of the different chapters is briefly as follows. In Chapter 1 we give sharp upper and lower bounds for orthonormal polynomials and their leading coefficients. Chapter 2 examines the location and asymptotic distribution of the zeros. An extremely important concept, *"regular (nth-root) asymptotic behavior"* (in symbols, $\mu \in \mathbf{Reg}$), is introduced and characterized in Chapter 3. Polynomials with this property are the natural analogue of classical orthogonal polynomials in the general case, and they have many applications and equivalent formulations in different subjects of approximation theory. To facilitate these applications one needs easy-to-use criteria for $\mu \in \mathbf{Reg}$, which are given in Chapter 4. In Chapter 5 a surprising phenomenon is investigated: The regularity is basically a local property. Finally, Chapter 6 contains several applications of $\mu \in \mathbf{Reg}$ of which we mention here only the results in Sections 6.1 and 6.2, where the classical connection between continued fractions and orthogonal polynomials is extended to rational interpolation and best rational approximation of Cauchy transforms of measures μ, and it is shown that $\mu \in \mathbf{Reg}$ is equivalent to an exact maximal rate of convergence for these rational interpolants or approximants.

In the Appendix we assemble those results from the theory of logarithmic potentials that are frequently used in the text. A list of symbols and an index helps the reader to keep track of notations.

We used formula numbers of the form $(a.b)$, where a denotes the section number within the actual chapter, and b the formula number within the section. If we want to refer to a formula in a different chapter, then we use the form $(c.a.b)$, where c denotes the chapter number and a and b are as above.

Acknowledgments

The research of the first author was supported in part by the Deutsche Forschungsgemeinschaft, Grant No. AZ:Sta 29914–1, and the research of the second author by the Hungarian National Science Foundation for Research, Grant No. 1157, and by the National Science Foundation, Grant No. DMS 9002794. Special thanks are due to Mrs. Ágnes Tillinkó for the fast and careful typing of the manuscript and to Mr. János Kincses and Mr. Károly Dévényi for their help in the editing process.

Berlin–Szeged

Symbols

$\partial\Omega$	boundary of Ω
$\mathbb{C}$	complex plane
$\overline{\mathbb{C}}$	extended complex plane, Riemann sphere
$\text{cap}(S)$	logarithmic capacity of S
$\text{cap}(V, S)$	condenser (or Green) capacity of the condenser (V, S)
$\text{card}(S)$	cardinality of S
$C(K)$	space of continuous functions on K
c_μ	minimal-carrier capacity corresponding to μ
$\text{Co}(S)$	convex hull of S
$\deg(P)$, $\deg P$	degree of the polynomial P
$D_\rho(x)$, $\Delta_\rho(x)$	disk of radius ρ with center at x
δ_x	Dirac measure at x
$\delta_{n,m}$	Kronecker's symbol
$\text{diam}(S)$	diameter of S
$\text{dist}(V, S)$	distance of V and S
$d(\mu, \nu)$	distance in the metric of the weak* topology
$d(z, E)$	distance of z and E
$\Gamma(\mu)$	set of carriers of μ
$\Gamma_0(\mu)$	set of minimal carriers of μ
$g_B(z; y)$	Green function of B with pole at y
$g(\alpha, \Omega; z)$	Green potential of α in Ω
$g_\mu(z; \infty)$	minimal-carrier Green function
$\gamma_n(\mu)$	leading coefficient of $p_n(\mu; \cdot)$
$\text{Im}(z)$	imaginary part of z
$I(\mu)$	logarithmic energy of μ
$\text{Int}(K)$	interior of K
$I(S)$	smallest interval containing $S \subseteq \mathbb{R}$
$\lambda(E)$, $\text{meas}(E)$	Lebesgue measure of E

$L(\mathbf{A})$	set of limit points of the interpolation scheme $\mathbf{A}$
$\mathcal{M}_\mu$	set of weak* limits of the normalized counting measures on the zeros of $p_n(\mu;\cdot)$
$\mathcal{M}[0,1]$	space of probability measures on $[0,1]$
$[m/n](z)$	Padé approximants
$\mathrm{MAX}p(\nu;\cdot)$	set of maximum points of $p(\nu;\cdot)$
$\vert\mu\vert$	total-variation measure corresponding to μ
$\Vert\mu\Vert$	total variation of μ, total mass
μ	positive Borel measure of compact support in $\mathbb{C}$
$\mu\sim\nu$	carrier related measures
μ–qu.e.	μ–a.e. except for a set of zero capacity
$\nu\ll\mu$	absolute continuity of ν with respect to μ
$\mathbb{N}$	set of natural numbers
$\nu_n \xrightarrow{*} \nu$, $\nu_n\to\nu$	weak* convergence of $\{\nu_n\}$ to ν
ν_S	counting measure on the set S
$\nu_{p_n(\mu;\cdot)}$	counting measure on the zeros of $p_n(\mu;\cdot)$
(ν,μ)	inner product with logarithmic kernel
ω_K	equilibrium measure of K
ω_μ	minimal-carrier equilibrium measure associated with μ
Ω	$\mathbb{C}\setminus\mathrm{Pc}(S(\mu))$, outer domain of $S(\mu)$
$\mathrm{Pc}(S)$	polynomial convex hull of S
$p_n(\mu;z)$	orthonormal polynomials with respect to μ
$p(\nu;z)$	logarithmic potential of ν
qu.e.	quasi everywhere
Π_n	set of polynomials of degree at most n
Π_n^*	set of monic polynomials of degree at most n
$q_n(\mu;z)$	monic orthogonal polynomials with respect to μ
$\mathbb{R}$	set of real numbers
$\mathrm{Re}(z)$	real part of z
Reg	set of regular measures
$\mathcal{R}_{m,n}\equiv\mathcal{R}_{mn}$	set of rational functions of numerator and denominator degrees at most m and n
$\mathcal{R}_n^{\mathbb{R}}$	set of rational functions with real coefficients of numerator and denominator degrees at most n
$r_n(f,A_n;z)$	interpolating rational function to f
$r_n^*=r_n^*(f,V;z)$	best rational approximant to f on V
$S(\mu)$, $\mathrm{supp}(\mu)$	support of μ
"$\subseteq$"	inclusion except for a set of zero capacity
$Z(p_n)$	set of zeros of p_n

1

Upper and Lower Bounds

This chapter is devoted to sharp upper and lower bounds for orthonormal polynomials with respect to general weights. The two bounds are given in terms of Green functions related to the carriers of the measure in question. As a corollary, sharp bounds are obtained for the leading coefficients. All subsequent chapters use both the notations and the results from the present one.

The chapter is organized as follows: Section 1.1 contains the statement of the main results; in Section 1.2 we prove some potential-theoretic preliminaries needed in the proofs. The actual proofs of the upper and lower estimates are carried out in Section 1.3, and the proof of their sharpness is given in Section 1.4. Finally, in Section 1.5 we construct some examples that illustrate the results.

1.1 Statement of the Main Results

The main results in this section are lower and upper asymptotic bounds for the nth root of the orthonormal polynomials $p_n(\mu; z)$ as $n \to \infty$, as well as their unimprovability.

In what follows $\text{cap}(S)$ denotes the (outer logarithmic) *capacity* of a bounded set $S \subseteq \mathbb{C}$, that is, $\text{cap}(S) = \inf_U \text{cap}(U)$, where the infimum extends over all open sets $U \supseteq S$ (see Chapter 11, Section 2 of [La] or Appendix I), and we say that a property holds *qu.e.* (*quasi everywhere*) on a set $S \subseteq \mathbb{C}$ if it holds on S with possible exceptions on a subset of capacity zero. The abbreviation "qu.e." will be also used for "quasi every." By $g_B(z; \infty)$ we denote the (generalized) *Green function* with logarithmic pole at infinity associated with a Borel set $B \subseteq \overline{\mathbb{C}}$ with bounded complement $\overline{\mathbb{C}} \setminus B$. For our investigation and especially for the definition of the minimal-carrier Green function given in (1.3), it is essential that the Green function is defined for Borel sets $B \subseteq \overline{\mathbb{C}}$, and not only for domains in $\overline{\mathbb{C}}$, as it is the case in most text books. More details about the definition of Green functions will be given at the beginning of the next section and in the Appendix (see Appendix V).

μ will always denote a finite Borel measure on $\mathbb{C}$ with compact support $S(\mu) := \operatorname{supp}(\mu)$. Let $\Omega = \Omega(\mu)$ denote the *outer domain* of $S(\mu)$, that is, the unbounded component of $\overline{\mathbb{C}} \setminus S(\mu)$, $\operatorname{Co}(S(\mu))$ the *convex hull*, and $\operatorname{Pc}(S(\mu))$ the *polynomial convex hull* of $S(\mu)$, that is, $\operatorname{Pc}(S(\mu)) = \overline{\mathbb{C}} \setminus \Omega$. (The name polynomial convex hull is derived from the fact that for any compact set $S \subseteq \mathbb{C}$ the set $\operatorname{Pc}(S)$ is the intersection of all sets $S_p := \{z \in \mathbb{C} \mid |p(z)| \leq \sup_{x \in S} |p(x)|\}$ with p a polynomial not identically zero.) The set $\partial\Omega \subseteq \partial S(\mu)$ is called the *outer boundary* of $S(\mu)$.

We shall always assume that the support of μ consists of infinitely many points. Then we can form the uniquely existing *orthonormal polynomials*

$$p_n(\mu; z) = \gamma_n(\mu) z^n + \cdots, \qquad \gamma_n(\mu) > 0,\ n \in \mathbb{N},$$

with respect to μ, which are defined by the orthogonality relations

$$\int p_n(\mu; z) \overline{p_m(\mu; z)} d\mu(z) = \delta_{n,m},$$

where $\delta_{n,m} = 1$ if $n = m$ and $\delta_{n,m} = 0$ otherwise. $\gamma_n(\mu)$ is called the *leading coefficient* of $p_n(\mu; \cdot)$.

Whereas $\operatorname{cap}(S(\mu))$ and $g_\Omega(z; \infty)$ depend only on the set $S(\mu)$, or more precisely, on the outer boundary $\partial\Omega$ of $S(\mu)$, we now introduce a type of capacity and Green function that depends on the carriers of the measure μ.

Definition 1.1.1. *Let $\Gamma(\mu)$ be the set of all carriers of the measure μ, that is,*

$$\Gamma(\mu) := \{C \subseteq \mathbb{C} \mid C \text{ a Borel set and } \mu(\mathbb{C} \setminus C) = 0\}, \tag{1.1}$$

then the minimal-carrier capacity *(of the measure μ) is defined as*

$$c_\mu := \inf\{\operatorname{cap}(C) \mid C \in \Gamma(\mu),\ C \text{ bounded}\} \tag{1.2}$$

and the minimal-carrier Green function *(of the measure μ) is defined as*

$$g_\mu(z; \infty) := \sup\{g_{\overline{\mathbb{C}} \setminus C}(z; \infty) \mid C \in \Gamma(\mu),\ C \text{ bounded}\}, \quad z \in \mathbb{C}. \tag{1.3}$$

Lemma 1.1.2. *We have*

$$c_\mu \leq \operatorname{cap}(S(\mu)) \tag{1.4}$$

and

$$g_\mu(z; \infty) \geq g_\Omega(z; \infty) \tag{1.5}$$

for all $z \in \mathbb{C}$.

Proof. The lemma immediately follows from $S(\mu) \in \Gamma(\mu)$. $\square$

The Examples 1.5.1 and 1.5.2 show that in (1.4) and (1.5) proper inequality as well as equality may hold true.

Before we come to the main theorems of this section, we state a lemma about the location of the zeros of the orthonormal polynomials $p_n(\mu; z)$, $n \in \mathbb{N}$. It will be proved together with related materials in Section 2.1.

Lemma 1.1.3. *All zeros of the orthonormal polynomials $p_n(\mu; z)$, $n \in \mathbb{N}$, are contained in the convex hull* $\mathrm{Co}(S(\mu))$, *and for any compact set $V \subseteq \Omega$ the number of zeros of $p_n(\mu; z)$, $n \in \mathbb{N}$, on V is bounded as $n \to \infty$.*

Remarks. (1) If $S(\mu) \subseteq \mathbb{R}$, then $\mathrm{Co}(S(\mu))$ is the smallest interval $I \subseteq \mathbb{R}$ containing $S(\mu)$. Lemma 1.1.3 shows that a well-known result from the theory of orthonormal polynomials associated with measures μ with support $S(\mu) \subseteq \mathbb{R}$, namely, that all zeros of $p_n(\mu; z)$, $n \in \mathbb{N}$, are contained in $\mathrm{Co}(S(\mu))$ (see Section 1.2.2 of [Fr]), carries over to weight measures μ with compact support in $\mathbb{C}$. However, the result that all zeros of $p_n(\mu; z)$, $n \in \mathbb{N}$, are simple, which holds in the case of measures μ with $S(\mu) \subseteq \mathbb{R}$, is no longer true in the general case. (See, for instance, the very simple Example 1.5.1.)

(2) In the case $S(\mu) \subseteq \mathbb{R}$ the orthonormal polynomials $p_n(\mu; z)$, $n \in \mathbb{N}$, can have zeros outside $\mathrm{Pc}(S(\mu))$ only if $S(\mu)$ is disconnected. The arc measure on the unit circle and Example 2.1.2 show that for measures μ with support in $\mathbb{C}$ the situation is different. There the support $S(\mu)$ can be a continuum (but not convex), and the orthonormal polynomials $p_n(\mu; z)$ can have zeros outside $\mathrm{Pc}(S(\mu))$. These zeros may even cluster outside of $\mathrm{Pc}(S(\mu))$ as $n \to \infty$. The phenomenon explains why in the next theorem the lower asymptotic bound requires different formulations for each of the three cases $z \notin \mathrm{Co}(S(\mu))$, $z \in \Omega \cap \mathrm{Co}(S(\mu))$, and $z \in \partial\Omega$.

For the formulation of our main result we introduce the following limit relation. We say that

$$\liminf_{n\to\infty} |f_n(z)| \geq h(z)$$

holds true *locally uniformly* in an open set D if for every $z \in D$ and $z_n \to z$ as $n \to \infty$ we have

$$\liminf_{n\to\infty} |f_n(z_n)| \geq h(z).$$

Thus, the two notions "locally uniformly in D" and "uniformly on compact subsets of D" are different. In general, the latter is stronger than the former one. However, if h is continuous, or merely upper semicontinuous (for a lim sup relation, h lower semicontinuous), then these two notions coincide.

Theorem 1.1.4. *We have*

$$\limsup_{n\to\infty} |p_n(\mu; z)|^{1/n} \leq e^{g_\mu(z;\infty)} \tag{1.6}$$

locally uniformly in $\mathbb{C}$, *and*

$$\liminf_{n\to\infty} |p_n(\mu; z)|^{1/n} \geq e^{g_\Omega(z;\infty)} \tag{1.7}$$

locally uniformly in $\mathbb{C} \setminus \operatorname{Co}(S(\mu))$. *In* $\operatorname{Co}(S(\mu)) \cap \Omega$ *the asymptotic lower bound* (1.7) *holds true only in capacity, that is, for every compact set* $V \subseteq \Omega$ *and every* $\varepsilon > 0$ *we have*

$$\lim_{n\to\infty} \operatorname{cap}(\{z \in V \mid |p_n(\mu; z)|^{1/n} < e^{g_\Omega(z;\infty)} - \varepsilon\}) = 0. \tag{1.8}$$

In $\operatorname{Co}(S(\mu)) \cap \Omega$ *the lower bound can also be given in the following form: For every infinite subsequence* $N \subseteq \mathbb{N}$ *we have*

$$\limsup_{n\to\infty,\ n\in N} |p_n(\mu; z)|^{1/n} \geq e^{g_\Omega(z;\infty)} \qquad \text{qu.e. in } \Omega \cap \operatorname{Co}(S(\mu)), \tag{1.9}$$

and on the outer boundary $\partial\Omega$ *of* $S(\mu)$ *we have*

$$\limsup_{n\to\infty,\ n\in N} |p_n(\mu; z)|^{1/n} \geq 1 \qquad \text{qu.e. on } \partial\Omega. \tag{1.10}$$

Remarks. (1) If the two Green functions $g_\Omega(z;\infty)$ and $g_\mu(z;\infty)$ are identical, then we have proper and identical limits in (1.6) and (1.7). (The existence of a proper limit and equality in (1.7) will be called regular asymptotic behavior in Section 3.1.) If the two Green functions are not identical for a weight measure μ, that is, if we have a proper inequality in (1.5) for some value $z \in \mathbb{C}$, then it will be shown in Theorem 1.1.9 that there exists a measure ν having the same set of carriers as μ, that is, $\Gamma(\nu) = \Gamma(\mu)$ and hence $g_\mu(z;\infty) \equiv g_\nu(z;\infty)$, such that we have equality in both (1.6) and (1.7) for $\{|p_n(\nu; z)|^{1/n} \mid n \in \mathbb{N}\}$. Hence, both bounds are sharp.

(2) Whereas the upper asymptotic bound (1.6) holds true locally uniformly on $\mathbb{C}$, the lower bound (1.7) holds true in this strong sense only on compact subsets of $\mathbb{C} \setminus \operatorname{Co}(S(\mu))$. This weakness of the lower asymptotic bound in $\operatorname{Co}(S(\mu))$ is caused by the zeros that the polynomials $p_n(\mu; z)$ may have everywhere in $\operatorname{Co}(S(\mu))$. In Corollary 1.1.5 the zeros in $\operatorname{Co}(S(\mu)) \cap \Omega$ will be factored out.

(3) The upper asymptotic bound (1.6) is not specific for orthonormal polynomials. It holds true for any sequence of polynomials normalized in $L^2(\mu)$, as will be shown in Theorem 1.1.8.

(4) On the outer boundary $\partial\Omega$ of $S(\mu)$ an asymptotic estimate in capacity such as the one in (1.8) cannot, in general, be true since there the asymptotic density of the zeros can be positive almost everywhere.

(5) The simple Example 1.5.1 shows that in the interior of the polynomial convex hull $\mathrm{Pc}(S(\mu))$ we cannot, in general, expect 1 as a lower asymptotic bound. The special case $\mathrm{Int}(\mathrm{Pc}(S(\mu))) = \emptyset$, which includes all weight measures μ on the real axis, is formulated separately in Corollary 1.1.6.

Corollary 1.1.5. *Let $U, V \subseteq \Omega$ be two compact sets, U containing V in its interior, and let $x_{n,1}, \ldots, x_{n,m(n)}$ be the zeros of $p_n(\mu; z)$, $n \in \mathbb{N}$, on U. Then we have*

$$\liminf_{n\to\infty} \left| \frac{p_n(\mu; z)}{\prod_{j=1}^{m(n)} (z - x_{nj})} \right|^{1/n} \geq e^{g_\Omega(z;\infty)} \tag{1.11}$$

uniformly on V.

Corollary 1.1.6. *If the interior of* $\mathrm{Pc}(S(\mu))$ *is empty, then for any infinite subsequence $N \subseteq \mathbb{N}$ we have*

$$\limsup_{n\to\infty,\ n\in N} |p_n(\mu; z)|^{1/n} \geq 1 \qquad \text{qu.e. on } S(\mu). \tag{1.12}$$

Note that the interior of $\mathrm{Pc}(S(\mu))$ is empty exactly when $S(\mu)$ has connected complement and empty interior.

If we consider the orthonormal polynomials $p_n(\mu; z)$, $n \in \mathbb{N}$, near infinity, then from the upper and lower asymptotic bounds (1.6) and (1.7) in Theorem 1.1.4 we immediately deduce upper and lower asymptotic bounds for the nth root of the *leading coefficient* $\gamma_n(\mu)$ of the orthonormal polynomials:

Corollary 1.1.7. *We have*

$$\frac{1}{\mathrm{cap}(S(\mu))} \leq \liminf_{n\to\infty} \gamma_n(\mu)^{1/n} \leq \limsup_{n\to\infty} \gamma_n(\mu)^{1/n} \leq \frac{1}{c_\mu}, \tag{1.13}$$

where $\mathrm{cap}(S(\mu)) = 0$ *or* $c_\mu = 0$ *is allowed.*

It has already been mentioned in Remark 3 to Theorem 1.1.4 that the upper asymptotic bound (1.6) is not specific for orthonormal polynomials. In the next theorem we state the result for arbitrary sequences of polynomials. The upper bound (1.6) then follows as a corollary.

Theorem 1.1.8. *For any sequence of polynomials P_n not identically zero and of degree at most $n \in \mathbb{N}$, we have*

$$\limsup_{n\to\infty} \left(\frac{|P_n(z)|}{\|P_n\|_{L^2(\mu)}} \right)^{1/n} \leq e^{g_\mu(z;\infty)} \tag{1.14}$$

locally uniformly in $\mathbb{C}$.

Finally, we show the sharpness of the estimates in Theorem 1.1.4. To do this let us call two measures ν and μ *carrier related* (denoted $\nu \sim \mu$) if ν and μ have the same carriers, that is, $\Gamma(\nu) = \Gamma(\mu)$. Obviously, this is the same as their mutual absolute continuity on each other. Since for carrier-related measures ν and μ the Green functions g_μ and g_ν are identical, and furthermore $S(\nu) = S(\mu)$, it follows that the upper and lower bounds (1.6) and (1.7) are the same for carrier-related measures. Now, from this point of view they are sharp.

Theorem 1.1.9. (a) *There is a* $\nu_1 \sim \mu$ *such that*

$$\lim_{n\to\infty} |p_n(\nu_1; z)|^{1/n} = e^{g_\mu(z;\infty)} \tag{1.15}$$

locally uniformly for $z \notin \mathrm{Co}(S(\mu))$.

(b) *There is a* $\nu_2 \sim \mu$ *such that*

$$\lim_{n\to\infty} |p_n(\nu_2; z)|^{1/n} = e^{g_\Omega(z;\infty)} \tag{1.16}$$

locally uniformly for $z \notin \mathrm{Co}(S(\mu))$.

(c) *There is a* $\nu_3 \sim \mu$ *such that*

$$\liminf_{n\to\infty} |p_n(\nu_3; z)|^{1/n} = e^{g_\Omega(z;\infty)} \tag{1.17}$$

and

$$\limsup_{n\to\infty} |p_n(\nu_3; z)|^{1/n} = e^{g_\mu(z;\infty)} \tag{1.18}$$

for every $z \notin \mathrm{Co}(S(\mu))$.

The unimprovability of the other estimates in Theorem 1.1.4 can also be verified. Without going into details we only remark concerning (1.8):

Corollary 1.1.10. *With* μ *and* ν_2 *as in Theorem 1.1.9, for every compact set* $V \subseteq \Omega$ *we have*

$$\limsup_{n\to\infty} |p_n(\nu_2; z)|^{1/n} \le e^{g_\Omega(z;\infty)} \tag{1.19}$$

uniformly on V.

1.2 Some Potential-theoretic Preliminaries

Minimal carriers will be introduced in this section. Representations for the minimal-carrier Green function $g_\mu(z;\infty)$ and a related principle of domination will be proved. All results will be used throughout the whole chapter. We start with some terminology.

A sequence of measures $\{\nu_n\}$ is said to *converge weakly* (or in the *weak** topology) to a measure ν (in $\overline{\mathbb{C}}$), written $\nu_n \xrightarrow{*} v$, if for every function f continuous in $\overline{\mathbb{C}}$ we have

$$\int f d\nu_n \to \int f d\nu \quad \text{as} \quad n \to \infty.$$

Since the unit ball of positive measures is weak* compact, from every sequence $\{\nu_n\}$ of probability measures we can select an infinite subsequence with index set $N \subseteq \mathbb{N}$ such that $\{\nu_n \mid n \in N\}$ is weak* convergent. This result is often called Helly's selection theorem.

For a (Borel) measure ν with compact support $S(\nu)$ in $\mathbb{C}$ we set

$$p(\nu; z) := \int \log \frac{1}{|z-x|} d\nu(x), \tag{2.1}$$

and call it the (*logarithmic*) *potential* of ν.

For a polynomial P we denote by ν_P the positive measure that has at every zero of P a mass equal to the multiplicity of that zero. This measure is called the *zero distribution* of P. Thus, $\|\nu_P\| = \deg(P)$ and for monic polynomials P we have $\log|P(z)| = -p(\nu_P; z)$.

The Green function $g_B(z;\infty)$ of a Borel set $B \subseteq \overline{\mathbb{C}}$ with bounded complement has already been mentioned in Section 1.1. In order to have a basis for proofs, we state three fundamental properties of $g_B(z;\infty)$, which can be taken as *defining properties*:

(i) $g_B(z;\infty)$ is nonnegative and subharmonic in $\mathbb{C}$, and harmonic in $\mathrm{Int}(B) \setminus \{\infty\}$.
(ii) $g_B(z;\infty) = \log|z| - \log(\mathrm{cap}(\mathbb{C} \setminus B)) + o(1)$ as $|z| \to \infty$, where $o(1)$ tends to 0 as $|z| \to \infty$.
(iii) $g_B(z;\infty) = 0$ for qu.e. z on $\mathbb{C} \setminus B$.

If $\mathrm{cap}(\mathbb{C} \setminus B) = 0$, then $g_B(z;\infty) \equiv \infty$. The function $g_B(z;\infty)$ exists and is uniquely determined by (i) to (iii) for every Borel set $B \subseteq \overline{\mathbb{C}}$ with bounded complement $\overline{\mathbb{C}} \setminus B$; see Appendix V. We note that we have defined $g_B(z;\infty)$ for all $z \in \mathbb{C}$. If the bounded complement $C := \mathbb{C} \setminus B$ is of positive capacity, that is, if

$$\mathrm{cap}(C) > 0, \tag{2.2}$$

then (see Appendixes IV and V) there uniquely exists a probability measure ω_C, called the *equilibrium distribution* of C, with $S(\omega_C) \subseteq \overline{C}$ such that the Green function $g_B(z;\infty)$ has the representation

$$g_B(z;\infty) \equiv -p(\omega_C; z) - \log(\mathrm{cap}(C)). \tag{2.3}$$

If C is a compact set, then $S(\omega_C)$ is contained in the outer boundary $\partial \mathrm{Pc}(C)$ of C; in general, however, we only know that $S(\omega_C) \subseteq \overline{C}$. Since

ω_C is in general not carried by C, it is sometimes useful to know that the set C can be exhausted from within by nested compact sets $C_n \subseteq C$ so the capacity, the associated Green functions, and the equilibrium distributions converge to the corresponding objects of the set C. More formally:

There exist compact sets $C_n \subseteq C$, $n \in \mathbb{N}$, with $C_n \subseteq C_{n+1}$ and $\text{cap}(C_n) > 0$ *such that*

$$\lim_{n\to\infty} \text{cap}(C_n) = \text{cap}(C), \tag{2.4}$$

$$\lim_{n\to\infty} g_{\overline{\mathbb{C}}\setminus C_n}(z;\infty) = g_{\overline{\mathbb{C}}\setminus C}(z;\infty) \tag{2.5}$$

for all $z \in \mathbb{C}$, and $\omega_{C_n} \xrightarrow{*} \omega_C$ as $n \to \infty$. The sequence $\{g_{\overline{\mathbb{C}}\setminus C_n}(z;\infty)\}$ is monotonically decreasing.

We will only sketch the proof: From the capacitability of Borel sets (see Appendix I) it follows that a sequence $\{C_n\}$ exists so that (2.4) holds true. The existence of limit (2.5) and the weak* limit of the equilibrium distributions follows from the monotonicity of the sequence $\{g_{\overline{\mathbb{C}}\setminus C_n}(z;\infty)\}$ (see Theorem 1.26 of [La]). That the limit (2.5) is identical with the Green function $g_B(z;\infty)$ can then be derived with the help of the three defining properties of Green functions by standard techniques (see also the reasonings applied in Appendix IV and V).

We note that all sets and functions to be subsequently defined are Borel measurable and therefore we do not have to care about inner and outer capacity.

After these general results from potential theory we turn to results related to the minimal-carrier Green function $g_\mu(z;\infty)$.

The elements of $\Gamma(\mu)$ are partially ordered by inclusion and the capacity is a monotone set function. It is therefore possible and often useful to consider elements in $\Gamma(\mu)$ which are minimal with respect to capacity.

Definition 1.2.1. *The elements of the set*

$$\Gamma_0(\mu) := \{C \in \Gamma(\mu) \mid \text{cap}(C) = c_\mu \quad \textit{and} \quad C \subseteq S(\mu)\} \tag{2.6}$$

are called minimal carriers.

Lemma 1.2.2. *The set $\Gamma_0(\mu)$ is not empty, and for every $C \in \Gamma(\mu)$ there exists an element $C_0 \in \Gamma_0(\mu)$ with $C_0 \subseteq C$.*

Remark. Contrary to the minimal-carrier capacity c_μ and the minimal-carrier Green function $g_\mu(z;\infty)$, which are both uniquely defined, there exist in general more than one minimal element in $\Gamma_0(\mu)$.

Proof. Let $C \in \Gamma(\mu)$ be arbitrary. By (1.2) there exists a sequence

$C_n \in \Gamma(\mu)$, $n \in \mathbb{N}$, with $\text{cap}(C_n) \to c_\mu$ as $n \to \infty$. If we set $C_{0n} := S(\mu) \cap C_n$, $n \in \mathbb{N}$, then we have $C_{0n} \in \Gamma(\mu)$ for all $n \in \mathbb{N}$, and therefore

$$C_0 := \bigcap_{n=1}^{\infty} C_{0n} \in \Gamma(\mu) \quad \text{and} \quad \text{cap}(C_0) = c_\mu. \tag{2.7}$$

Hence, $C_0 \in \Gamma_0(\mu)$ and $C_0 \subseteq C$. □

The minimal carriers of $\Gamma_0(\mu)$ provide us with a representation of the minimal-carrier Green function $g_\mu(z;\infty)$ as an ordinary Green function.

Lemma 1.2.3. *For any $C \in \Gamma_0(\mu)$ we have*

$$g_{\overline{\mathbb{C}}\setminus C}(z;\infty) \equiv g_\mu(z;\infty), \tag{2.8}$$

and

$$g_\mu(z;\infty) = 0 \qquad \textit{qu.e. on } C. \tag{2.9}$$

Proof. If $c_\mu = 0$, then $g_{\overline{\mathbb{C}}\setminus C}(z;\infty) \equiv \infty$ and $\text{cap}(C) = 0$ for all $C \in \Gamma_0(\mu)$. Since in this case we also have $g_\mu(z;\infty) \equiv \infty$, the identities (2.8) and (2.9) hold true.

Let us now assume $c_\mu > 0$, and let C_1 and C_2 be two arbitrary elements of $\Gamma_0(\mu)$. $C_0 := C_1 \cap C_2$ also belongs to $\Gamma_0(\mu)$, and therefore $\text{cap}(C_j) = c_\mu$ for $j = 0, 1, 2$. By checking the three defining properties (i) to (iii) of a Green function, it is easy to verify that both Green functions $g_{\overline{\mathbb{C}}\setminus C_j}(z;\infty)$, $j = 1, 2$, are at the same time the Green function of the set $\overline{\mathbb{C}} \setminus C_0$. Hence, by the uniqueness of Green functions we have

$$g_{\overline{\mathbb{C}}\setminus C_1}(z;\infty) \equiv g_{\overline{\mathbb{C}}\setminus C_2}(z;\infty), \tag{2.10}$$

which shows that all Green functions $g_{\overline{\mathbb{C}}\setminus C}(z;\infty)$, $C \in \Gamma_0(\mu)$, are identical.

Since for two bounded Borel sets $C_1, C_2 \subseteq \mathbb{C}$ the inclusion $C_1 \subseteq C_2$ implies

$$g_{\overline{\mathbb{C}}\setminus C_1}(z;\infty) \geq g_{\overline{\mathbb{C}}\setminus C_2}(z;\infty) \quad \text{for all} \quad z \in \mathbb{C},$$

and since every carrier $C \in \Gamma(\mu)$ contains a minimal carrier $C_0 \in \Gamma_0(\mu)$, identity (2.9) follows from identity (2.10) and definition (1.3) of the minimal-carrier Green function $g_\mu(z;\infty)$. Equality (2.9) then follows from the corresponding property of $g_{\overline{\mathbb{C}}\setminus C}(z;\infty)$. □

The next lemma immediately follows from Lemma 1.2.3 together with representation (2.3) for ordinary Green functions.

Lemma 1.2.4. *If $c_\mu > 0$, then all $C \in \Gamma_0(\mu)$ possess the same equilibrium distribution, which we denote by ω_μ, and for $g_\mu(z;\infty)$ we have the representation*

$$g_\mu(z;\infty) \equiv -p(\omega_\mu; z) - \log c_\mu. \tag{2.11}$$

Definition 1.2.5. *The probability measure* ω_μ *is called the* minimal-carrier equilibrium distribution *associated with* μ.

Lemma 1.2.6. *We have*

$$g_\Omega(z;\infty) \equiv g_\mu(z;\infty) \tag{2.12}$$

if and only if

$$c_\mu = \text{cap}(S(\mu)). \tag{2.13}$$

Proof. Let us assume (2.12). Then (2.13) follows from considering both Green functions in (2.12) near infinity.

Let us now assume (2.13). Then $S(\mu)$ is a minimal carrier and (2.12) is a consequence of (2.8). □

Next we investigate the special case when the weight measure μ is equal to the equilibrium distribution $\omega_{S(\mu)}$ of $S(\mu)$. We shall need the results later. But besides that, the lemma shows that the equilibrium distribution is an instructive example of a weight measure μ that satisfies (2.13).

Lemma 1.2.7. *Let* $S \subseteq \mathbb{C}$ *be a compact set with* $\text{cap}(S) > 0$, *and set* $\omega := \omega_S$. *Then we have*

$$c_\omega = \text{cap}(S) = \text{cap}(S(\omega)), \tag{2.14}$$

$$g_\omega(z;\infty) \equiv g_{\overline{\mathbb{C}}\setminus S}(z;\infty), \tag{2.15}$$

and

$$\omega_\omega = \omega. \tag{2.16}$$

Proof. If we represent $g_\Omega(z;\infty)$, $\Omega := \mathbb{C} \setminus \text{Pc}(S(\omega))$, and $g_\omega(z;\infty)$ as in (2.3) and (2.11), respectively, by logarithmic potentials and apply Fubini's theorem, then we get

$$\int g_\omega(z;\infty)d\omega(z) = \int g_\Omega(z;\infty)d\omega_\omega(z) + \log\frac{\text{cap}(S(\omega))}{c_\omega}. \tag{2.17}$$

(For the case $\text{cap}(S(\omega)) = c_\omega = 1$, equality (2.17) is known in potential theory as the formula of reciprocity.)

Let C_0 be the set of all $z \in \mathbb{C}$ with $g_\omega(z;\infty) = 0$. Then from Lemma 1.2.3 we know that $\text{cap}(C \setminus C_0) = 0$ for any $C \in \Gamma_0(\omega)$. Since $\omega = \omega_S$ is of finite energy it follows that $\omega(C \setminus C_0) = 0$ and therefore $C_0 \in \Gamma(\omega)$. Hence, the integral on the left-hand side of (2.17) is zero. The integral on the right-hand side of (2.17) is nonnegative, which implies $c_\omega \geq \text{cap}(S(\omega))$, and with inequality (1.4), formula (2.14) follows.

Identity (2.15) then follows from Lemma 1.2.6, and (2.16) follows from (2.15) together with representation (2.11). □

The last result in this section is a domination principle for the minimal-carrier Green function $g_\mu(z;\infty)$. It turns out that a combination of the measure-theoretic "μ–almost everywhere" and the potential-theoretic "quasi everywhere," which we call "μ–quasi everywhere," is very appropriate in this connection. It is also very helpful for the formulation of polynomial extremality problems in $L^2(\mu)$.

Definition 1.2.8. *We say that a property holds* μ–quasi everywhere (μ-qu.e.) *on a set* $S \subseteq \overline{\mathbb{C}}$ *if there exists a Borel set* $S_0 \subseteq S$ *with* $\text{cap}(S_0) = 0$ *and the property holds* μ*–almost everywhere on* $S \setminus S_0$. *If* $S = \mathbb{C}$ *or* $= \overline{\mathbb{C}}$, *the reference to* S *may be dropped.*

From Lemma 1.2.3 we know, for instance, that the minimal-carrier Green function $g_\mu(z;\infty)$ is equal to zero μ–quasi everywhere.

Lemma 1.2.9. *Let* ν *be a probability measure with compact support* $S(\nu) \subseteq \mathbb{C}$, *and let* c *be a constant such that*

(2.18) $$p(\nu;z) + c \geq 0 \quad \mu\text{–quasi everywhere in } \mathbb{C}.$$

Then we have

(2.19) $$p(\nu;z) + c \geq -g_\mu(z;\infty) \quad \text{for all} \quad z \in \mathbb{C},$$

and

(2.20) $$c \geq \log c_\mu.$$

Remark. The case $c_\mu = 0$ has not been excluded in Lemma 1.2.9. The assumption that $S(\nu)$ is compact is not essential; it has been made here in order to safeguard the existence of $p(\nu;z)$.

Proof. If $c_\mu = 0$, then $g_\mu(z;\infty) \equiv \infty$, and (2.19) and (2.20) are trivially satisfied. Let us now assume $c_\mu > 0$, and let C' be the set of all points $z \in \mathbb{C}$ where the inequality in (2.18) holds true. From Definition 1.2.8 it follows that there exists a Borel set $C'' \subseteq S(\mu)$ of capacity zero such that $C' \cup C'' \in \Gamma(\mu)$. From Lemma 1.2.2 we know that there exists a minimal carrier $C \in \Gamma_0(\mu)$ with $C \subseteq C' \cup C''$. As in (2.4) and (2.5) there exist compact sets $C_n \subseteq C$, $n \in \mathbb{N}$, with $C_n \subseteq C_{n+1}$, $\text{cap}(C_n) > 0$, such that the limits (2.4) and (2.5) hold true. Let ω_n be the equilibrium distribution of the set C_n. Because $\text{cap}(C_n) > 0$, the measures ω_n are of finite logarithmic energy and therefore $\omega_n(C'') = 0$ and so $C' \in \Gamma(\omega_n)$ for all $n \in \mathbb{N}$.

From $C' \in \Gamma(\omega_n)$ and from the definition of the set C' it then follows that

$$(2.21) \qquad p(\nu; z) + c \geq 0 \geq -g_{\overline{\mathbb{C}}\setminus C_n}(z;\infty) \equiv p(\omega_n; z) + \log(\mathrm{cap}(C_n))$$

ω_n–almost everywhere. Since ω_n is of finite energy, it follows from (2.21) and from the principle of domination of potential theory (see Appendix III) that

$$(2.22) \qquad p(\nu; z) + c \geq -g_{\overline{\mathbb{C}}\setminus C_n}(z;\infty) \quad \text{for all} \quad z \in \mathbb{C}.$$

Using the limit (2.5) and representation (2.8) for $g_\mu(z;\infty)$, we derive (2.19) from (2.22). The inequality (2.20) follows from (2.19) by considering both sides in (2.19) near infinity. □

1.3 Proof of the Upper and Lower Bounds

The positive results of Section 1.1 will be proved here. There are two groups of problems which require separate treatment: On the one hand, the upper asymptotic bound in Theorem 1.1.4 and in Theorem 1.1.8, and on the other hand, Lemma 1.1.3, the lower asymptotic bounds (1.7) to (1.10) in Theorem 1.1.4, and the lower asymptotic bound (1.11) in Corollary 1.1.5. We start with the first group.

The proof of Theorem 1.1.8 is a combination of the principle of domination for $g_\mu(z;\infty)$, which has just been proved in Lemma 1.2.9, and the next lemma.

Lemma 1.3.1. *For any infinite sequence of functions $h_n \in L^2(\mu)$, $n \in N$, $N \subseteq \mathbb{N}$, with*

$$(3.1) \qquad \limsup_{n\to\infty,\ n\in N} \|h_n\|_{L^2(\mu)}^{1/n} \leq 1$$

there exists an infinite subsequence $N_0 \subseteq N$ such that

$$(3.2) \qquad \limsup_{n\to\infty,\ n\in N_0} |h_n(z)|^{1/n} \leq 1 \quad \mu\text{–almost everywhere.}$$

Proof. Set $f_n := \{|h_n|/(n\|h_n\|_{L^2(\mu)})\}^2$, $n \in N$. Then $f_n \to 0$ in $L^1(\mu)$, and therefore there exists an infinite subsequence $N_0 \subseteq N$ such that

$$(3.3) \qquad \lim_{n\to\infty,\ n\in N_0} f_n(z) = 0 \quad \mu\text{–almost everywhere.}$$

Using $|h_n|^2 = f_n n^2 \|h_n\|_{L^2(\mu)}^2$, we get (3.2) from (3.3) and (3.1). □

Proof of Theorem 1.1.8. Since the polynomials P_n, $n \in \mathbb{N}$, in Theorem 1.1.8 can be multiplied by a nonzero constant without changing the value of the terms on the left-hand side of (1.14), we can assume without loss of generality that all polynomials P_n, $n \in \mathbb{N}$, are monic.

(a) In a first stage we still assume that all zeros of the polynomials P_n are contained in a disc $\{z \mid |z| \leq R\}$ with a fixed radius $R > 0$. From Helly's selection theorem and an ordinary compactness argument it follows that we can select an infinite subsequence $N \subseteq \mathbb{N}$ so that the following two limits exist:

$$(3.4) \qquad \begin{aligned} &\frac{1}{n}\nu_{P_n} \xrightarrow{*} \nu, \\ &\frac{1}{n}\log \|P_n\|_{L^2(\mu)} \to c \in \mathbb{R} \cup \{-\infty\} \end{aligned}$$

as $n \to \infty$, $n \in N$. It is easy to verify that the measure ν is positive, $S(\nu) \subseteq \{z \mid |z| \leq R\}$, and $\|\nu\| \leq 1$. From (3.4), the principle of descent and the lower envelope theorem of potential theory (see Appendix III) we deduce that

$$(3.5) \qquad \limsup_{n\to\infty,\, n\in N} \frac{1}{n} \log \frac{|P_n(z)|}{\|P_n\|_{L^2(\mu)}} \leq -p(\nu; z) - c$$

locally uniformly for all $z \in \mathbb{C}$, and equality holds true in (3.5) for qu.e. z in $\mathbb{C}$. If we now apply Lemma 1.3.1 to the sequence $h_n := P_n/\|P_n\|_{L^2(\mu)}$, $n \in \mathbb{N}$, then (3.1) is trivially satisfied. From (3.2), from the fact that we have equality quasi everywhere in (3.5), and from the definition of μ–quasi everywhere (see Definition 1.2.8) it follows that

$$(3.6) \qquad -p(\nu; z) - c \leq 0 \quad \mu\text{–quasi everywhere.}$$

From Lemma 1.2.9 it then follows that

$$(3.7) \qquad -p(\nu; z) - c \leq g_\mu(z; \infty) \quad \text{for all} \quad z \in \mathbb{C}.$$

This together with (3.5) proves (1.14) for the subsequence N.

From (3.7) the proof that (1.14) holds true locally uniformly in $\mathbb{C}$ is easy: Suppose on the contrary that the latter one fails to hold. Then there are a subsequence N of the natural numbers, a point $z \in \mathbb{C}$, and for every $n \in N$ a point $z_n \in \mathbb{C}$ such that $z_n \to z$ as $n \to \infty$, $n \in N$, but

$$\lim_{n\to\infty} \left(\frac{|P_n(z_n)|}{\|P_n\|_{L^2(\mu)}} \right)^{1/n} =: \gamma > e^{g_\mu(z;\infty)}.$$

From N we can select a subsequence, which we continue to denote by N, for which the limits in (3.4) exist. Now the preceding relation implies via the principle of descent that

$$\exp(-p(\nu; z) - c) \geq \gamma > \exp(g_\mu(z; \infty)),$$

but this contradicts (3.7), and this contradiction fully proves that (1.14) holds true locally uniformly.

(b) In the second stage of the proof we drop the temporary assumption that all zeros of the polynomials P_n are contained in a disc $\{z \mid |z| \leq R\}$. Let $V \subseteq \mathbb{C}$ be a compact set and let $R > 0$ be so large that $S(\mu) \cup V \subseteq \{z \mid |z| \leq R\}$. Furthermore let $k > 1$ be a fixed constant. We factorize the polynomials P_n into two monic polynomials h_n and q_n so that

$$P_n = h_n q_n, \quad n \in \mathbb{N}, \tag{3.8}$$

$h_n(z) \neq 0$ for all $|z| \leq kR$, and $q_n(z) \neq 0$ for all $|z| > kR$. By elementary calculations we can show that

$$\left[\frac{|P_n(z)|}{\|P_n\|_{L^2(\mu)}}\right]^{1/n} \leq \frac{k+1}{k-1}\left[\frac{|q_n(z)|}{\|q_n\|_{L^2(\mu)}}\right]^{1/2} \tag{3.9}$$

for all $z \in \{z \mid |z| \leq R\}$. Since the sequence of polynomials q_n, $n \in \mathbb{N}$, satisfies the assumption of the first part of the proof, and since k can be chosen arbitrarily large, from (3.9) the theorem follows in full generality.

□

Together with Theorem 1.1.8 we have also proved the upper asymptotic bound (1.6) of Theorem 1.1.4.

Next we come to the second group of problems: the proofs of the lower asymptotic bounds (1.7) to (1.10) of Theorem 1.1.4, and the lower asymptotic bound (1.11) of Corollary 1.1.5. These proofs are based on the following well-known and easily provable extremality property of the leading coefficient $\gamma_n(\mu)$ of the orthonormal polynomial $p_n(\mu; z)$, $n \in \mathbb{N}$, in $L^2(\mu)$: We have

$$\frac{1}{\gamma_n(\mu)^2} = \left\|\frac{p_n(\mu; z)}{\gamma_n(\mu)}\right\|^2_{L^2(\mu)} \leq \int |q_n|^2 d\mu \tag{3.10}$$

for every monic polynomial $q_n(z) = z^n + \cdots$ of degree n, and equality holds in (3.10) if and only if $q_n = p_n(\mu; z)/\gamma_n(\mu)$. In fact, it is enough to write q_n as a linear combination of the $p_k(\mu; \cdot)$s, and to apply orthogonality.

In the sequel we denote the monic orthogonal polynomial with respect to the measure μ by

$$q_n(\mu; z) := \frac{p_n(\mu; z)}{\gamma_n(\mu)}. \tag{3.11}$$

Besides minimality (3.10) the next lemma plays a major role.

Lemma 1.3.2. *Let $V, S \subseteq \mathbb{C}$ be two compact sets.*

(a) *If V and $\operatorname{Pc}(S)$ are disjoint, then there exist $a < 1$ and $m \in \mathbb{N}$ such that* *for arbitrary m points $x_1, \dots, x_m \in V$ there exist m points $y_1, \dots, y_m \in \mathbb{C}$ for which the rational function*

$$r_m(z) := \prod_{j=1}^{m} \frac{z - y_j}{z - x_j} \tag{3.12}$$

has on S a supremum norm satisfying

$$\|r_m\|_S \le a. \tag{3.13}$$

(b) *If V and* $\mathrm{Co}(S)$ *are disjoint, then we can assume in part* (a) *that* $m = 1$.

Proof. Without loss of generality we may assume that $\mathrm{cap}(S) > 0$ and that $\mathrm{Pc}(S)$ is regular with respect to the Dirichlet problem in $\overline{\mathbb{C}} \setminus \mathrm{Pc}(S)$. For otherwise we can replace S by $\{z \in \mathbb{C} \mid \mathrm{dist}(z; S) \le \varepsilon\}$ with $\varepsilon > 0$ sufficiently small.

(a) Let us assume that V and $\mathrm{Pc}(S)$ are disjoint, and let us further assume that part (a) of the lemma is wrong. Then there exists $N_0 \subseteq \mathbb{N}$ such that for every $m \in N_0$ there exists a set $X_m := \{x_1, \dots, x_m\} \subseteq V$ of m points so that for any choice of a set $Y_m := \{y_1, \dots, y_m\} \subseteq \mathbb{C}$ of m points we have

$$\liminf_{n \to \infty, m \in N_0} \|r_m\|_S \ge 1, \tag{3.14}$$

where r_m is defined as in (3.12). We will disprove (3.14).

By Helly's selection theorem there exists an infinite subsequence $N \subseteq N_0$ with

$$\frac{1}{m} \nu_{X_m} \xrightarrow{*} \nu \quad \text{as} \quad m \to \infty, \ m \in N, \tag{3.15}$$

where ν is a probability measure with $S(\nu) \subseteq V$. We consider the Green potential

$$q(z) := \int g_{\overline{\mathbb{C}} \setminus \mathrm{Pc}(S)}(z; x) d\nu(x), \tag{3.16}$$

where $g_{\overline{\mathbb{C}} \setminus \mathrm{Pc}(S)}(z; x)$ is the Green function of the domain $\overline{\mathbb{C}} \setminus \mathrm{Pc}(S)$ with logarithmic pole at $x \notin \mathrm{Pc}(S)$. Since $\mathrm{Pc}(S)$ has been assumed to be regular, we have $g_{\overline{\mathbb{C}} \setminus \mathrm{Pc}(S)}(z; x) = 0$ for all $z \in \mathrm{Pc}(S)$ and $x \notin \mathrm{Pc}(S)$ (see Theorem 4.7 of [La] or Appendix VI). As a consequence we have $q(z) = 0$ for all $z \in \mathrm{Pc}(S)$. As an analogue to representation (2.3) of the Green function $g_{\overline{\mathbb{C}} \setminus \mathrm{Pc}(S)}(z; \infty)$, we have for $x \in \mathbb{C} \setminus \mathrm{Pc}(S)$ the representation

$$g_{\overline{\mathbb{C}} \setminus \mathrm{Pc}(S)}(z; x) \equiv p(\delta_x - \omega_x; z) + g_{\overline{\mathbb{C}} \setminus \mathrm{Pc}(S)}(x; \infty),$$

where ω_x is the balayage of the point mass δ_x on $\partial \mathrm{Pc}(S)$ (see (A.18) in

Appendix VII or Chapter IV, Section 3, p. 217 of [La]) and δ_x is Dirac's measure of the point x. This representation yields the identity

$$q(z) \equiv \int p(\delta_x - \omega_x; z) d\nu(x) + \int g_{\overline{\mathbb{C}} \setminus \mathrm{Pc}(S)}(x; \infty) d\nu(x),$$

from which we see that

$$p(\nu - \omega_\nu; z) \equiv q(z) - q(\infty), \tag{3.17}$$

where ω_ν is the probability measure on $\partial \mathrm{Pc}(S)$ defined by

$$\omega_\nu := \int \omega_x d\nu(x), \tag{3.18}$$

and

$$q(\infty) = \int g_{\overline{\mathbb{C}} \setminus \mathrm{Pc}(S)}(x; \infty) d\nu(x) > 0 \tag{3.19}$$

is a positive constant since $g_{\overline{\mathbb{C}} \setminus \mathrm{Pc}(S)}(x; \infty) > 0$ for all $x \in V$. It may be of some help for the reader to remark that (3.18) is nothing else than the balayage of ν onto $\partial \mathrm{Pc}(S)$ (see Appendix VII).

Since the discrete measures are weak* dense in the unit ball of positive measures, we can select a sequence of sets $Y_m := \{y_1, \ldots, y_m\} \subseteq \partial \mathrm{Pc}(S)$, $m \in N$, with

$$\frac{1}{m} \nu_{Y_m} \xrightarrow{*} \omega_\nu \quad \text{as} \quad m \to \infty,\ m \in N. \tag{3.20}$$

From (3.15) it follows that

$$\lim_{m \to \infty,\ m \in N} \frac{1}{m} \log \prod_{j=1}^{m} \frac{1}{|z - x_j|} = p(\nu; z) \tag{3.21}$$

uniformly on S, and from (3.20) together with the principle of descent of potential theory (Appendix III) it follows that

$$\liminf_{m \to \infty,\ m \in N} \frac{1}{m} \log \prod_{j=1}^{m} \frac{1}{|z - y_j|} \geq p(\omega_\nu; z) \tag{3.22}$$

locally uniformly on S. Since the regularity of $\mathrm{Pc}(S)$ implies that $p(\omega_\nu; z)$ is continuous throughout $\mathbb{C}$, it follows that (3.22) is true uniformly on S. Hence, by (3.17) and the fact that $q(z) = 0$ for all $z \in S$, we have shown that

$$\limsup_{m \to \infty,\ n \in N} \|r_m\|_S^{1/m} \leq e^{-q(\infty)} < 1, \tag{3.23}$$

where the rational function r_m is defined as in (3.12) by the points of the two sets X_m and Y_m. The asymptotic estimate (3.23) contradicts (3.14), and therefore part (a) of the lemma is proved.

(b) Let us now assume that V and $\mathrm{Co}(S)$ are disjoint. Set $m = 1$ and

$$a := 1 \Big/ \sqrt{1 + \left(\frac{\mathrm{dist}(V,S)}{\mathrm{diam}(S)}\right)^2} < 1. \tag{3.24}$$

For an arbitrary $x_1 \in V$ define $y_1 \in \mathrm{Co}(S)$ as the nearest point to x_1 in $\mathrm{Co}(S)$. By elementary geometrical considerations we can easily show that the convexity of $\mathrm{Co}(S)$ implies that

$$\left|\frac{z-y_1}{z-x_1}\right| \le a \quad \text{for all} \quad z \in \mathrm{Co}(S). \tag{3.25}$$

Thus, part (b) of the lemma is also proved. □

Proof of Theorem 1.1.4 and Corollary 1.1.5. Since the upper asymptotic bound (1.6) in Theorem 1.1.4 follows immediately from Theorem 1.1.8, only the different forms of the lower asymptotic bounds (1.7) to (1.11) remain to be proved. We start with the proof of (1.10), which turns out to be the key for the proof of all other lower bounds.

(a) By Helly's selection theorem and an ordinary compactness argument we know that any infinite subsequence $N \subseteq \mathbb{N}$ contains an infinite subsequence, which we continue to denote by N, such that the two limits

(3.26)

$$\frac{1}{n}\nu_{p_n(\mu;\cdot)} \xrightarrow{*} \nu \text{ and } \frac{1}{n}\log\gamma_n(\mu) \to c \in \mathbb{R}\cup\{\infty\} \quad \text{as } n\to\infty,\ n\in N,$$

exist (cf. also Lemma 1.1.3 which easily follows from Lemma 1.3.2 and will be proved in Theorem 2.1.1). From the lower envelope theorem of potential theory it then follows that

$$\limsup_{n\to\infty,\ n\in N} \frac{1}{n}\log|p_n(\mu;z)| = -p(\nu;z) + c \tag{3.27}$$

qu.e. in $\mathbb{C}$. Note that the right-hand side of (3.27) is defined quasi everywhere even if $c = \infty$, for a potential can be infinite only on a set of zero capacity (see, e.g., Theorem 7.33 of [He]). From Lemma 1.1.3 we know that $p_n(\mu;z)$ has all its zeros in $\mathrm{Co}(S(\mu))$. Therefore, (3.26) implies

$$\lim_{n\to\infty,\ n\in N} \frac{1}{n}\log|p_n(\mu;z)| = -p(\nu;z) + c \tag{3.28}$$

locally uniformly for $z \in \mathbb{C} \setminus \mathrm{Co}(S(\mu))$. Furthermore, from (3.26), Lemma 1.1.3, and from the fact that $\deg p_n(\mu;\cdot) = n$, it follows that ν is a probability measure with

$$S(\nu) \subseteq \mathrm{Pc}(S(\mu)). \tag{3.29}$$

Because of (3.27), the lower asymptotic bound (1.10) is proved if for any infinite subsequence $N \subseteq \mathbb{N}$, for which the two limits in (3.26) exist, we can show that

$$-p(\nu; z) + c \geq 0 \tag{3.30}$$

for qu.e. z on $\partial\Omega$. Actually, we shall prove that (3.30) holds true for all $z \in \partial\Omega$. The proof will be carried out indirectly.

Let us assume that (3.30) does not hold true for all $z \in \partial\Omega$. Then

$$\partial\Omega \cap A_0 \neq \emptyset \quad \text{for the set} \quad A_0 := \{z \in \mathbb{C} \mid -p(\nu; z) + c < 0\}. \tag{3.31}$$

This implies that there exists $\delta > 0$ such that

$$\partial\Omega \cap A_\delta \neq \emptyset \quad \text{for} \quad A_\delta := \{z \in \mathbb{C} \mid -p(\nu; z) + c < -\delta\}. \tag{3.32}$$

From the upper semicontinuity of $-p(\nu; z)$ it follows that A_δ is open. The subharmonicity of $-p(\nu; z)$ implies that every component of A_δ is simply connected. Every such component A_δ^* has the same boundary ∂A_δ^* in the fine topology and in the ordinary planar topology (see Appendix II). This combined with that fact that potentials are continuous in the fine topology implies that on ∂A_δ^* we must have

$$-p(\nu; z) + c \equiv -\delta.$$

In particular, if $\eta > \delta$ and A_η^* is a component of A_η lying in A_δ^*, then we have $\overline{A_\eta^*} \subseteq A_\delta^*$. By the minimum principle of harmonic functions this is possible only if $A_\delta^* \cap S(\nu) \neq \emptyset$. Hence every component of the set A_δ has a nonempty intersection with $S(\nu)$. Therefore, there exists a simply connected domain B with

$$B \cap S(\nu) \neq \emptyset, \qquad B \cap \Omega \neq \emptyset, \qquad \text{and} \qquad \overline{B} \subseteq A_\delta. \tag{3.33}$$

Since B is simply connected, the set

$$S := \mathrm{Pc}(S(\mu)) \setminus B \tag{3.34}$$

is polynomially convex. Now let $V \subseteq B$ be a compact set with

$$\mathrm{Int}(V) \cap S(\nu) \neq \emptyset. \tag{3.35}$$

Since S is polynomially convex and S and V are disjoint, part (a) of Lemma 1.3.2 can be applied. Let $a < 1$ and $m \in \mathbb{N}$ be the two numbers introduced in part (a) of Lemma 1.3.2. Because of (3.35) and the first limit of (3.26), for sufficiently large $n \in \mathbb{N}$ there exist at least m zeros $x_{1,n}, \ldots, x_{m,n}$ of the polynomial $p_n(\mu; z)$ on V. From Lemma 1.3.2 we know that there also exist m points $y_{1,n}, \ldots, y_{m,n} \in \mathbb{C}$ so that the rational function r_m defined in (3.12) by the points $x_{1,n}, \ldots, x_{m,n}$ and $y_{1,n}, \ldots, y_{m,n}$ satisfies the inequality (3.13).

Using the rational function r_m, we construct the modified monic polynomial

$$\tilde{q}_n(z) := r_m(z) q_n(\mu; z) = z^n + \cdots, \tag{3.36}$$

where $q_n(\mu; \cdot)$ are the monic orthogonal polynomials corresponding to μ, and derive an upper estimate for the $L^2(\mu)$ norm of $\tilde{q}_n$.

Since the polynomials $\tilde{q}_n$ and $q_n(\mu; z)$, $n \in N$, differ only in m zeros, both sequences of polynomials have the same asymptotic zero distribution, that is, we have

$$\frac{1}{n} \nu_{\tilde{q}_n} \xrightarrow{*} \nu \quad \text{as} \quad n \to \infty,\ n \in N, \tag{3.37}$$

where ν is the same probability measure as in (3.26). From (3.37), (3.32), and the principle of descent of potential theory (see Appendix III) we derive that

$$\limsup_{n \to \infty,\ n \in N} \frac{1}{n} \log |\tilde{q}_n(z)| \leq -p(\nu; z) < -\delta - c \tag{3.38}$$

locally uniformly for $z \in A_\delta$ and so the inequality between the left- and right-hand sides holds true uniformly for $z \in B$. From (3.38) and the second limit in (3.26) it follows that there exists a zero sequence $\{\varepsilon_n > 0\}$ so that the supremum norm of $\tilde{q}_n$ on B satisfies

$$\|\tilde{q}_n\|_B \leq \frac{1}{\gamma_n(\mu)} e^{-n\delta/2} \leq \frac{\varepsilon_n}{\gamma_n(\mu)} \tag{3.39}$$

for all $n \in N$ sufficiently large. We consider now the $L^2(\mu)$ norm of $\tilde{q}_n$ and use (3.34), (3.36), (3.39), and the estimate (3.13) for the supremum norm of the rational function r_m, which give us the following sequence of inequalities

$$\begin{aligned} \|\tilde{q}_n\|^2_{L^2(\mu)} &= \int_B |\tilde{q}_n|^2 d\mu + \int_S |\tilde{q}_n|^2 d\mu \\ &\leq \mu(B) \|\tilde{q}_n\|^2_B + \|r_m\|^2_S \int_S |q_n(\mu; \cdot)|^2 d\mu \\ &\leq \frac{\mu(B)\varepsilon_n^2}{\gamma_n(\mu)^2} + \frac{a^2}{\gamma_n(\mu)^2} < \frac{1}{\gamma_n(\mu)^2} \end{aligned} \tag{3.40}$$

for all $n \in N$ sufficiently large. The estimate (3.40) contradicts the minimality (3.10) of the monic orthogonal polynomials $q_n(\mu; z)$. Hence, (3.30) and thereby also (1.10) is proved.

(b) In the next stage we prove the lower asymptotic bounds (1.7) and (1.9). If $\mathrm{cap}(S(\mu)) = 0$, then there are polynomials, say, the Chebyshev polynomials of $S(\mu)$, that tend to zero on $S(\mu)$ faster than any geometric progression. This implies $\lim_{n\to\infty} \gamma_n(\mu)^{1/n} = \infty$ and hence the estimates (1.7) and (1.9) are trivially true (use also Lemma 1.1.3 and for (1.9) also the fact, see, e.g., Chapter II, Section 4, p. 173 of [La], that if $p_n(z) = z^n + \cdots$ then the lemniscate set $\{z \mid |p_n(z)| \le a^n\}$ has capacity a). Therefore we can assume without loss of generality that $\mathrm{cap}(S(\mu)) > 0$.

Let us assume that $N \subseteq \mathbb{N}$ is an infinite subsequence such that the two limits in (3.26) exist. If $c = \infty$ then exactly as before we are ready. Thus, let $c < \infty$. Since $g_\Omega(z;\infty) = 0$ for all regular points of $\partial\Omega$ (see Appendix VI), it follows from (3.30), which has been proved in part (a) for all $z \in \partial\Omega$, that

$$-p(\nu; z) + c - g_\Omega(z;\infty) \ge 0 \quad \text{for all regular points on } \partial\Omega. \tag{3.41}$$

Seeing that quasi every point of $\partial\Omega$ is regular (see Appendix VI or Kellogg's lemma in [La, Chapter IV, Section 10, p. 232]) we get the inequality in (3.41) for quasi every $z \in \partial\Omega$. We claim that

$$-p(\nu; z) + c \ge g_\Omega(z;\infty) \quad \text{for all} \quad z \in \Omega. \tag{3.42}$$

Suppose that on the contrary there is an $\varepsilon > 0$ such that the set

$$A_\varepsilon = \{z \in \Omega \mid -p(\nu; z) + c - g_\Omega(z;\infty) < -\varepsilon\}$$

is not empty. Then by a variant of the minimum principle for harmonic functions (see, e.g., Theorem III.28 of [Ts]) for every component A_ε^* of A_ε we must have

$$\mathrm{cap}(\partial A_\varepsilon^* \cap \partial\Omega) > 0.$$

Thus, there is a point $z_0 \in \partial A_\varepsilon^* \cap \partial\Omega$ for which (3.41) holds with z replaced by z_0. Since z_0 is also a boundary point of A_ε^* in the fine topology, and the left-hand side of (3.41) is continuous everywhere in this topology, we have

$$-p(\nu; z_0) + c - g_\Omega(z_0;\infty) \le -\varepsilon,$$

which contradicts the choice of z_0 (namely, that (3.41) holds true at z_0). This contradiction proves (3.42).

From (3.42) together with (3.27) the estimate (1.9) follows for any infinite subsequence $N \subseteq \mathbb{N}$ for which the two limits in (3.26) exist. But since any infinite subsequence $N \subseteq \mathbb{N}$ contains an infinite subsequence for which the two limits in (3.26) exist, and since the right-hand side of (1.9) is independent of the subsequence N, (1.9) is generally proved.

The lower asymptotic bound (1.7) follows for any infinite subsequence $N \subseteq \mathbb{N}$ for which the two limits in (3.26) exist from limit (3.28) and inequality (3.42). Since the right-hand side of (1.7) is independent of the sequence N, (1.7) is also generally proved.

(c) Next we prove the lower asymptotic estimate (1.11) of Corollary 1.1.5. Let $U, V \subseteq \Omega$ be two arbitrary compact sets with U containing V in its interior. Consider the factorization

$$p_n(\mu; z) = h_n(z) s_n(z), \quad n \in \mathbb{N}, \tag{3.43}$$

where $h_n(z) \neq 0$ for all $z \notin U$, h_n a monic polynomial, and $s_n(z) \neq 0$ for all $z \in U$, that is,

$$h_n(z) = z^{m(n)} + \cdots, \qquad s_n(z) = \gamma_n(\mu) z^{n-m(n)} + \cdots, \tag{3.44}$$

with $m(n)$ the number of zeros that the polynomial h_n has on the set U. By Lemma 1.1.3 this number is bounded as $n \to \infty$. Therefore, we have

$$\lim_{n\to\infty} \frac{1}{n} m(n) = 0. \tag{3.45}$$

Let now $N \subseteq \mathbb{N}$ be an infinite subsequence such that the two limits in (3.26) exist. Since the polynomials s_n and $p_n(\mu; z)$, $n \in \mathbb{N}$, differ only in $m(n)$ zeros, both sequences of polynomials have the same asymptotic zero distribution, and as in (3.37) we have

$$\frac{1}{n} \nu_{s_n} \xrightarrow{*} \nu \quad \text{as} \quad n \to \infty,\ n \in N, \tag{3.46}$$

where ν is the same probability measure as that in (3.26). Since the zero distributions ν_{s_n} of the polynomials s_n have no mass on U, that is, $S(\nu_{s_n}) \cap U = \emptyset$ for all $n \in \mathbb{N}$, it follows from (3.46) and the second limit of (3.26) that

$$\lim_{n\to\infty,\ n\in N} \frac{1}{n} \log|s_n(z)| = -p(\nu; z) + c \tag{3.47}$$

uniformly for $z \in V$, which is an analogue of (3.28). From (3.47) and inequality (3.42) we deduce (1.11) for the subsequence N. Since the right-hand side of (1.11) is independent of the subsequence N, (1.11) is generally proved.

(d) We now come to the last stage of the proof: the verification of the lower asymptotic bound (1.8), which holds only in capacity. We assume that the two sets V and U are the same as in part (c) and that for the infinite subsequence $N \subseteq \mathbb{N}$ the two limits in (3.26) exist. Let $\varepsilon > 0$ be arbitrary with $\varepsilon < 1$. There exists δ $(0 < \delta < \infty)$ such that

$$-\delta > \log \frac{e^{g_\Omega(z;\infty)} - \varepsilon}{e^{g_\Omega(z;\infty)} - \frac{\varepsilon}{2}} \quad \text{for all} \quad z \in V. \tag{3.48}$$

Set

$$A_n := \left\{ z \in \mathbb{C} \mid \frac{1}{n} \log|h_n(z)| \leq -\delta \right\} \quad \text{for} \quad n \in \mathbb{N}. \tag{3.49}$$

Since h_n is a polynomial of degree $m(n)$, we can exactly calculate the capacity of the lemniscate sets A_n (see Chapter II, Section 4, p. 173 of [La]) and we have

$$\text{cap}(A_n) = e^{-\frac{n}{m(n)}\delta} \quad \text{for} \quad n \in \mathbb{N}. \tag{3.50}$$

Equations (3.42), (3.47), (3.48), and (3.49) imply for sufficiently large $n \in N$ that

(3.51)

$$|p_n(\mu; z)|^{1/n} = |h_n(z)|^{1/n} |s_n(z)|^{1/n} \geq e^{-\delta} \left(e^{g_\Omega(z;\infty)} - \frac{\varepsilon}{2} \right) \geq e^{g_\Omega(z;\infty)} - \varepsilon$$

for all $z \in V \setminus A_n$. With (3.50) this proves (1.8). □

1.4 Proof of the Sharpness of the Upper and Lower Bounds

In this section we verify Theorem 1.1.9 and Corollary 1.1.10.

Proof of Theorem 1.1.9. We may assume $\|\mu\| = 1$. First of all we remark that it is enough to show the existence of measures ν_1, ν_2, ν_3 with the properties that they are carrier related to μ and

$$\lim_{n\to\infty} \gamma_n(\nu_1)^{1/n} = \frac{1}{c_\mu}, \tag{4.1}$$

$$\lim_{n\to\infty} \gamma_n(\nu_2)^{1/n} = \frac{1}{\text{cap}(S(\mu))}, \tag{4.2}$$

and

$$\limsup_{n\to\infty} \gamma_n(\nu_3)^{1/n} = \frac{1}{c_\mu}; \qquad \liminf_{n\to\infty} \gamma_n(\nu_3)^{1/n} = \frac{1}{\text{cap}(S(\mu))}, \tag{4.3}$$

respectively.

In fact, the verification of, for example, (1.15) then runs as follows: Let $\varepsilon > 0$. By (1.6) for large n the functions

$$h_{n,\varepsilon}(z) = g_\mu(z; \infty) - \frac{1}{n} \log |p_n(\mu; z)| + \varepsilon$$

are nonnegative and harmonic in $\overline{\mathbb{C}} \setminus \text{Co}(S(\mu))$, and since (4.1) implies

$$\lim_{n\to\infty} h_{n,\varepsilon}(\infty) = \varepsilon,$$

(1.15) follows from Harnack's inequality (see Appendix III, cf. also (1.6)).

In what follows let $\nu \ll \mu$ denote the absolute continuity of the measure ν with respect to μ. First we prove the following lemma.

Lemma 1.4.1. *For each n there is a $\mu_n \ll \mu$ such that $\|\mu_n\| < 2n^{-2}$ and $\gamma_n(\mu_n) \le \sqrt{2}n\gamma_n(\omega_{S(\mu)})$.*

Proof. It follows from Lemma 1.2.7 and Corollary 1.1.7 that

$$\lim_{n\to\infty} \gamma_n(\omega_{S(\mu)})^{1/n} = \frac{1}{\mathrm{cap}(S(\mu))}. \tag{4.4}$$

Now keep the natural number n fixed throughout the proof. Let Π_n^* be the set of monic polynomials $p_n(z) = z^n + \cdots$, and let

$$\Pi_n^{**} = \{p_n \in \Pi_n^* \mid \int |p_n|^2 d\mu \le 2 \max_{z\in S(\mu)} |z|^{2n}\}.$$

We will set $\mu_n = \frac{1}{n^2}(\mu + \rho)$, where $\rho \ll \mu$, $\|\rho\| \le 1$ $(= \|\mu\|)$. In estimating $\gamma_n(\mu_n)$ from above we have to estimate the infimum of $\int |p_n|^2 d\mu_n$ for $p_n \in \Pi_n^*$ from below. Now in minimizing $\int |p_n|^2 d\mu_n$ for $p_n \in \Pi_n^*$ we may assume $p_n \in \Pi_n^{**}$ (otherwise $p_n^*(z) := z^n$ is a better choice). But Π_n^{**} is already compact in $C(S(\mu))$, hence for some $\varepsilon > 0$

$$\int_{z:|p_n(z)|\ge\varepsilon} |p_n|^2 d\omega_{S(\mu)} \ge \frac{3}{4} \int |p_n|^2 d\omega_{S(\mu)}, \quad p_n \in \Pi_n^{**}.$$

This and the uniform equicontinuity of the p_ns in Π_n^{**} easily imply the existence of a $\delta > 0$ such that if $\partial S(\mu) = \cup_1^M S_j$, $S_j \cap S_{j'} = \emptyset$ for $j \ne j'$, is a decomposition of $\partial S(\mu)$ into measurable subsets $S_1, \ldots, S_M$ in such a way that $\omega_{S(\mu)}(S_j) < \delta$, $\mathrm{diam}(S_j) < \delta$, and $A_1, \ldots, A_M$ are arbitrary measurable sets with $\mathrm{diam}(A_j) < \delta$, $\mathrm{dist}(S_j, A_j) < \delta$, $j = 1, \ldots, M$, then for any measure ρ for which $\rho(A_j) = \omega_{S(\mu)}(S_j)$, $j = 1, \ldots, M$, we have

$$\int |p_n|^2 d\rho > \frac{1}{2} \int |p_n|^2 d\omega_{S(\mu)}. \tag{4.5}$$

Take such a decomposition $\{S_j\}$ of $\partial S(\mu)$. Since $\mathrm{supp}(\mu) = S(\mu)$, we can choose $\{A_j\}$ with the aforementioned properties such that $\mu(A_j) > 0$ also holds for each $j = 1, \ldots, M$. Now if

$$\rho := \sum_{j=1}^{M} (\omega_{S(\mu)}(S_j)/\mu(A_j))\mu_{|A_j} \tag{4.6}$$

and $\mu_n = \frac{1}{n^2}(\mu + \rho)$, then for $p_n \in \Pi_n^{**}$

$$\int |p_n|^2 d\mu_n \ge \frac{1}{n^2} \int |p_n|^2 d\rho \ge \frac{1}{2n^2} \int |p_n|^2 d\omega_{S(\mu)}.$$

Since the infimum of the left-hand side for $p_n \in \Pi_n^{**}$ gives $\gamma_n(\mu_n)^{-2}$, whereas that of the right-hand side for $p_n \in \Pi_n^* \supseteq \Pi_n^{**}$ gives $\gamma_n(\omega_{S(\mu)})^{-2}$, Lemma 1.4.1 is proved. □

Now we return to the construction of ν_1, ν_2, ν_3 with properties (4.1)–(4.3). Let

$$\nu_2 := \mu + \sum_{n=1}^{\infty} \mu_n,$$

where the μ_ns are the measures from Lemma 1.4.1. Since $\mu_n \ll \mu$ for every n (actually we have $\mu_n \sim \mu$), ν_2 and μ are carrier related. Now the estimate

$$\gamma_n(\nu_2) \leq \gamma_n(\mu_n) \leq \sqrt{2}n\gamma_n(\omega_{S(\mu)})$$

together with (4.4) and (1.13) yields (4.2).

Next we consider (4.1). Take a carrier C_0 of μ of minimal capacity. There are then compact sets $K_1 \subseteq K_2 \subseteq \cdots \subseteq C_0$ with $\mathrm{cap}(K_m) \nearrow c_\mu$ and $\mu(K_m) \nearrow \|\mu\|$ as $m \to \infty$. Then $C := \cup_1^\infty K_m$ is again a carrier of μ.

Corollary 1.1.7 applied to $\mu|_{K_m}$ implies that there are numbers N_m and polynomials $p_n^{(m)} \in \Pi_n^*$, $n \geq N_m$, such that

$$\int_{K_m} |p_n^{(m)}|^2 d\mu < \left(\mathrm{cap}(K_m) + \frac{1}{m}\right)^n.$$

We may assume $N_1 < N_2 < \cdots$. We can choose positive constants c_m such that for $N_m \leq n < N_{m+1}$ the inequality

$$c_m \int |p_n^{(m)}|^2 d\mu < (\mathrm{cap}(K_m))^n$$

holds, and again we may assume $1 \geq c_1 > c_2 > \cdots$.

Set

$$\nu_1 := \sum_{j=1}^{\infty} \frac{c_j}{2^j} \mu|_{K_j}$$

and

$$q_n := p_n^{(m)} \quad \text{if} \quad N_m \leq n < N_{m+1}.$$

Using the fact that

$$\sum_{j=m+1}^{\infty} \frac{c_j}{2^j} \leq c_m \quad \text{and} \quad \sum_{j=1}^{\infty} \frac{c_j}{2^j} \leq 1,$$

we get for $N_m \le n < N_{m+1},\ m = 1, 2, \ldots,$

$$\int |q_n|^2 d\nu_1 = \int |p_n^{(m)}|^2 \left(d\left(\sum_1^m \frac{c_j}{2^j} \mu|_{K_j} \right) + d\left(\sum_{j=m+1}^{\infty} \frac{c_j}{2^j} \mu|_{K_j} \right) \right)$$

$$\le \int_{K_m} |p_n^{(m)}|^2 d\mu + c_m \int |p_n^{(m)}|^2 d\mu < 2 \left(\operatorname{cap}(K_m) + \frac{1}{m} \right)^n,$$

and so

$$\limsup_{n\to\infty} \frac{1}{\gamma_n(\nu_1)^{1/n}} \le c_\mu.$$

This and (1.13) prove (4.1), because $\nu_j \sim \mu$ easily follows from our construction. Thus, the existence of ν_1 has been established.

Finally, we construct a $\nu_3 \sim \mu$ that satisfies (4.3).

Let K_m be as in the preceding proof, and consider the measures μ_n from Lemma 1.4.1. Easy modification of the proof of Lemma 1.4.1 shows that μ_n can be chosen in such a way that it is supported in some K_{m_n} (this is why we wrote $\mu_n \ll \mu$ in the lemma and not $\mu_n \sim \mu$) and $\mu_n \ge n^{-2}\mu|_{K_n}$. In fact, assuming $\mu(K_1) > 0$, consider the set

$$\left\{ p_n \in \Pi_n^* \mid \int_{K_n} |p_n|^2 d\mu \le 2 \max_{z\in S(\mu)} |z|^{2n} \right\}$$

instead of Π_n^{**}, and with $\mu_n = \frac{1}{n^2}(\mu|_{K_n} + \rho)$ argue as in the proof of Lemma 1.4.1. If $A_1, \ldots, A_M$ are the sets constructed just before (4.6) then we must have $\mu(K_{m_n} \cap A_j) > 0$ for all j provided m_n is large enough, and so replacing in (4.6) each A_j by $K_{m_n} \cap A_j$ our claim ($\operatorname{supp}(\mu_n) \subseteq K_{m_n}$ and $\mu_n \ge n^{-2}\mu|_{K_n}$) follows.

We set

$$\nu_3 := \sum_{j=1}^{\infty} \mu_{l_j}$$

for some $\{l_j\}$ to be chosen subsequently. Let $l_1 = 1$, and suppose that $l_1, \ldots, l_{j-1}$ have already been selected. Since we may suppose that $m_1 < m_2 < \cdots$, the measure $\sum_{k=1}^{j-1} \mu_{l_k}$ is supported on $K_{m_{l_{j-1}}}$. Thus, by Corollary 1.1.7, there is an L_{j-1} such that

$$\gamma_{L_{j-1}} \left(\sum_{k=1}^{j-1} \mu_{l_k} \right)^{1/L_{j-1}} > \frac{1}{\operatorname{cap}(K_{m_{l_{j-1}}})} - \frac{1}{j}.$$

But then, if l_j is sufficiently large compared to L_{j-1}, we have independently of the choice of $l_{j+1}, l_{j+2}, \ldots$ that the inequality

$$\gamma_{L_{j-1}}(\nu_3)^{1/L_{j-1}} > \frac{1}{\operatorname{cap}(K_{m_{l_{j-1}}})} - \frac{1}{j} \tag{4.7}$$

holds (note that $\|\mu_n\| < 2n^{-2}$). Choose such an $l_j > L_{j-1}$, and this completes the definition of the indexes $\{l_j\}$.

By Lemma 1.4.1

$$\gamma_{l_j}(\nu_3)^{1/l_j} \le \gamma_{l_j}(\mu_{l_j})^{1/l_j} \le (\sqrt{2}l_j)^{1/l_j}\gamma_{l_j}(\omega_{S(\mu)})^{1/l_j},$$

which, together with (4.4), yields

$$\liminf_{n\to\infty} \gamma_n(\nu_3)^{1/n} \le \frac{1}{\operatorname{cap}(S(\mu))}.$$

Since (4.7) implies

$$\limsup_{n\to\infty} \gamma_n(\nu_3)^{1/n} \ge \frac{1}{c_\mu},$$

the limits in (4.3) follow by Corollary 1.1.7 once we verify the carrier relatedness of ν_3 and μ. But this is easy: $\nu_3 \ll \mu$ follows immediately from the construction, and $\mu \ll \nu_3$ follows from the facts that each μ_{l_j}, and hence ν_3 itself, is at least as large as $l_j^{-2}\mu|_{K_{l_j}}$, and $\cup_{j=1}^{\infty} K_{l_j}$ is a carrier of μ. □

Proof of Corollary 1.1.10. By letting $z \to \infty$ in (1.16) we get (4.2). Now arguing as in the beginning of the proof of Theorem 1.1.9 but with the functions

$$h_{n,\varepsilon}^*(z) := \frac{1}{n}\log\left|\frac{p_n(\nu_2, z)}{\prod_{j=1}^{m(n)}(z - x_{nj})}\right| - g_\Omega(z;\infty) + \varepsilon$$

instead of $h_{n,\varepsilon}$, where $x_{n1}, \ldots, x_{nm(n)}$ are the zeros of $p_n(\nu_2, z)$ contained in some suitable fixed compact subset of $\overline{\mathbb{C}} \setminus \operatorname{Pc}(S(\mu))$ containing V in its interior (cf. Corollary 1.1.5), we can derive from (4.2) and Corollary 1.1.5 that

$$\lim_{n\to\infty}\left|\frac{p_n(\nu_2; z)}{\prod_{j=1}^{m(n)}(z - x_{n,n})}\right|^{1/n} = e^{g_\Omega(z;\infty)}$$

uniformly on V. Now since $m(n) \le D_V$ for some constant D_V (see Lemma 1.1.3), this implies (1.19). □

1.5 Examples

Three examples will be given, which have been used in Section 1.1 to illustrate results or to prove their sharpness. The last two of the three examples could actually be saved in view of Theorem 1.1.9, but their simplicity may show more insight into the nature of minimal carries than the more general construction of Theorem 1.1.9.

Example 1.5.1. Let μ be the planar Lebesgue measure on the open disc $D := \{z \mid |z| < 1\}$.

The support $S(\mu) = \overline{D}$ is convex and its complement is connected. Hence, we have $\mathrm{Co}(S(\mu)) = \mathrm{Pc}(S(\mu)) = S(\mu)$ and $\Omega = \{z \mid |z| > 1\}$. The orthonormal polynomials (and the leading coefficients) are given by

$$(5.1) \qquad p_n(\mu; z) = \sqrt{\frac{n+1}{\pi}} z^n \qquad \left(\gamma_n(\mu) = \sqrt{\frac{n+1}{\pi}}\right) \quad \text{for} \quad n \in \mathbb{N}.$$

Hence, we have

$$(5.2) \qquad \lim_{n\to\infty} |p_n(\mu; z)|^{1/n} = |z| \quad \text{locally uniformly for} \quad z \in \mathbb{C}.$$

With the help of Criterion Λ (cf. Theorem 4.2.1) we can easily derive that in this example

$$(5.3) \qquad g_\mu(z;\infty) = g_\Omega(z;\infty) = \begin{cases} \log|z| & \text{for } |z| \geq 1; \\ 0 & \text{for } |z| < 1, \end{cases}$$

and furthermore that $\mathrm{cap}(S(\mu)) = c_\mu = 1$. Hence, in this example the lower and upper asymptotic bounds (1.7) and (1.6) of Theorem 1.1.4 are identical. □

With respect to the value of c_μ and the function $g_\mu(z;\infty)$ we shall observe the extreme opposite situation in the next example.

Example 1.5.2. We construct a weight measure μ on the closed unit disc $\overline{D}$ with $S(\mu) = \overline{D}$ so that $c_\mu = 0$, $\mathrm{cap}(S(\mu)) = 1$, and

$$(5.4) \qquad g_\Omega(z;\infty) \equiv \max(0, \log|z|) < g_\mu(z;\infty) \equiv \infty \quad \text{for all } z \in \mathbb{C}.$$

In this example the nth root of the orthonormal polynomials $p_n(\mu; z)$, $n \in \mathbb{N}$, asymptotically assume the upper bound (1.6) in Theorem 1.1.4.

Let $D_Q := Q \cap \{z \mid |z| \leq 1\}$ be the set of all rational points in the unit disc, and let $\{r_n \mid n \in \mathbb{N}\} = D_Q$ be an enumeration of these points. The measure μ is defined as

$$(5.5) \qquad \mu := \sum_{n=1}^{\infty} 2^{-n^2} \delta_{r_n},$$

where δ_x is Dirac's measure of the point $x \in \mathbb{C}$. We have $\mathrm{Co}(S(\mu)) = \mathrm{Pc}(S(\mu)) = S(\mu) = \overline{D}$, and $\mathrm{cap}(D_Q) = 0$ since D_Q is a countable set. As $D_Q \in \Gamma(\mu)$, we have $c_\mu = 0$ and therefore $g_\mu(z;\infty) \equiv \infty$. From the extremality property (3.10) of the monic orthogonal polynomials $q_n(\mu; z)$, we derive

$$(5.6) \qquad \frac{1}{\gamma_n(\mu)^2} \leq \int |q_n|^2 d\mu \leq 4^n \sum_{j=n+1}^{\infty} 2^{-j^2} \leq 4^n 2^{-n^2}, \quad n \in \mathbb{N},$$

where in (5.6) q_n is the monic polynomial of degree n having zeros at the first n points $r_1, \dots, r_n$. From (5.6) it follows that $\gamma_n(\mu)^{1/n} \to \infty$ as $n \to \infty$. Together with the fact that all zeros of $p_n(\mu; z)$, $n \in \mathbb{N}$, are contained in $\{z \mid |z| \le 1\}$ (see Lemma 1.1.3) this implies

$$\lim_{n\to\infty} |p_n(\mu; z)|^{1/n} = \infty \quad \text{locally uniformly for } |z| > 1. \tag{5.7}$$

Hence, in this example the upper asymptotic bound (1.6) in Theorem 1.1.4 is assumed, whereas the lower bound (1.7) stays far away. □

For upper estimates of the leading coefficient $\gamma_n(\mu)$ the following lemma is often very helpful.

Lemma 1.5.3. *Let μ and ν be positive measures with compact support in $\mathbb{C}$, and $c > 0$. Then*

$$\gamma_n(c\mu + \nu) \le \sqrt{\frac{1}{c}}\gamma_n(\mu), \qquad n \in \mathbb{N}. \tag{5.8}$$

Proof. From the extremality property (3.10) of the monic orthogonal polynomials $q_n(\mu; z)$ it follows that

$$\begin{aligned} \frac{1}{\gamma_n(c\mu+\nu)^2} &= \int |q_n(c\mu+\nu; x)|^2 d(c\mu+\nu)(x) \\ &\ge c\int |q_n(c\mu+\nu; x)|^2 d\mu(x) \ge \frac{c}{\gamma_n(\mu)^2}. \end{aligned} \tag{5.9}$$

□

In the next example we construct a weight measure μ on ∂D, for which (5.4) holds true as in Example 1.5.2, but now the nth roots of the polynomials $|p_n(\mu; z)|$, $n \in \mathbb{N}$, behave like the lower bound in (1.7).

Example 1.5.4. We construct a weight measure μ on ∂D with $c_\mu = 0$, $\operatorname{cap}(S(\mu)) = 1$, such that $\lim_{n\to\infty} |p_n(\mu; z)|^{1/n} = e^{g_\Omega(z;\infty)}$ for $|z| > 1$.

Let $S_n := \{z \in \partial D \mid z^{n+1} = 1\}$ be the set of the $(n+1)$th roots of unity. For every $n \in \mathbb{N}$ we define the probability measure

$$\mu_n := \frac{1}{n+1} \sum_{x \in S_n} \delta_x. \tag{5.10}$$

It is easy to verify that $p_k(\mu_n; z) = z^k$, and therefore $\gamma_k(\mu_n) = 1$ for all $0 \le k \le n$. The weight measure μ is defined by

$$\mu := \sum_{n=1}^{\infty} \frac{1}{n^2}\mu_n. \tag{5.11}$$

We have $S(\mu) = \partial D$ and $\mathrm{Co}(S(\mu)) = \mathrm{Pc}(S(\mu)) = \overline{D}$. Since μ has countable carriers, $c_\mu = 0$. For the leading coefficient $\gamma_n(\mu)$ we can derive from Lemma 1.5.3 the upper estimate

$$(5.12) \qquad \gamma_n(\mu) \leq n\gamma_n(\mu_n) = n \quad \text{for} \quad n \in \mathbb{N},$$

which implies with Corollary 1.1.7 that $\gamma_n(\mu)^{1/n} \to 1$ as $n \to \infty$. Now this yields (cf. the beginning of the proof of Theorem 1.1.9 or Theorem 3.1.1)

$$(5.13) \quad \lim_{n\to\infty} |p_n(\mu; z)|^{1/n} = |z| = e^{g_\Omega(z;\infty)} \quad \text{locally uniformly for } |z| > 1,$$

which shows that here the lower asymptotic bound (1.7) of Theorem 1.1.4 is assumed. □

2

Zero Distribution of Orthogonal Polynomials

This chapter is devoted to the investigation of the asymptotic zero distribution of general orthogonal polynomials. In Section 2.1 we characterize all possible weak* limits of the zeros and show that they can actually occur at some orthogonal polynomial systems. In Section 2.2 the relation between zero asymptotics and asymptotics for the leading coefficients (called norm asymptotics) is explored. It turns out that in the case when $S(\mu)$ has empty interior and connected complement, the existence of the asymptotic zero distribution implies norm asymptotics and in the two extreme cases appearing in Corollary 1.1.7 the converse implication also holds: Norm asymptotics implies the existence of the asymptotic zero distribution. In general, no other implication is possible in between these concepts, which is demonstrated by several nontrivial constructions. Finally, in Section 2.3 we discuss the more difficult problem of characterizing weak* limits when $c_\mu > 0$.

2.1 Zeros of Orthogonal Polynomials

Recall that for a polynomial P by ν_P we denote the measure that places mass 1 to every zero of P (counting multiplicity). Then $\|\nu_P\| = \deg(P)$ and $\nu_P/\|\nu_P\| = \frac{1}{\deg(P)}\nu_P$ is often referred to as the *normalized counting measure* on the zeros of P.

We are primarily interested in the location of the zeros of orthogonal polynomials and in their distribution. Of course, by the existence of the (asymptotic) distribution of the zeros we mean that the sequence of normalized counting measures

$$\left\{\frac{1}{n}\nu_{p_n(\mu;\cdot)}\right\} \tag{1.1}$$

converges in the weak* topology of Borel measures. Similarly, by a weak* limit of the zeros we mean a weak* limit point of the measures (1.1) (we prefer the short but somewhat inaccurate expression "weak* limit of the zeros" to the more precise "weak* limit of the normalized counting measures on the zeros").

Our first result, which is an immediate consequence of Lemma 1.3.2, is basically the repetition of Lemma 1.1.3.

Theorem 2.1.1. *All zeros of the orthonormal polynomials* $p_n(\mu; z)$, $n \in \mathbb{N}$, *are contained in the convex hull* $\mathrm{Co}(S(\mu))$ *of* $S(\mu)$, *and for any compact set* $V \subseteq \Omega$ *the number of zeros of* $p_n(\mu; z)$, $n \in \mathbb{N}$, *on* V *is bounded as* $n \to \infty$. *Consequently, every weak* limit point of the zeros is supported on the polynomial convex hull* $\mathrm{Pc}(S(\mu))$ *of* $S(\mu)$.

Recall that Ω is the unbounded component of the complement of the support of μ. We also mention (see [Sa2]) that the zeros lie strictly inside the convex hull of $S(\mu)$ unless this support is on a line.

Proof of Theorem 2.1.1. (a) Let us first assume that V is a compact set contained in Ω. Then part (a) of Lemma 1.3.2 can be applied with $S = S(\mu)$. Let $a < 1$ and $m \in \mathbb{N}$ be the two numbers from part (a) of Lemma 1.3.2, and let us assume that $p_n(\mu; z)$ has at least m zeros $x_1, \ldots, x_m$ on V. By Lemma 1.3.2 there exist m points $y_1, \ldots, y_m \in \mathbb{C}$ such that the rational function r_m defined as in (1.3.12) by the points $x_1, \ldots, x_m$ and $y_1, \ldots, y_m$ satisfies the inequality (1.3.13). With r_m we define the modified monic polynomial

$$\tilde{q}_n(z) := r_m(z) q_n(\mu; z) = z^n + \cdots,$$

For the $L^2(\mu)$ norm of this polynomial we have the estimate

$$\|\tilde{q}_n\|_{L^2(\mu)} \leq \|r_m\|_{S(\mu)} \|q_n(\mu; \cdot)\|_{L^2(\mu)} < \|q_n(\mu; \cdot)\|_{L^2(\mu)},$$

which contradicts the minimality (1.3.10) of the monic orthogonal polynomial $q_n(\mu; z)$. Hence, we have proved that $p_n(\mu; z)$ has at most $m-1$ zeros on V.

(b) Now let V be a compact set contained in $\mathbb{C} \setminus \mathrm{Co}(S(\mu))$, and let $p_n(\mu; z)$ have at least one zero on V. Then it follows from part (b) of Lemma 1.3.2 that we can repeat the analysis of part (a) of the proof, but now with $m = 1$. This proves that $p_n(\mu; z)$, $n \in \mathbb{N}$, cannot have any zero in $\mathbb{C} \setminus \mathrm{Co}(S(\mu))$. □

In general, nothing more can be said about the zeros in Ω, as is shown by the following example.

Example 2.1.2. A weight measure μ with a nonconvex simply connected support $S(\mu)$ is constructed. For this measure μ it is shown that infinitely many of the orthonormal polynomials $p_n(\mu; z)$, $n \in \mathbb{N}$, have at least one zero outside of $S(\mu)$, and these zeros cluster at every point of a prescribed set in $\mathrm{Co}(S(\mu)) \setminus S(\mu)$.

Let S be the semicircle $\{z = e^{it} \mid t \in [0, \pi]\}$,

$$\varphi : [0,1] \to S \quad \text{with} \quad \varphi(t) = e^{i\pi t} \tag{1.2}$$

a parametrization, and

$$S_n := \left\{ \varphi(t) \mid t = \frac{2j-1}{2^{n+1}},\ j = 1, \dots, 2^n \right\} \subseteq S, \quad n = 0, 1, \dots, \tag{1.3}$$

a finite, equally spaced set of 2^n points. For $m \neq n$ the two sets S_n and S_m are disjoint. Set

$$T_n := S_{n-1} \cup \cdots \cup S_1, \quad n = 1, 2, \dots. \tag{1.4}$$

This is again an equally spaced set in S, it contains $2^n - 1$ points, and for $m \geq n$ the two sets T_n and S_m are disjoint. The measure μ is constructed as an infinite sum

$$\mu = \mu_0 + \mu_1 + \cdots, \tag{1.5}$$

in which the single terms μ_j with support on S_j, $j = 1, 2, \dots$, will be recursively defined.

We assume that the first n measures $\mu_0, \dots, \mu_{n-1}$ have been defined in such a way that

$$\|\mu_j\| > 0 \quad \text{and} \quad S(\mu_j) = S_j \quad \text{for} \quad j = 0, \dots, n-1 \tag{1.6}$$

(μ_0 is the unit measure at $z = i$). Let now y_1 and y_2 be two arbitrarily selected distinct points from S_n, and distribute to these points two masses $b_1 > 0$ and $b_2 > 0$ with $b_1 + b_2 \leq \|\mu_{n-1}\|/4$. If $b_1 + b_2$ is sufficiently small, then the orthonormal polynomial

$$p_m(\mu_1 + \cdots + \mu_{n-1} + b_1\delta_{y_1} + b_2\delta_{y_2}; z), \quad m = 2^n, \tag{1.7}$$

has $2^n - 1$ zeros close to the $2^n - 1$ points of the set T_n and exactly one zero close to the line connecting the two points y_1 and y_2. Depending on the ratio b_1/b_2, this last zero will be located closer to y_1 or y_2.

Since the last two sentences are very important for the understanding of the whole example, we will elaborate on them in more detail: Let

$$\mu^{(n)} := \mu_1 + \cdots + \mu_{n-1} + b_1\delta_{y_1} + b_2\delta_{y_2} \tag{1.8}$$

be the measure used in (1.7), $q_m = q_m(\mu^{(n)}; \cdot) := p_m(\mu^{(n)}; \cdot)/\gamma_m(\mu^n)$ the monic orthogonal polynomial with respect to the measure $\mu^{(n)}$, and $q(z) = z^m + \cdots$ the monic polynomial having a zero at every point of T_n and one zero at $z = 0$. From the minimality property (1.3.10) of orthogonal polynomials we derive

$$\int |q_m|^2 d\mu^{(n)} \leq \int |q|^2 d\mu^{(n)} = O(b_1 + b_2) \quad \text{as} \quad b_1 + b_2 \to 0. \tag{1.9}$$

From (1.9) it follows that the orthogonal polynomial q_m has $m-1$ zeros near the $m-1$ points of the set T_n and that these zeros converge to the points of T_n if $b_1 + b_2$ tends to zero. Let the first $m-1$ zeros of the polynomial q_m in an appropriate order be denoted by $z_1, \ldots, z_{m-1}$, the points of T_n by $x_1, \ldots, x_{m-1}$, and the differences between corresponding pairs by $\Delta z_j := z_j - x_j$, $j = 1, \ldots, m-1$. In order to have more precise information about the magnitude of the differences Δz_j, $j = 1, \ldots, m-1$, we expand the integral on the left-hand side of (1.9) with respect to the differences Δz_j :

(1.10)

$$\int |q_m|^2 d\mu^{(n)} = \sum_{j=1}^{m-1} \mu^{(n)}(\{x_j\})|q_m'(x_j)|^2|\Delta z_j|^2 + b_1|q_m(y_1)|^2$$
$$+ b_2|q_m(y_2)|^2 + +O\left(\sum_{j=1}^{m-1} |\Delta z_j|^3\right) \quad \text{as} \quad b_1 + b_2 \to 0.$$

This immediately implies the preliminary estimate $|\Delta z_j| = O((b_1+b_2)^{1/2})$. Because of the minimality (1.3.10) of q_m in $L^2(\mu^{(n)})$, the directional derivative of the left-hand side of (1.10) with respect to z_k in any direction must vanish. Taking this directional derivative of both sides in (1.10) in the direction $\overrightarrow{x_k z_k}$ leads to

$$0 = \mu^{(n)}(\{x_k\})|q_m'(x_k)|^2 2|\Delta z_k| + O\left(\sum_{j=1}^{m-1} |\Delta z_j|^2\right) + O(b_1 + b_2),$$

which, together with the preceding preliminary estimates gives $|\Delta z_k| = O(b_1 + b_2)$ provided $|q_m'(x_k)|$ is not small, that is, provided that the last zero z_m of q_m is not near to any point of T_n. This will be justified in the sequel, hence we can conclude

$$|\Delta z_j| = O(b_1 + b_2) \quad \text{as} \quad b_1 + b_2 \to 0 \tag{1.11}$$

for $j = 1, \ldots, m-1$.

Now let the polynomial Q_{m-1} be defined by $q_m(z) = Q_{m-1}(z)(z - z_m)$. From (1.10) and (1.11) we derive

(1.12)

$$\int |q_m|^2 d\mu^{(n)} = b_1|Q_{m-1}(y_1)|^2|y_1 - z_m|^2 + b_2|Q_{m-1}(y_2)|^2|y_2 - z_m|^2$$
$$+ O((b_1 + b_2)^2) \quad \text{as} \quad b_1 + b_2 \to 0.$$

If we neglect terms of higher order in $b_1 + b_2$ then we see that the integral in (1.12) is minimal if

$$(1.13) \qquad z_m = \frac{b_1|Q_{m-1}(y_1)|^2 y_1 + b_2|Q_{m-1}(y_2)|^2 y_2}{b_1|Q_{m-1}(y_1)|^2 + b_2|Q_{m-1}(y_2)|^2},$$

which is a point on the line connecting the two points y_1 and y_2. The position on this line depends on the ratio of b_1 and b_2. If we do not neglect the terms of higher order in $b_1 + b_2$, then the point z_m may move away from the position given in (1.13), but the movement will be small if we keep $b_1 + b_2$ sufficiently small. Finally, to justify the aforementioned claim that the last zero z_m cannot lie close to T_m we have only to observe that the integral in (1.12) obtained by the value (1.13) is smaller than the smallest value of the analogous integral if the zero z_m of q_m is close to a point of T_n.

With this last remark our investigation of the location of the zeros of the orthonormal polynomial (1.7) is completed. We continue with the construction of the single terms in (1.5).

If we add further terms to the measure (1.8), then the location of the zeros of the polynomial (1.7) will move. However, the movement can be kept small if the total mass of the added terms is kept sufficiently small. Having this connection in mind, we distribute a small total mass $b > 0$ uniformly to all 2^n points of S_n. The measure μ_n is then defined as

$$(1.14) \qquad \mu_n := b_1\delta_{y_1} + b_2\delta_{y_2} + \frac{b}{2^n}\sum_{y \in S_n} \delta_y.$$

In this way we can go on and define all further terms in (1.5). At every step we have to take care that the zeros of the orthonormal polynomials of degree $m = 2^j$, $j = 1, \ldots, n$, do not change too much. Ultimately we get a measure μ with the property that the associated orthonormal polynomials $p_m(\mu; z)$ have at least one zero outside of $S = S(\mu)$ for $m \in \mathcal{N} := \{m = 2^n \mid n \in \mathbb{N}\} \subseteq \mathbb{N}$, and, as $m \to \infty$, $m \in \mathcal{N}$, this zero will cluster at prescribed points in $\mathrm{Co}(S) \setminus S$ if at every step the points y_1 and y_2 and the corresponding masses b_1 and b_2 have been appropriately selected.

It is not difficult to see that the construction of the example can be refined so that more than one zero of any $p_m(\mu; z)$, $m \in \mathcal{N}$, lies outside of S. However, because of Theorem 2.1.1, there exists an upper bound for the number of zeros that any polynomial $p_m(\mu; z)$, $m \in \mathcal{N}$, may have on a given compact subset of $\mathrm{Co}(S) \setminus S$. □

P. Turán [Tu] raised the following problem. Suppose that the measure μ is supported on the unit circumference. Can it happen that the set of all zeros of all orthogonal polynomials $p_n(\mu; \cdot)$, $n = 1, 2, \ldots$, is dense in the unit disk? The affirmative answer was given in [AlVi]. With obvious modifications in the construction given in Example 2.1.2 we can get the following stronger statement.

Theorem 2.1.3. *Let S be any compact subset of $\mathbf{C}$. Then there is a measure μ such that its support is S and every point of the convex hull* $\mathrm{Co}(S)$ *of S is a limit point of the zeros of $p_n(\mu;\cdot)$, $n = 1, 2, \ldots$.*

Next we characterize the sets of measures that arise as weak* limit points of zeros of orthogonal polynomials corresponding to measures μ in the important special case when $S(\mu) = [0,1]$, although our method works in many other cases as well. In other words, we want to determine the sets

$$\mathcal{M}_\mu = \left\{ \nu \mid \nu \text{ is a weak* limit point of the measures } \frac{1}{n}\nu_{p_n(\mu;\cdot)},\right.$$
$$\left. n = 1, 2, \ldots \right\}$$

when the support of μ is contained in $[0,1]$. Then, of course, $\mathcal{M}_\mu \subseteq \mathcal{M}[0,1]$, where $\mathcal{M}[0,1]$ denotes the set of unit Borel measures on $[0,1]$, and we always think of the latter space as equipped with the weak* topology.

Theorem 2.1.4. *$\mathcal{M}_\mu$ is a closed and connected subset of $\mathcal{M}[0,1]$, and if $\mathcal{M}^*$ is any closed and connected subset of $\mathcal{M}[0,1]$, then there is a μ such that $\mathcal{M}_\mu = \mathcal{M}^*$.*

Proof of Theorem 2.1.4. Let $\{f_k\}$ be a countable dense set in the unit ball of $C[0,1]$ and set

$$d(\mu,\nu) = \sum_{k=1}^{\infty} 2^{-k} \left| \int f_k d\mu - \int f_k d\nu \right|.$$

Then the topology on $\mathcal{M}[0,1]$ associated with the metric d is exactly the weak* topology.

Obviously, $\mathcal{M}_\mu$ is closed. To prove its connectedness we mention first of all that the interlacing property of the zeros of consecutive orthogonal polynomials implies that $d(\frac{1}{n}\nu_{p_n(\mu;\cdot)}, \frac{1}{n+1}\nu_{p_{n+1}(\mu;\cdot)}) \to 0$ as $n \to \infty$. Now suppose that $\mathcal{M}_\mu$ is not connected, and there are two nonempty disjoint closed subsets K_1 and K_2 of $\mathcal{M}_\mu$ such that their union is $\mathcal{M}_\mu$. Since $\mathcal{M}[0,1]$ with the metric d is compact, the distance of K_1 and K_2 is positive, say d_0. Now the points $\frac{1}{n}\nu_{p_n(\mu;\cdot)}$ wander from points close to K_1 to points close to K_2 and back, while their distance tends to zero, so it is obvious that there is a sequence $\{\frac{1}{n_k}\nu_{p_{n_k}(\mu;\cdot)}\}_{k=1}^{\infty}$ with the property $d(K_j, \frac{1}{n_k}\nu_{p_{n_k}(\mu;\cdot)}) \geq d_0/3$ for all k and $j = 1, 2$. But then no weak* limit of $\{\frac{1}{n_k}\nu_{p_{n_k}(\mu;\cdot)}\}_{k=1}^{\infty}$ can belong to $K_1 \cup K_2$, which is a contradiction and hence the connectedness has been proved.

Suppose now that $\mathcal{M}^*$ is closed and connected. Then we can choose a sequence $\{\rho_n\} \subset \mathcal{M}^*$ such that its closure is $\mathcal{M}^*$ and $d(\rho_n, \rho_{n+1}) \to 0$ as $n \to 0$, the existence of which is guaranteed by the connectedness of $\mathcal{M}^*$. We choose a sequence $Y = \{y_k\}_{k=1}^{\infty} \subset [0,1]$ of distinct points such that with

$$\sigma(m;Y) = \frac{1}{m}\sum_{k=1}^{m} \delta_{y_k}$$

for some sequence $\{N_n\}$ we have

$$\lim_{\substack{m\to\infty \\ N_n \le m < N_{n+1}}} d(\sigma(m;Y), \rho_n) = 0. \tag{1.15}$$

That this is possible is standard, but for completeness we give a sketch here.

For each n there are sequences $Y^{(n)} = \{y_k^{(n)}\}_{k=1}^{\infty} \subseteq [0,1]$ such that

$$\lim_{m\to\infty} d(\sigma(m;Y^{(n)}), \rho_n) = 0$$

(see, e.g., [Tol]). Without loss of generality we may assume that all the points $\{y_k^{(n)}\}$ are different. Now with some fast increasing sequence $\{N_n\}$, $N_1 = 1$, to be chosen subsequently, let

$$y_k = y_{k-N_n+1}^{(n)} \quad \text{for} \quad N_n \le k < N_{n+1}.$$

If $N_n \le m < N_{n+1}$ then $\sigma(m;Y)$ is a convex linear combination of

$$\sigma(N_2 - N_1, Y^{(1)}),\ \sigma(N_3 - N_2, Y^{(2)}),\ \dots,\ \sigma(N_n - N_{n-1}, Y^{(n-1)}),$$
$$\sigma(m - N_n + 1, Y^{(n)})$$

with corresponding coefficients

$$\frac{N_2 - N_1}{m},\ \frac{N_3 - N_2}{m},\ \dots,\ \frac{N_n - N_{n-1}}{m},\ \frac{m - N_n + 1}{m},$$

and if $N_n > N_{n-1}n^2$ is large compared to N_{n-1} then only the last two terms will not be negligible, that is,

$$d\left(\sigma(m;Y),\ \frac{N_n - N_{n-1}}{m}\sigma(N_n - N_{n-1}, Y^{(n-1)}) + \frac{m - N_n + 1}{m}\sigma(m - N_n + 1, Y^{(n)})\right) < \frac{1}{n}.$$

Since for large N_n either

$$\frac{m - N_n + 1}{m} < \frac{1}{n}$$

or

$$d(\sigma(m - N_n + 1, Y^{(n)}), \rho_n) < \frac{1}{n},$$

and furthermore

$$d(\sigma(N_n - N_{n-1}, Y^{(n-1)}), \rho_{n-1}) < \frac{1}{n},$$

we get

$$d\left(\sigma(m; Y), \ \frac{N_n - N_{n-1}}{m}\rho_{n-1} + \frac{m - N_n + 1}{m}\rho_n\right) < \frac{4}{n}.$$

Finally, since $d(\rho_{n-1}, \rho_n) \to 0$ and

$$\frac{N_n - N_{n-1}}{m} + \frac{m - N_n + 1}{m} > 1 - \frac{1}{n^2},$$

we obtain

$$d(\sigma(m; Y), \rho_n) \leq \frac{4}{n} + o(1), \quad N_n \leq m < N_{n+1},$$

and (1.15) is proved.

What we have done so far implies that if $\{x_k^{(n)}\}_{k=1}^n$, $n = 1, 2, \ldots$, is an arbitrary point system such that

$$\max_{1 \leq k \leq n} |x_k^{(n)} - y_k| \to 0 \tag{1.16}$$

as $n \to \infty$, then the set of the weak* limit points of the sequence

$$\left\{\frac{1}{n}\sum_{k=1}^{n} \delta_{x_k^{(n)}}\right\}$$

coincides with $\mathcal{M}^*$. Thus, it is enough to show a μ such that the zeros of the corresponding orthogonal polynomials in a suitable numbering satisfy (1.16). But this is easy: Set

$$\mu = \sum_{j=1}^{\infty} c_j \delta_{y_j},$$

where the positive coefficients c_j will be chosen in a moment.

We claim that if $c_1, \ldots, c_n$ have already been chosen, then there is a number $\alpha_n > 0$ such that if

$$\rho = \sum_{j=1}^{n} c_j \delta_{y_j} + \rho_0$$

is any measure with $\|\rho_0\| < \alpha_n$, then the zeros $x_1^{(n)}, \dots, x_n^{(n)}$ of $P_n(\rho; \cdot)$ (in a suitable ordering) satisfy

$$|y_j - x_j^{(n)}| < \frac{1}{n}, \quad 1 \le j \le n.$$

In fact, let $\beta_n = \min_{\substack{1\le j,l\le n\\ j\ne l}} |y_j - y_l|/n$, and set $p_n(x) = \prod_{j=1}^n (x - y_j)$. For this we have

$$\int p_n^2 d\rho \le \|\rho_0\| < \alpha_n,$$

but for any polynomial P_n of degree n with real zeros and leading coefficient 1, if P_n does not have a zero in $(y_j - \beta_n, y_j + \beta_n)$ for some $1 \le j \le n$, then we have

$$\int P_n^2 d\rho \ge P_n^2(y_j)c_j \ge \beta_n^n c_j.$$

Therefore, if $\alpha_n < \beta_n^n \min_{1\le j\le n} c_j$, then $p_n(\rho;\cdot)$, having minimal $L^2(\rho)$ norm, must have a zero in each of the intervals $(y_j - \beta_n, y_j + \beta_n)$, which proves the claim.

Now using this claim we can choose the coefficients one after the other to satisfy $0 < c_{n+1}$ and $\sum_{k=n+1}^\infty c_k < \min_{1\le j\le n} \alpha_j$ for the corresponding α_js and the theorem is proved. □

In the proof above the measures that we constructed were discrete ones, that is, for them $c_\mu = 0$. The analogous problem when $c_\mu > 0$ is much more difficult and will be discussed in Section 2.3.

Our last result in this section shows that weak* limits need not be equilibrium distributions even if $c_\mu > 0$ is assumed (see the historical notes at the end of the book). Actually this follows from the more difficult result of Section 2.3 but we found it instructive to give here the following concrete example.

Example 2.1.5. Let $\omega_{[a,b]}$ be the equilibrium (or Chebyshev) distribution of the interval $[a, b]$, that is, $d\omega_{[a,b]}(x) = \pi^{-1}((x-a)(b-x))^{-1/2}dx$ for $x \in [a, b]$. We set $\kappa = \frac{1}{2}(\omega_{[-1,1]} + \omega_{[-2,2]})$ and show a μ with support in $[-2, 2]$ and $c_\mu \ge 1/2$ such that κ is a weak* limit of the zeros of the orthogonal polynomials $p_n(\mu; \cdot)$. This shows the aforementioned phenomenon, since κ is not the equilibrium distribution of any set.

Let

$$v(x) = \exp(p(\kappa; x) - \tfrac{1}{2}\log 2) = |x + \sqrt{x^2 - 1}|^{-1/2}, \quad x \in [-2, 2],$$

with the branch of the square root that is positive for positive x, and first we establish the following properties of these weights. For simpler notation let $\prod_n^*$ be the set of all polynomials of the form $z^n + \cdots$.

(a) There are polynomials $Q_n \in \prod_n^*$ with zeros in $[-2,2]$ and

$$\|Q_n v^n\|_{L^\infty[-2,2]} \leq 8 \cdot 2^{-n/2}.$$

(b) If for some polynomials $P_n \in \prod_n^*$, $n \in N \subset \mathbb{N}$,

$$\limsup_{n\to\infty,\ n\in N} \|P_n v^n\|_{L^\infty[-2,2]}^{1/n} \leq 2^{-1/2},$$

then $\nu_{P_n} \xrightarrow{*} \kappa$ as $n \to \infty$, $n \in N$.

Property (a) follows from the fact that for the Chebyshev polynomials

$$\begin{aligned} T_n(x) &= 2^{-n+1}\cos(n \cdot \arccos x) \\ &= 2^{-n}((x+\sqrt{x^2-1})^n + (x-\sqrt{x^2-1})^n) \end{aligned}$$

we have $\|T_n\|_{L^\infty[-1,1]} = 2^{1-n}$, and so, for example, for even n we can put $Q_n(x) = 2^{n/2}T_{n/2}(x)T_{n/2}(x/2)$ (when n is odd replace the second $n/2$ by $(n+1)/2$ and the other two ones by $[n/2]$).

Property (b) obviously follows from the following somewhat more general lemma that we will also need in subsequent sections.

Lemma 2.1.6. *Let σ be an arbitrary probability measure with support in $[a,b]$ and suppose that its potential $p(\sigma;\cdot)$ is continuous. Set $w(x) = \exp(p(\sigma;x) - \max p(\sigma;\cdot))$, where $\max p(\sigma;\cdot)$ denotes the maximum of $p(\sigma;\cdot)$ on $[a,b]$. If for some polynomials $P_n \in \prod_n^*$, $n \in N \subset \mathbb{N}$, with zeros in some interval $[A,B]$ containing $[a,b]$*

$$\limsup_{n\to\infty,\ n\in N} \|P_n w^n\|_{L^\infty[a,b]}^{1/n} \leq \exp(-\max p(\sigma;\cdot)),$$

then $\nu_{P_n} \xrightarrow{*} \sigma$ as $n \to \infty$, $n \in N$.

Proof of Lemma 2.1.6. Let ω be a weak* limit point of the measures $\{\frac{1}{n}\nu_{P_n}\}_{n\in N}$. We have to prove that $\omega = \sigma$. Without loss of generality we may assume that the whole sequence $\{\frac{1}{n}\nu_{P_n}\}_{n\in N}$ converges to ω. For $\varepsilon > 0$ and large $n \in N$ we have

$$\frac{1}{n}\log\frac{1}{|P_n(x)|} + \epsilon \geq p(\sigma;x) \tag{1.17}$$

for all $x \in [a,b]$. Since the latter set contains the support of σ and σ has finite logarithmic energy (this follows from the continuity of the potential $p(\sigma;\cdot)$ by integration with respect to σ), we can conclude from the principle of domination of potential theory (see Appendix III) that (1.17) holds for all $x \in \mathbb{C}$. Letting n tend to infinity through N we get for $x \in \mathbb{C}\setminus[A,B]$ the inequality $p(\omega;x)(x)+\epsilon \geq p(\sigma;x)$, and finally for $\epsilon \to 0$ we get $p(\omega;x) - p(\sigma;x) \geq 0$ outside $[A,B]$. But the left-hand side is harmonic in $\overline{\mathbb{C}} \setminus [A,B]$ and vanishes at infinity, so therefore the two potentials $p(\sigma;\cdot)$ and $p(\omega;\cdot)$ coincide outside $[A,B]$. But then they coincide everywhere and we get $\omega = \sigma$. □

Before we can turn to the actual construction of the measure μ we need one more lemma.

Lemma 2.1.7. *If w is a positive continuous weight on $[a,b]$ then*

$$\sup_{\substack{P_n \not\equiv 0 \\ \deg P_n = n}} \left(\|P_n w^n\|_{L^\infty[a,b]} / \|P_n w^n\|_{L^2[a,b]}\right)^{1/n} \to 1$$

as $n \to \infty$.

Proof of Lemma 2.1.7. The proof is an easy consequence of the well-known Nikolskii inequality [Ti, 4.9.6] according to which

$$\|P_n\|_{L^\infty[\alpha,\beta]} \leq 3(\beta-\alpha)^{-1/2} n \|P_n\|_{L^2[\alpha,\beta]}, \quad \deg P_n = n.$$

In fact, if $\epsilon > 0$ is given, then using the continuity of w we can decompose $[a,b]$ into finitely many intervals I_k, $1 \leq k \leq l$, such that $w(x)/w(y) \leq 1+\epsilon$ if x and y belong to the same interval I_k. Let $\tau_k \in I_k$ be a fixed point in I_k. Since

$$\begin{aligned} \|P_n w^n\|_{L^\infty[I_k]} &\leq (1+\epsilon)^n w(\tau_k)^n \|P_n\|_{L^\infty[I_k]} \\ &\leq (1+\epsilon)^n w(\tau_k)^n 3|I_k|^{-1/2} n \|P_n\|_{L^2[I_k]} \\ &\leq 3(1+\epsilon)^{2n} |I_k|^{-1/2} n \|P_n w^n\|_{L^2[a,b]}, \end{aligned}$$

we get the statement. □

After these preliminaries let μ_0 be the Lebesgue measure on $[-1,1]$ and

$$\mu_n = \frac{1}{n^2 N_n} \sum_{k=1}^{N_n} v(x_k^{(n)})^{2n} \delta_{x_k^{(n)}},$$

where $\{x_k^{(n)}\}_{k=1}^{N_n}$ are the left endpoints of the intervals obtained by dividing $[-2,2]$ into N_n equal parts, and where N_n are chosen in such a way that for polynomials P_n of degree at most n we have

$$\text{(1.18)} \qquad \frac{1}{2}\int P_n^2 d\mu_n \leq n^{-2} \int_{-2}^{2} P_n^2(x) v^{2n}(x) dx \leq 2 \int P_n^2 d\mu_n.$$

We show that with an appropriate sequence $\{n_k\}$ the measure $\mu = \mu_0 + \sum_{k=1}^{\infty} \mu_{n_k}$ satisfies the requirements.

$c_\mu \geq \frac{1}{2}$ is obvious because $\mu \geq \mu_0$. Let $m_k = \sum_{s=1}^{k-1} N_{n_s}$, $L_k = n_k - m_k$, and choose $\{n_k\}$ so that m_k/n_k tends to zero as k tends to infinity. With

$$P_{n_k}(x) = Q_{L_k}(x) \prod_{1\leq t\leq N_{n_s}, 1\leq s\leq k-1} (x - x_t^{(n_s)}),$$

where Q_{L_k} are the polynomials from (a), we have $\deg(P_{n_k}) = n_k$ and

$$\int P_{n_k}^2 d\mu_{n_s} = 0, \quad s = 1, 2, \ldots, k-1.$$

When $s \geq k$ then

$$\begin{aligned} \int P_{n_k}^2 d\mu_{n_s} &\leq 2n_s^{-2} \int_{-2}^{2} P_{n_k}^2(x) v^{2n_s}(x) dx \\ &\leq 2n_s^{-2} \int_{-2}^{2} P_{n_k}^2(x) v^{2L_k}(x) dx \\ &\leq 2 \cdot 4^{m_k} n_s^{-2} \|Q_{L_k} v^{L_k}\|_{L^\infty[-2,2]}^2 \end{aligned}$$

and

$$\begin{aligned} \int P_{n_k}^2 d\mu_0 &= \int_{-1}^{1} P_{n_k}^2(x) dx \\ &= \int_{-1}^{1} P_{n_k}^2(x) v^{2n_k}(x) dx \leq 4^{m_k} \|Q_{L_k} v^{L_k}\|_{L^\infty[-2,2]}^2, \end{aligned}$$

from which we have

$$\int P_{n_k}^2 d\mu \leq 4^{m_k+1} \|Q_{L_k} v^{L_k}\|_{L^\infty[-2;2]}^2 \leq 4^{m_k+7} 2^{-L_k}.$$

But

$$\int q_{n_k}^2(\mu; x) d\mu(x) \leq \int P_{n_k}^2 d\mu,$$

and from (1.18) and Lemma 2.1.7 we have for any $C > 1$ and large k

$$\|q_{n_k}(\mu; \cdot) v^{n_k}\|_{L^\infty[-2,2]}^2 \leq C^{2n_k} \int q_{n_k}^2(\mu; x) v^{2n_k}(x) dx$$

$$\leq 2 \cdot C^{2n_k} n_k^2 \int q_{n_k}^2(\mu; x) d\mu_{n_k}(x) \leq 2 \cdot C^{2n_k} n_k^2 \int q_{n_k}^2(\mu; x) d\mu(x),$$

which, together with the previous estimates and $L_k/n_k \to 1$, yields

$$\limsup_{k\to\infty} \|q_{n_k}(\mu; \cdot) v^{n_k}\|_{L^\infty[-2,2]}^{1/n_k} \leq 2^{-1/2}.$$

Thus, $\nu_{p_{n_k}(\mu;\cdot)} \xrightarrow{*} \kappa$ as $k \to \infty$ follows from proposition (b). □

2.2 Norm Asymptotics and Zero Distribution

Since $1/\gamma_n(\mu)$ equals the $L^2(\mu)$ norm of the monic orthogonal polynomial $p_n(\mu;z)/\gamma_n(\mu)$, asymptotics for $\{\gamma_n^{1/n}(\mu)\}$ are often referred to as *norm asymptotics* for orthogonal polynomials. As we shall see they are closely related to zero asymptotics.

To avoid pathological cases (cf. Theorem 2.1.4 and Examples 1.5.1, 3.5.2, and 2.2.5), in this section we will assume that $c_\mu > 0$ and $\mathrm{Int}(\mathrm{Pc}(S(\mu))) = \emptyset$, that is, that μ does not have a carrier of zero capacity and $S(\mu) = \mathrm{supp}(\mu)$ has connected complement and empty interior. Recall also that $\nu_{p_n(\mu,\cdot)}$ denotes the normalized counting measure on the zeros of $p_n(\mu,\cdot)$.

Theorem 2.2.1. *We assume that* $c_\mu > 0$ *and* $\mathrm{Int}(\mathrm{Pc}(S(\mu))) = \emptyset$.
(a) *The existence of the weak* limit*

$$\frac{1}{n}\nu_{p_n(\mu;\cdot)} \to \nu \quad \text{as} \quad n \to \infty,\ n \in N, \tag{2.1}$$

for an infinite subsequence $N \subseteq \mathbb{N}$ *implies the existence of the limit*

$$\lim_{n\to\infty,\ n\in N} \gamma_n(\mu)^{1/n}. \tag{2.2}$$

Furthermore, if for each of two infinite subsequences $N_1,\ N_2 \subseteq \mathbb{N}$ *the limit* (2.1) *exists, and the two limits are identical, then also the two corresponding limits* (2.2) *are equal.*
(b) *If*

$$\lim_{\substack{n\to\infty\\ n\in N}} \gamma_n(\mu)^{1/n} = \frac{1}{\mathrm{cap}(S(\mu))} \tag{2.3}$$

for some subsequence $N \subseteq \mathbb{N}$, *then we also have*

$$\frac{1}{n}\nu_{p_n(\mu;\cdot)} \to \omega_{S(\mu)} \quad \text{as} \quad n \to \infty,\ n \in N \tag{2.4}$$

in the weak sense.*
(c) *If*

$$\lim_{\substack{n\to\infty\\ n\in N}} \gamma_n(\mu)^{1/n} = \frac{1}{c_\mu}, \tag{2.5}$$

then

$$\frac{1}{n}\nu_{p_n(\mu;\cdot)} \to \omega_\mu \quad \text{as} \quad n \to \infty,\ n \in N \tag{2.6}$$

in the weak sense.*

Recall that $\omega_{S(\mu)}$ and ω_μ are the equilibrium measure of $S(\mu)$ and the minimal-carrier equilibrium distribution associated with μ (see Definition 1.2.5), respectively.

Remarks. (1) Note that (see Corollary 1.1.7) $1/\text{cap}(S(\mu))$ and $1/c_\mu$ are the smallest and largest possible limit points of the sequence $\{\gamma_n(\mu)^{1/n}\}$, thus (b) and (c) provide the converse of (a) in these two extreme cases. In Example 2.2.7 we will show that nothing similar can be true in between $1/\text{cap}(S(\mu))$ and $1/c_\mu$.

(2) Both assumptions $c_\mu > 0$ and $\text{Int}(\text{Pc}(S(\mu))) = \emptyset$ are really necessary. The necessity of $c_\mu > 0$ will be shown in Example 3.5.2. For the necessity of $\text{Int}(\text{Pc}(S(\mu))) = \emptyset$ see Example 2.2.5.

(3) Part (a) of the theorem is true together with its proof if we merely assume $S(\mu) = \partial\Omega$, that is, that the support of μ lies on the boundary of its polynomial convex hull.

Next we list some corollaries of Theorem 2.2.1. In each of them we keep the assumptions

(2.7) $$c_\mu > 0, \quad \text{Int}(\text{Pc}(S(\mu))) = \emptyset.$$

Corollary 2.2.2. *With assumption* (2.7) *the existence of the limit distribution of zeros implies norm asymptotics* (*i.e., the existence of the limit of* $\{\gamma_n(\mu)^{1/n}\}$) *for the orthonormal polynomials.*

Corollary 2.2.3. *Assuming* (2.7), *the relations*

(2.8) $$\lim_{n\to\infty} \gamma_n(\mu)^{1/n} = \frac{1}{\text{cap}(S(\mu))}$$

and

(2.9) $$\lim_{n\to\infty} \frac{1}{n}\nu_{p_n(\mu;\cdot)} = \omega_{S(\mu)}$$

are equivalent.

Corollary 2.2.4. *Assuming* (2.7), *the relations*

$$\lim_{n\to\infty} \gamma_n(\mu)^{1/n} = \frac{1}{c_\mu}$$

and

$$\lim_{n\to\infty} \frac{1}{n}\nu_{p_n(\mu;\cdot)} = \omega_\mu$$

are equivalent.

Proof of Theorem 2.2.1. (a) Let us assume that for an infinite subsequence $N \subseteq \mathbb{N}$ the limit (2.1) exists, but the limit (2.2) does not. From

Corollary 1.1.7 we know that the sequence in (2.2) is bounded, and therefore we can select two infinite subsequences $N_1,\ N_2 \subseteq N$ so that the two limits

$$\lim_{n\to\infty,\ n\in N_j} \gamma_n(\mu)^{1/n} = e^{c_j}, \quad j=1,2, \tag{2.10}$$

exist and $c_1 \neq c_2$. If we apply, as in (1.3.27), the lower envelope theorem of potential theory (Appendix III) to the two sequences

$$\{(1/n)\log|p_n(\mu;\cdot)| \mid n \in \mathbb{N}_j\},\ j=1,2,$$

we derive from the upper and lower asymptotic bounds (1.1.6) and (1.1.10) of Theorem 1.1.4 that

$$0 \leq -p(\nu;z) + c_j \leq g_\mu(z;\infty), \quad j=1,2, \tag{2.11}$$

holds true qu.e. on $\partial\Omega$. From (2.11) and Lemma 1.2.3 it follows that for every minimal carrier $C \in \Gamma_0(\mu)$ we have

$$-p(\nu;z) + c_j = 0 \qquad \text{qu.e. on } C. \tag{2.12}$$

Since $\operatorname{cap}(C) = c_\mu > 0$, we get from (2.12) that $c_1 = c_2$, which disproves our assumption that the limit (2.2) does not exist.

If for two infinite subsequences $N_1,\ N_2 \subseteq \mathbb{N}$ the two limits (2.1) exist and are identical, then we can merge these sequences, and for the new sequence we have convergence again. Therefore, the limit (2.2) also exists for the merged sequence, which proves that the two original limits (2.2) have to be equal. Thus part (a) is proved.

(b) Let us now assume (2.3). It has been proved in the beginning of the proof of Theorem 1.1.9 that (2.3) implies

$$\lim_{n\to\infty, n\in N} |p_n(\mu;z)|^{1/n} = e^{g_\Omega(z;\infty)}$$

for $z \in \mathbb{C} \setminus \operatorname{Co}(S(\mu))$, where $\operatorname{Co}(S(\mu))$ denotes the convex hull of $S(\mu)$. Thus, if σ is any weak* limit point of $\{\frac{1}{n}\nu_{p_n(\mu;\cdot)}\}_{n\in N}$ then for its potential $p(\sigma;z)$ in $\mathbb{C} \setminus \operatorname{Co}(S(\mu))$ we have

$$p(\sigma;z) = -g_\Omega(z;\infty) + \log\frac{1}{\operatorname{cap}(S(\mu))} = p(\omega_{S(\mu)};z). \tag{2.13}$$

Here σ is supported on $S(\mu)$ (see Theorem 2.1.1), hence (2.13) holds true in Ω, as well, because the two sides are harmonic in Ω. Since both sides are continuous in the fine topology and the boundary of Ω in this topology contains $\partial\Omega = S(\mu)$, (2.13) extends to all z and so $\sigma = \omega_{S(\mu)}$ by the principle of domination and the one-to-one correspondence between measures and their potentials. This proves (2.4) because σ was an arbitrary weak* limit point of $\{\frac{1}{n}\nu_{p_n(\mu;\cdot)}\}_{n\in N}$.

(c) Exactly as in the proof of (b), condition (2.5) implies

$$\lim_{\substack{n\to\infty \\ n\in N}} |p_n(\mu;z)|^{1/n} = e^{g_\mu(z;\infty)},$$

and, on applying

$$g_\mu(z;\infty) = -p(\omega_{S(\mu)};z) + \log\frac{1}{c_\mu}$$

(see (1.2.11)), we can complete the proof as before. □

Corollary 2.2.2 is an immediate consequence of Theorem 2.2.1.

Proof of Corollary 2.2.3. We only have to show that (2.9) implies (2.8). If we assume (2.9), then part (a) of Theorem 2.2.1 ensures the existence of the limit

$$\lim_{n\to\infty} \gamma_n(\mu)^{1/n} = c,$$

and we have to prove $c = 1/\operatorname{cap}(S(\mu))$. Now with the lower envelope theorem of potential theory (Appendix III) it follows from these assumptions that

$$\limsup_{n\to\infty} \frac{1}{n}\log|p_n(\mu;z)| = -p(\omega_{S(\mu)};z) + \log c \tag{2.14}$$

qu.e. in $\mathbb{C}$. If we compare (2.14) with the upper asymptotic bound (1.1.6) of Theorem 1.1.4, then we derive with representation (A.10) in Appendix V of the Green function $g_\Omega(z;\infty)$ the inequality

$$g_\Omega(z;\infty) + \log\operatorname{cap}(S(\mu)) + \log c \le g_\mu(z;\infty) \qquad \text{qu.e. on } \mathbb{C}. \tag{2.15}$$

From Corollary 1.1.7 we know that $\log c \ge -\log\operatorname{cap}(S(\mu))$. It follows from Lemma 1.2.3 that $c_\mu > 0$ implies $g_\mu(z;\infty) = 0$ on a set of positive capacity, hence in (2.15) we must have $\log c = -\log\operatorname{cap}(S(\mu))$. □

Proof of Corollary 2.2.4. As in the preceding proof we only have to show that

$$\lim_{n\to\infty}\frac{1}{n}\nu_{p_n(\mu;\cdot)} = \omega_\mu, \qquad \lim_{n\to\infty}\gamma_n(\mu)^{1/n} = c$$

imply $c = 1/c_\mu$. Instead of (2.14) now we have

$$\limsup_{n\to\infty}\frac{1}{n}\log|p_n(\mu;z)| = -p(\omega_\mu;z) + \log c,$$

and taking into account (1.1.10) and the representation (1.2.11) we get for qu.e. $z \in \partial\Omega = S(\mu)$

$$0 \le -p(\omega_\mu;z) + \log c = g_\mu(z;\infty) + \log c_\mu + \log c,$$

and so $c_\mu > 0$ implies again $c \geq 1/c_\mu$. Finally, here the equality must hold because of Corollary 1.1.7. □

We promised the following example in Remark 2 after Theorem 2.2.1.

Example 2.2.5. We construct a measure μ with $c_\mu > 0$ such that for some $N \subseteq \mathbb{N}$

$$\lim_{\substack{n\to\infty \\ n\in N}} \frac{1}{n}\nu_{p_n(\mu;\cdot)} \tag{2.16}$$

exists, but $\{\gamma_n(\mu)^{1/n}\}_{n\in N}$ does not converge. Of course, for this μ we must have by Theorem 2.2.1(a) that $\mathrm{Int}(\mathrm{Pc}(S(\mu))) \neq \emptyset$.

The construction is based on the following lemma.

Lemma 2.2.6. *Let $q_n(z) = z^n + \cdots$ be any monic polynomial of degree n and let*

$$A_L(q_n) = A_L := \{z \mid |q_n(z)| = L\}$$

be a level set of q_n. Then $p_n(\omega_{A_L}; z)/\gamma_n(\omega_{A_L}) \equiv q_n(z)$, that is, the nth monic orthogonal polynomial with respect to the equilibrium distribution of A_L is exactly q_n.

Proof of Lemma 2.2.6. It is known that $\mathrm{cap}(A_L) = L^{1/n}$ (see [La, Chapter II, Section 17, p. 173]) and so (see [Ts, Theorem III.12])

$$p(\omega_{A_L}; z) \leq -\log L^{1/n}, \quad z \in \mathbb{C}.$$

Now if $R_n(z) = (z - z_1)\cdots(z - z_n)$ is any monic polynomial of degree n then from the inequality between the geometric and quadratic means we have

$$\begin{aligned} \|R_n\|_{L^2(\omega_{A_L})} &\geq \exp\left(\int \log|R_n(z)| d\omega_{A_L}(z)\right) \\ &= \exp\left(\sum_{j=1}^{n} -p(\omega_{A_L}; z_j)\right) \geq (L^{1/n})^n = L, \end{aligned}$$

and since for q_n we have

$$\|q_n\|_{L^2(\omega_{A_L})} = \left(\int L^2 d\omega_{A_L}\right)^{1/2} = L,$$

our claim follows from the extremality property and uniqueness of monic orthogonal polynomials with respect to minimal L^2 norm. □

Returning now to Example 2.2.5 we set

$$\mu = \mu_0 + \sum_{k=1}^{\infty} \mu_k, \tag{2.17}$$

where μ_0 is the Lebesgue measure on $[-1,1]$ and μ_k, $k \geq 1$, will be some discrete measures on some level sets of certain polynomials. In any case $c_\mu \geq c_{\mu_0} = \frac{1}{2}$.

We set

$$\mu^{(m)} = \mu_0 + \sum_{k=1}^{m} \mu_k.$$

Each μ_k, $k \geq 1$, will have finite support; therefore, for $\mu^{(m)}$ the lower and upper bounds in (1.1.13) are identical to 2, and on applying Corollary 2.2.3 we can conclude that the limit zero distribution of the orthogonal polynomials corresponding to $\mu^{(m)}$ coincides with $\omega_{[-1,1]}$. Using the metric d introduced in the proof of Theorem 2.1.4 for weak* topology, we can thus guarantee a large number N_m, which is then fixed in the rest of the proof, such that

$$d\left(\frac{1}{N_m}\nu_{p_{N_m}(\mu^{(m)};\cdot)};\omega_{[-1;1]}\right) < \frac{1}{m}, \tag{2.18}$$

and at the same time

$$\gamma_{N_m}(\mu^{(m)})^{1/N_m} < 3. \tag{2.19}$$

In (2.17) we require that each μ_k has norm $\|\mu_k\| < \varepsilon_k$, where the positive sequence $\{\varepsilon_k\}$ will also be defined during our construction. If $\varepsilon_{k+1} < \varepsilon_k/2$ is satisfied, then we have in any case

$$\left\| \sum_{k=m+1}^{\infty} \mu_k \right\| < 2\varepsilon_{m+1}.$$

Therefore if ε_{m+1} is small enough, then independently of the actual choice of $\mu_{m+1}, \mu_{m+2}, \ldots$ (2.18) and (2.19) imply

$$d\left(\frac{1}{N_m}\nu_{p_{N_m}(\mu;\cdot)};\omega_{[-1;1]}\right) < \frac{1}{m} \tag{2.20}$$

and

$$\gamma_{N_m}(\mu)^{1/N_m} < 3. \tag{2.21}$$

Fix such an $\varepsilon_{m+1} > 0$ (which we can also choose smaller than $\varepsilon_m/2$ provided ε_m has already been chosen).

For large n the level set of $|p_n(\mu^{(m)};\cdot)|$ corresponding to $L = 10^n$ is contained in the disk $D_{10} := \{z||z| \le 10\}$ because outside $\text{supp}(\mu^{(m)})$ we have locally uniformly

$$\lim_{n\to\infty} |p_n(\mu^{(m)};\cdot)|^{1/n} = e^{g_{\mathbf{C}\setminus[-1,1]}(z;\infty)}$$

and

$$g_{\mathbf{C}\setminus[-1,1]}(z;\infty) = \log|z + \sqrt{z^2-1}|.$$

Now set

$$q_n(z) \equiv q_{n,m}(z) = p_n(\mu^{(m)};z)/\gamma_n(\mu^{(m)})$$

and $L = L(m,n) = 10^n/\gamma_n(\mu^{(m)})$ in Lemma 2.2.6. This lemma then asserts that with

$$\omega_{n,m} := \omega_{A_{L(m,n)}} = \omega_{A_{L(m,n)}(q_{n,m})}$$

the nth monic orthogonal polynomial corresponding to $\omega_{n,m}$ is $q_{n,m}$. Since the same was true for the measure $\mu^{(m)}$, we can conclude that the monic orthogonal polynomial with respect to $\mu^{(m)} + \varepsilon_{m+1}\omega_{n,m}$ is again $q_{n,m}$, and so for large n

$$\begin{aligned}(2.22)\qquad & 1/\gamma_n^2(\mu^{(m)} + \varepsilon_{m+1}\omega_{n,m}) \\ &= \int |q_{n,m}(z)|^2 d\mu^{(m)}(z) + \varepsilon_{m+1}\int |q_{n,m}(z)|^2 d\omega_{n,m}(z) \\ &= (1+\varepsilon_{m+1}10^{2n})/\gamma_n(\mu^{(m)})^2 \le 2\varepsilon_{m+1}(10^n/\gamma_n(\mu^{(m)}))^2,\end{aligned}$$

because $|p_n(\mu^{(m)};z)| = 10^n$ on the support of $\omega_{n,m}$. Thus, we can choose a large number $n = M_m$ such that

$$(2.23)\qquad \gamma_{M_m}(\mu^{(m)} + \varepsilon_{m+1}\omega_{M_m,m}) > 4^{M_m}$$

$$\begin{aligned}(2.24)\qquad & d\left(\frac{1}{M_m}\nu_{p_{M_m}(\mu^{(m)}+\varepsilon_{m+1}\omega_{M_m,m};\cdot)};\omega_{[-1,1]}\right) \\ &= d\left(\frac{1}{M_m}\nu_{p_{M_m}(\mu^{(m)};\cdot)};\omega_{[-1;1]}\right) < \frac{1}{m}\end{aligned}$$

(for (2.23) consider (2.22) and

$$\lim_{n\to\infty}\gamma_n(\mu^{(m)})^{1/n} = 2$$

established previously). Fix an $M_m > N_m$ with properties (2.23) and (2.24). If μ_{m+1} is a sufficiently good discrete approximation with finite support in D_{10} of the measure $\varepsilon_{m+1}\omega_{M_m}$ such that $\|\mu_{m+1}\| = \varepsilon_{m+1}$, then we have with (2.23)–(2.24)

$$(2.25)\qquad \gamma_{M_m}(\mu^{(m)} + \mu_{m+1})^{1/M_m} = \gamma_{M_m}(\mu^{(m+1)})^{1/M_m} > 4,$$

(2.26) $$d\left(\frac{1}{M_m}\nu_{p_{M_m}(\mu^{(m+1)};\cdot)};\omega_{[-1,1]}\right)<\frac{1}{m}.$$

Now if ε_{m+2} is sufficiently small (compared to $1/M_m$, which has already been fixed) then independently of the actual choice of $\mu_{m+2},\mu_{m+3},\ldots$ we will have from (2.25) and (2.26)

(2.27) $$\gamma_{M_m}(\mu)^{1/M_m}>4,$$

and

(2.28) $$d\left(\frac{1}{M_m}\nu_{p_{M_m}(\mu;\cdot)};\omega_{[-1,1]}\right)<\frac{1}{m}.$$

This finishes the induction step during which we choose the numbers $\{\varepsilon_m\}$, $\{N_m\},\{M_m\}$ and the measures $\{\mu_m\}$ in the order $\ldots$, N_m, ε_{m+1}, M_m, μ_{m+1}, N_{m+1}, ε_{m+2}, $\ldots$ depending on the terms that have already been chosen. The construction yields a μ with $\operatorname{supp}(\mu)\subseteq D_{10}$, $c_\mu\geq 1/2$, and properties (2.20), (2.21), (2.27), and (2.28). Setting N to be the sequence with terms N_m and M_m, $m=1,2,\ldots$, we can see that the limit (2.16) exists ($=\omega_{[-1,1]}$) but $\{\gamma_n(\mu)^{1/n}\}_{n\in N}$ does not have a limit. □

Our final result shows that, contrary to Corollaries 2.2.3 and 2.2.4, norm asymptotics, in general, does not imply zero asymptotics even if $c_\mu>0$ and $\operatorname{Int}(\operatorname{Pc}(S(\mu)))=\emptyset$.

Example 2.2.7. We construct a weight measure μ with support on $[0,3]$ such that $c_\mu\geq\frac{1}{4}$ and

(2.29) $$\lim_{n\to\infty}\gamma_n(\mu)^{1/n}=2,$$

but every equilibrium measure $\omega_{[\tau,\tau+2]}$, $\tau\in[0,1]$, is a weak* limit point of the zeros of orthogonal polynomials corresponding to μ.

Let μ_0 be the Lebesgue measure on [1,2], and with a sequence $\{a_n\}\subseteq[0,1]$ to be chosen subsequently let μ_n be the measure that assigns mass n^{-5} to each of the points of the set $S_n=\{a_n+2tn^{-3}\}_{t=1}^{n^3}$. Choose $\{a_n\}\subset[0,1]$ in the following way: For an n and $\varepsilon>0$ let $n_\varepsilon^-\leq n\leq n_\varepsilon^+$ be the smallest and the largest integers, respectively, with the property that for $n_\varepsilon^-\leq k\leq n_\varepsilon^+$ we have $|a_n-a_k|\leq\varepsilon$. Now it is easy to construct a sequence $\{a_n\}\subset[0,1]$ in such a way that each point of $[0,1]$ is a limit point of $\{a_n\}$ and for every $\varepsilon>0$

$$(n_\varepsilon^-)^4/n\to 0\quad\text{and}\quad 6^{2n}/n_\varepsilon^+\to 0$$

hold as $n\to\infty$.

Let

$$T_n(x;a,b)=(b-a)^n2^{-2n+1}\cos\left(n\cdot\arccos\left(\frac{2(x-a)}{b-a}-1\right)\right)$$

be the classical Chebyshev polynomials on $[a, b]$ with leading coefficient 1. The estimates

$$|T_n(x; a, b)| \le (b-a)^n 2^{-2n+1}, \quad x \in [a, b], \tag{2.30}$$

and

$$|T_n(x; a, b)| \le 3^n, \quad x \in [0, 3], \ a, b \in [0, 3], \tag{2.31}$$

follow immediately from the definitions.

Now for an n and $\varepsilon > 0$ set $m_n = n - |\cup_{k<n_\varepsilon^-} S_k|$, $\alpha_n = \max(a_n - \varepsilon, 0)$, $\beta_n = \min(a_n + 2 + \varepsilon, 3)$, and

$$P_n(x) = T_{m_n}(x; \alpha_n, \beta_n) \prod_{\tau \in \cup_{k<n_\varepsilon^-} S_k} (x - \tau).$$

Since $|\cup_{k<n_\varepsilon^-} S_k| \le (n_\varepsilon^-)^4$, we have $m_n \ge n - (n_\varepsilon^-)^4$, $\deg(P_n) = n$, and P_n has leading coefficient 1. For $x \in [a_n - \varepsilon, a_n + 2 + \varepsilon] \cap [0, 3]$ we get from (2.30) that

$$|P_n(x)| \le (1+\varepsilon)^{m_n} 2^{-m_n+1} 3^{(n_\varepsilon^-)^4},$$

and this combined with (2.31) and the fact that P_n vanishes on the support of μ_k for $1 \le k < n_\varepsilon^-$ yields

$$\begin{aligned}
\gamma_n^{-2}(\mu) &\le \int P_n^2 d\mu = \int P_n^2 d\mu_0 + \sum_{k=n_\varepsilon^-}^{n_\varepsilon^+} \int P_n^2 d\mu_k + \sum_{k=n_\varepsilon^+ + 1}^{\infty} \int P_n^2 d\mu_k \\
&\le (1+\varepsilon)^{2m_n} 2^{-2m_n+2} 3^{2(n_\varepsilon^-)^4} \left(1 + \sum_{k=n_\varepsilon^-}^{n_\varepsilon^+} k^{-2}\right) + 3^{2n} \sum_{k=n_\varepsilon^+ + 1}^{\infty} k^{-2} \\
&\le 4(1+\varepsilon)^{2m_n} 2^{-2m_n+2} 3^{2(n_\varepsilon^-)^4} + 3^{2n}/n_\varepsilon^+ \le \exp(o(n))(1+\varepsilon)^{2n} 2^{-2n}
\end{aligned}$$

from which

$$\liminf_{n\to\infty} \gamma_n^{1/n}(\mu) \ge 2 \tag{2.32}$$

follows.

On the other hand, we easily get from Markov's inequality

$$\|q_n'(\mu; \cdot)\|_{L^\infty[a_n, a_n+2]} \le n^2 \|q_n(\mu; \cdot)\|_{L^\infty[a_n, a_n+2]}$$

that

$$\begin{aligned}
\gamma_n^{-2} &= \int q_n^2(\mu; \cdot) d\mu \ge \int q_n^2(\mu; \cdot) d\mu_n \ge n^{-5} \max_{x \in S_n} q_n^2(\mu; x) \\
&\ge \tfrac{1}{2} n^{-5} \|q_n(\mu)\|^2_{L^\infty[a_n, a_n+2]} \ge 2^{-2n} n^{-5},
\end{aligned}$$

where, in the last step, we used the well-known fact that the supremum norm of any polynomial of degree n with leading coefficient 1 over an interval of length 2 is at least as large as 2^{-n+1}. This together with (2.32) proves (2.29).

Now let $\tau \in [0,1]$ be arbitrary. There is a sequence $\{a_{n_k}\}_{k=1}^{\infty}$ converging to τ. Before, we verified that

$$\lim_{k\to\infty} \|q_{n_k}(\mu;\cdot)\|_{L^\infty[a_{n_k},a_{n_k}+2]}^{1/n_k} = \tfrac{1}{2},$$

which combined with the fact that

$$\limsup_{k\to\infty} \|q_{n_k}(\mu;\cdot)\|_{L^\infty[\tau,\tau+2]}^{1/n_k} \le \limsup_{k\to\infty} \|q_{n_k}(\mu;\cdot)\|_{L^\infty[a_{n_k},a_{n_k}+2]}^{1/n_k}$$

(this immediately follows from the well-known Bernstein–Walsh lemma; see [Wa, p. 77] or (A.31) in Appendix IX for $w \equiv 1$) yields

$$\limsup_{k\to\infty} \|q_{n_k}(\mu;\cdot)\|_{L^\infty[\tau,\tau+2]}^{1/n_k} \le \tfrac{1}{2}.$$

Now

$$\lim_{k\to\infty} \nu_{p_{n_k}(\mu\cdot)} = \kappa_{[\tau,\tau+2]}$$

is an immediate consequence of this and Lemma 2.1.6 (applied to $\sigma = \omega_{[\tau,\tau+2]}$, $[a,b] = [\tau,\tau+2]$). □

2.3 Asymptotic Behavior of Zeros when $c_\mu > 0$

Now we return to the question of characterizing the sets

$$M_\mu = \{\nu \mid \nu \text{ is a weak* limit point of the measures } \frac{1}{n}\nu_{p_n(\mu;\cdot)}\}.$$

For arbitrary weight measure μ with support in $[0,1]$ this was done in Theorem 2.1.3, but the question for $c_\mu > 0$ was left open (note that different μs may have the same M_μ classes).

In the sequel we will encounter potentials $p(\nu;\cdot)$ that are bounded above and at some point they take on their supremum. We shall write MAX$p(\mu;\cdot)$ for the *set of maximum points of* $p(\nu;\cdot)$ with the agreement that MAX$p(\nu;\cdot)$ is empty if the potential is not bounded above or if its supremum is not attained. With this concept we prove the following theorem.

Theorem 2.3.1. *Let $S(\mu) \subseteq [0,1]$. If $c_\mu > 0$ and C is a minimal carrier of μ then any weak* limit ν of the zero distributions $\{\frac{1}{n}\nu_{p_n(\mu;\cdot)}\}$ satisfies C "$\subseteq$"MAX$p(\nu;\cdot)$ and supp$(\nu) \subseteq \overline{C}$, where "$\subseteq$" means inclusion except for a set of zero capacity. Conversely, if $C \subseteq [0,1]$ is of positive capacity and*

$\mathcal{M}_C$ is the set of probability measures ν satisfying C"$\subseteq$"MAX$p(\nu;\cdot)$ and $\text{supp}(\nu) \subseteq \overline{C}$, then there is a measure μ such that C is a minimal carrier of μ and $\mathcal{M}_\mu = \mathcal{M}_C$.

This theorem says that all possible weak* limits can occur among the limit distributions of the zeros of orthogonal polynomials even in the case $c_\mu > 0$.

Proof of Theorem 2.3.1. Let $c_\mu > 0$ and $\nu \in \mathcal{M}_\mu$. $\text{supp}(\nu) \subseteq \text{supp}(\mu) \subseteq \overline{C}$ follows from the well-known fact that in any interval contiguous to $\text{supp}(\mu)$ each of the polynomials $p_n(\mu;\cdot)$ can have at most one zero (otherwise there would be a polynomial of degree at most $n-1$ having the same sign as $p_n(\mu;\cdot)$ on $\text{supp}(\mu)$, contradicting orthogonality). Furthermore, if the subsequence $\{\frac{1}{n}\nu_{p_n(\mu;\cdot)}\}_{n\in N}$ converges to ν then $\{\gamma_{n(\mu;\cdot)}^{1/n}\}_{n\in N}$ also converges to some number α (see Theorem 2.2.1.(a)) and by the lower envelope theorem (see Appendix III) $p(\nu;x) - \log\alpha$ coincides quasi everywhere with the lim inf of the sequence

$$\left\{\frac{1}{n}\log(1/|p_n(\mu;x)|)\right\}_{n\in N}$$

But

$$\limsup_{\substack{n\to\infty \\ n\in N}} |p_n(\mu;x)|^{1/n} = 1$$

quasi everywhere on C (see (1.1.6), (1.1.10) of Theorem 1.1.4, and Lemma 1.2.3) and

$$\limsup_{\substack{n\to\infty \\ n\in N}} |p_n(\mu;x)|^{1/n} \geq 1$$

quasi everywhere (see Theorem 1.1.4), thus the potential $p(\nu;\cdot)$ satisfies $p(\nu;x) \leq \log\alpha$ quasi everywhere, and equality holds for quasi every $x \in C$. From the lower semicontinuity of $p(\nu;\cdot)$ we finally obtain $p(\nu;x) \leq \log\alpha$ everywhere, and C "$\subseteq$"MAX$p(\nu;\cdot)$ has been verified.

The second half of the theorem requires a more sophisticated argument and we divide the construction into several steps.

(I) Choose a dense sequence $\{\nu_k\}_{k=1}^\infty$ in $\mathcal{M}_C$. Then there is a $C_0 \subseteq C$ such that $\text{cap}(C_0) = \text{cap}(C)$ and $C_0 \subseteq$ MAX$p(\nu_k;\cdot)$ for all k. C_0 is not necessarily compact, but we can choose an increasing sequence $C_1 \subseteq C_2 \subseteq \cdots$ of compact subsets of C_0 such that $\text{cap}(C_n) \to \text{cap}(C)$ as n tends to infinity.

(II) Fix a k. By [La, Theorem 1.2.7] there is an increasing sequence $\{\nu_k^i\}_{i=1}^\infty$ of measures (with total mass at most one) such that the support of ν_k^i is contained in the support of ν_k, $p(\nu_k^i;\cdot)$ is continuous, $\nu_k^i \xrightarrow{*} \nu_k$ as $i \to \infty$, and for all x we have

$$p(\nu_k^i;x) \to p(\nu_k;x), \quad i \to \infty. \tag{3.1}$$

Since then $\{p(\nu_k^i;\cdot)\}_{i=1}^{\infty}$ must be increasing on $[0,1]$, and on the compact set C_k it converges to the constant $\max p(\nu_k;\cdot)$, on C_k we have uniform convergence in (3.1). These imply that there is a probability measure $\sigma_k (= \nu_k^i/\|\nu_k^i\|$ for sufficiently large i) such that

(i) $d(\sigma_k;\nu_k) < 1/k$,
(ii) $\mathrm{supp}(\sigma_k) \subseteq \mathrm{supp}(\nu_k) \subseteq \overline{C}$,
(iii) $p(\sigma_k;\cdot)$ is continuous,
(iv) $\max p(\sigma_k;\cdot) - \frac{1}{k} \le p(\sigma_k;x) \le \max p(\sigma_k;\cdot)$ for all $x \in C_k$.

In (i) d is the metric for the weak* topology introduced in the proof of Theorem 2.1.4.

(III) We set

$$w_k(x) = \exp(p(\sigma_k;x) - \max p(\sigma_k;\cdot)).$$

Then w_k is continuous,

$$\exp(-\max p(\sigma_k;\cdot)) \le w_k(x) \le 1$$

for $x \in [0,1]$, and

$$e^{-1/k} \le w_k(x) \le 1$$

for $x \in C_k$. For any polynomial P_n of degree at most n we have

$$\|P_n w_k^n\|_{[0,1]} = \|P_n w_k^n\|_{\mathrm{supp}(\sigma_k)}. \tag{3.2}$$

In fact, if M is the right-hand side, then

$$\frac{1}{n}\log\frac{1}{|P_n(x)|} \ge p(\sigma_k;x) - \max p(\sigma_k;\cdot) - (\log M)/n \tag{3.3}$$

for all $x \in \mathrm{supp}(\sigma_k)$, and since σ_k has finite logarithmic energy, the principle of domination implies that (3.3) holds true for all $x \in \mathbb{C}$ and this implies (3.2).

(IV) Now we select a certain finite subset of $\mathrm{supp}(\sigma_k)$ on which the maximum of every weighted polynomial $|P_n w_k^n|$ is comparable with the norm of $P_n w_k^n$. Let

$$\{x_0^{(k,n)},\ x_1^{(k,n)},\ \ldots,\ x_n^{(k,n)}\} = \{x_0,\ x_1,\ \ldots,\ x_n\}$$

be $n+1$ points from $\mathrm{supp}(\sigma_k)$ maximizing the determinant

$$D(\{x_i\}) := |x_i^j w_k^n(x_i)|_{0\le i,j\le n}$$

among all such systems. If $P_n(x) = \sum_{j=0}^n a_j x^j$ is any polynomial and $x \in \mathrm{supp}(\sigma_k)$ is fixed, then the vector $v_x := (x^j w_k^n(x))_{0\le j\le n} \in \mathbb{R}^{n+1}$ is a linear combination of the vectors $v_{x_i} := (x_i^j w_k^n(x_i))_{0\le j\le n}$, $0 \le i \le n$ (note that $n+2$ vectors in $\mathbb{R}^{n+1}$ are always linearly dependent whereas the latter $n+1$ vectors form a linearly independent system by the choice of the points $\{x_i\}$):

$$v_x = \sum_{i=0}^n \beta_i v_{x_i},$$

where the coefficients β_i also depend on x. Recalling the way the β_is are computed (Cramer's rule) and the maximality of the determinant $|x_i^j w_k^n(x_i)|_{0\le i,j\le n}$ we get that each β_i is at most 1 in absolute value. But

$$P_n(x)w_k^n(x) = \sum_{i=0}^n \beta_i P_n(x_i) w_k^n(x_i),$$

from which we get

$$\|P_n w_k^n\|_{\mathrm{supp}(\sigma_k)} \le (n+1)\|P_n w_k^n\|_{\{x_i\}_{0\le i\le n}}.$$

Finally, if we also take into account that $\mathrm{supp}(\sigma_k) \subseteq \overline{C}$ and that w_k is continuous, a simple compactness argument yields an $(n+1)$-point set $S_{k,n} \subseteq C$ such that for all polynomials P_n of degree at most n we have

$$\|P_n w_k^n\|_{[0,1]} = \|P_n w_k^n\|_{\mathrm{supp}(\sigma_k)} \le (n+2)\|P_n w_k^n\|_{S_{n,k}}$$

(use the fact that the set of polynomials P_n, $\deg(P_n) \le n$, with $\|P_n w_k^n\|_{\mathrm{supp}(\sigma_k)} \le 1$ is compact).

(V) Next we need the existence of polynomials $P_{n,k}$ of the form $x^n + \cdots$ such that

$$\lim_{n\to\infty} \|P_{n,k} w_k^n\|_{[0,1]}^{1/n} = \exp(-\max p(\sigma_k;\cdot)).$$

The existence of such polynomials follows from the theory of weighted potentials sketched in Appendix IX (cf. (A.30) in that appendix), but for completeness we present a direct construction using the points from the previous step.

Consider the points $\{x_i\}_{0\le i\le n} = \{x_i^{(k,n)}\}_{0\le i\le n}$ from the preceding step, and without loss of generality we may assume that

$$\left|\prod_{j\ne 0}(x_0 - x_j)w_k^n(x_0)\right| \le \left|\prod_{j\ne i}(x_i - x_j)w_k^n(x_i)\right| \tag{3.4}$$

for all i. If we factor out $w_k^n(x_j)$ from the jth row of the determinant

$D(\{x_i\})$, then what remains is a Vandermonde determinant; hence the product of the right-hand sides of (3.4) for $i = 0, 1, \ldots, n$ equals

$$D^2(\{x_i\}) \left(\prod_{i=0}^{n} w_k^n(x_i) \right)^{-1}$$

Now if $P_{n,k}(x) := \prod_{i=1}^{n}(x - x_i)$, then the extremality of the system $\{x_i\}_{0 \le i \le n}$ implies

$$|P_{n,k}(x)| w_k^n(x) \le |P_{n,k}(x_0)| w_k^n(x_0)$$

for $x \in \operatorname{supp}(\sigma_k)$ and so by the choice of x_0 and by what we have just said (cf. also step IV)

$$\text{(3.5)} \quad \|P_{n,k} w_k^n\|_{[0,1]}^{1/n} \le D^{2/n(n+1)}(\{x_i\}) \left(\prod_{i=0}^{n} w_k(x_i) \right)^{-1/(n+1)} =: R_{n,k}.$$

We recall that the points $\{x_i\}_{0 \le i \le n} = \{x_i^{(k,n)}\}_{0 \le i \le n}$ depend on both w_k and n. We claim that the corresponding counting measures

$$\rho_{k,n} := \frac{1}{n+1} \sum_{i=0}^{n} \delta_{x_i^{(k,n)}}$$

converge to σ_k in the weak* topology. In fact, let $\{\rho_{k,n}\}_{n \in N}$ converge to some σ. Then with

$$(\nu, \mu) = \int\int \log \frac{1}{|z - t|} d\nu(t) d\mu(z)$$

we have

$$\begin{aligned}
&(\sigma, \sigma) - 2(\sigma, \sigma_k) + 2 \max p(\sigma_k; \cdot) \\
&= \int\int \left(\log \frac{1}{|z - t|} - p(\sigma_k; z) - p(\sigma_k; t) + 2 \max p(\sigma_k; \cdot) \right) d\sigma(t) d\sigma(z) \\
&= \lim_{M \to \infty} \lim_{\substack{n \to \infty \\ n \in N}} \int\int \left(\min(M, \log \frac{1}{|z - t|} \right) - p(\sigma_k; z) - p(\sigma_k; t) \\
&\quad + 2 \max p(\sigma_k; \cdot)) d\rho_{k,n}(t) d\rho_{k,n}(z) \\
&\le \lim_{M \to \infty} \limsup_{\substack{n \to \infty \\ n \in N}} \Big(\frac{M}{n+1} - \frac{2}{(n+1)^2} \Big(\sum_{i=0}^{n} (p(\sigma_k; x_i^{(k,n)}) - \max p(\sigma_k; \cdot) \Big) \Big) \\
&\quad + \frac{1}{(n+1)^2} \log(D^{-2}(\{x_i^{(k,n)}\}))) = \limsup_{\substack{n \to \infty \\ n \in N}} \log(D^{-2/n(n+1)}(\{x_i^{(k,n)}\})).
\end{aligned}$$

But by the definition of the extremal points $\{x_i^{(k,n)}\}$ we have for any points $\{y_i\}_{i=0}^n \subset \mathrm{supp}(\sigma_k)$

$$\log(D^{-2}(\{x_i^{(k,n)}\})) \le \log(D^{-2}(\{y_i\})),$$

and integrating this inequality with respect to $d\sigma_k(y_0)\cdots d\sigma_k(y_n)$ we obtain

$$\log(D^{-2}(\{x_i^{(k,n)}\}) \le n(n+1)$$

$$\times \int\int \left(\log\frac{1}{|z-t|} - p(\sigma_k; z) - p(\sigma_k; t) + 2\max p(\sigma_k; \cdot)\right) d\sigma_k(t) d\sigma_k(z)$$

$$= n(n+1)(-(\sigma_k, \sigma_k) + 2\max p(\sigma_k; \cdot)).$$

Collecting the previous estimates we can finally conclude that

$$0 \le -(\sigma, \sigma) + 2(\sigma, \sigma_k) - (\sigma_k, \sigma_k) = -(\sigma - \sigma_k, \sigma - \sigma_k),$$

which is possible only when $\sigma = \sigma_k$ (see [La, Theorem 1.16]).

Thus we have proved $\sigma = \sigma_k$ and as a side product of our computations we get that the limit of $\log(R_{k,n})$ from (3.5) as n tends to infinity is

$$(\sigma_k, \sigma_k) - 2\max p(\sigma_k; \cdot) - \int (p(\sigma_k; t) - \max p(\sigma_k; \cdot)) d\sigma_k(t) = -\max p(\sigma_k; \cdot),$$

by which the existence of the polynomials $P_{n,k}$ with the requested property is proved.

(VI) With a sequence $\{n_k\}$ to be chosen subsequently we set

$$\mu_k := \frac{1}{(n_k+1)^2}\omega_{C_k} + \frac{1}{(n_k+1)^3} \sum_{\tau \in S_{n_k,k}} w_k^{2n_k}(\tau)\delta_\tau =: \mu_k' + \mu_k'',$$

where ω_{C_k} is the equilibrium measure of the compact set C_k (cf. step I) and $S_{n_k,k}$ is the subset of C appearing in step IV. Finally, let

$$\mu := \sum_{k=1}^{\infty} \mu_k.$$

We will show that $\mathcal{M}_C = \mathcal{M}_\mu$ and C is a minimal carrier of μ, by which the proof will be complete.

That

$$\mathrm{cap}(C) \ge c_\mu \ge \sup_k \mathrm{cap}(C_k) = \mathrm{cap}(C)$$

is obvious because $\mathrm{supp}(\mu_k) \subseteq C_k \subseteq C$, $\mu \ge \omega_{C_k}$ for all k, and it is easy to show that C_k is a minimal carrier of ω_{C_k}. Thus, C is a minimal carrier for μ. Then, of course, by the first part of our theorem, we must have

$\mathcal{M}_\mu \subseteq \mathcal{M}_C$; therefore we only have to show the converse inclusion. This will be done in step VIII after the preparation to be done in step VII.

(VII) Set

$$T_{n_k}(x) := P_{m_k,k}(x) \prod_{\substack{\tau \in \cup S_{n_r,r} \\ r<k}} (x-\tau),$$

where the $P_{m_k,k}$ are the polynomials from step V and $S_{n_r,r}$ are the sets from step IV, and where

$$m_k = n_k - |\cup_{r=1}^{k-1} S_{n_r,r}| \geq n_k - \sum_{r=1}^{k-1}(n_r+1).$$

Thus, $\deg(T_{n_k}) = n_k$ and T_{n_k} has leading coefficient 1. Clearly, T_{n_k} vanishes on the support of μ_r'' for $1 \leq r < k$, and since on the support of μ_r' for $1 \leq r \leq k$, which are all contained in C_k, we have by step III the inequality $1 \geq w_k(x) \geq e^{-1/k}$, we get

$$\int T_{n_k}^2 d\left(\sum_{r=1}^{k} \mu_r'\right)$$

$$\leq \|T_{n_k}\|_{C_k}^2 \leq e^{2n_k/k}\|T_{n_k} w_k^{m_k}\|_{[0,1]}^2 \leq e^{2n_k/k}\|P_{m_k,k} w_k^{m_k}\|_{[0,1]}^2$$

$$\leq e^{2n_k/k} \exp\left(2m_k\left(-\max p(\sigma_k;\cdot)+\frac{1}{k}\right)\right)$$

$$\leq e^{7n_k/k}\exp(2n_k(-\max p(\sigma_k;\cdot))),$$

provided n_k is large enough compared to n_{k-1}. By the definition of μ_k'' we have

$$\int T_{n_k}^2 d\mu_k'' \leq \|T_{n_k} w_k^{n_k}\|_{[0,1]}^2,$$

for which the preceding estimate can be applied. Finally,

$$\int T_{n_k}^2 d\left(\sum_{r=k+1}^{\infty} \mu_r\right) \leq \sum_{r=k+1}^{\infty} \|\mu_r\| \leq 2\sum_{r=k+1}^{\infty}(n_r+1)^{-2}$$

$$\leq \frac{2}{n_{k+1}} \leq 2\exp(2n_k(-\max p(\sigma_k;\cdot)))$$

provided n_{k+1} is large compared to n_k.

Until now we have not said anything about the sequence $\{n_k\}$. Let us choose it to be so fast increasing that all the estimates in this step hold. From these we get

$$\int T_{n_k}^2 d\mu \leq e^{8n_k/k}\exp(2n_k(-\max p(\sigma_k;\cdot))),$$

and

$$\int q_{n_k}^2(\mu;\cdot)d\mu \le e^{8n_k/k}\exp(2n_k(-\max p(\sigma_k;\cdot))),$$

where $q_{n_k}(\mu;\cdot)$ denotes the monic orthogonal polynomials belonging to μ. But the left-hand side is at least as large as

$$\int q_{n_k}^2(\mu;\cdot)d\mu_k \ge (n_k+1)^{-3}\|q_{n_k}(\mu;\cdot)w_k^{n_k}\|_{S_{n_k,k}}^2$$
$$\ge (n_k+2)^{-4}\|q_{n_k}(\mu;\cdot)w_k^{n_k}\|_{[0,1]}^2,$$

where we used the property of the sets $S_{n_k,k}$ established in step IV. Thus, finally we can conclude

$$\|q_{n_k}(\mu;\cdot)w_k^{n_k}\|_{[0,1]} \le (n_k+2)^2 e^{4n_k/k}\exp(n_k(-\max p(\sigma_k;\cdot))). \tag{3.6}$$

(VIII) Now we complete the proof. Let $\sigma \in \mathcal{M}_C$ be arbitrary. By the choice of the measures ν_k (step I) and σ_k (step II) there is a sequence $N \subseteq \mathbb{N}$ such that σ_k converges in the weak* topology to σ as k tends to infinity through N. By (3.6) we have for all $x \in [0,1]$

$$\frac{1}{n_k}\log\frac{1}{|q_{n_k}(\mu;x)|} \ge p(\sigma_k;x) - \frac{4}{k} - \frac{\log((n_k+2)^2)}{n_k}, \qquad k \in N. \tag{3.7}$$

By the principle of domination then this holds for all $x \in \mathbb{C}$. Now if ϑ is a weak* limit point of the measures $\nu_{q_{n_k}(\mu;\cdot)}$, $k \in N$, then going to infinity through a suitable subsequence of N we get from (3.7) for $x \notin [0,1]$ (note that all the zeros of q_{n_k} are contained in $[0,1]$)

$$p(\vartheta;x) \ge p(\sigma;x),$$

from which $\vartheta = \sigma$ follows as at the end of the proof of Lemma 2.1.6. □

It is an open problem to determine which subsets of $\mathcal{M}[0,1]$, where $\mathcal{M}[0,1]$ denotes the set of probability measures on $[0,1]$, arise as $\mathcal{M}_\mu$ for some μ with $c_\mu > 0$. That this class is quite extensive is indicated by Theorem 2.3.1 and also by the following result that can be verified by combining the methods applied in the proofs of Example 2.2.7 and Theorem 2.3.1. Suppose that $\mathcal{K}$ is a set of compact subsets of $[0,1]$ such that the set $\mathcal{M}_\mathcal{K}$ of the corresponding equilibrium measures is a closed subset of $\mathcal{M}[0,1]$ and all $K \in \mathcal{K}$ contain a common set C of positive capacity. Suppose further that each pair K, K' from $\mathcal{K}$ can be joined for every $\varepsilon > 0$ by a sequence $K = K_0, K_1, \ldots, K_n = K'$ from $\mathcal{K}$ in the sense that $\mathrm{cap}((K_i \setminus K_{i+1}) \cup (K_{i+1} \setminus K_i)) < \varepsilon$. Then there is a μ with $c_\mu \ge \mathrm{cap}(C) > 0$ such that $\mathcal{M}_\mathcal{K} = \mathcal{M}_\mu$.

3

Regular nth-root Asymptotic Behavior of Orthonormal Polynomials

If the weight measure μ is sufficiently dense on its support, then in Theorem 1.1.4 the lower bound (1.1.7) is asymptotically assumed. If this is the case, then the measure and the corresponding orthonormal polynomials are said to have regular (nth-root) asymptotic behavior. Orthogonal polynomials of regular asymptotic behavior are the natural generalization of classical orthogonal polynomials if we only want to preserve the asymptotic behavior in the nth-root sense, and they appear in many applications (see, e.g., Chapter 6). There exist several definitions of regular asymptotic behavior, which will be discussed together with their immediate consequences in this chapter.

3.1 Regular Asymptotic Behavior

Different definitions of regular nth-root asymptotic behavior of orthonormal polynomials will be reviewed and relationships among them will be studied.

The sequence $\{p_n(\mu; z) \mid n \in \mathbb{N}\}$ is said to have regular *exterior* asymptotic behavior if

$$\lim_{n\to\infty} |p_n(\mu; z)|^{1/n} = e^{g_\Omega(z;\infty)} \tag{1.1}$$

locally uniformly for $z \in \mathbb{C} \setminus \operatorname{Co}(S(\mu))$. This terminology has been introduced by J. Ullman (cf. [Ul2]) for the special case of weight measures μ with $S(\mu) = [-1, 1]$, in which case (1.1) takes the form

$$\lim_{n\to\infty} |p_n(\mu; z)|^{1/n} = |z + \sqrt{z^2 - 1}|, \tag{1.2}$$

where that branch of the square root is taken which is positive for positive z. The meaning of (1.2) is that $p_n(\mu; z)$ asymptotically behaves like the classical orthogonal polynomials, say, Chebyshev or Jacobi polynomials.

If $S(\mu) = [-1,1]$, then the outer boundary $\partial\Omega$ of $S(\mu)$ is identical with $[-1,1]$. For this case in [Ul2] it has been said that the sequence $\{p_n(\mu;z) \mid n \in \mathbb{N}\}$ has regular *interior* asymptotic behavior if

$$(1.3) \qquad \limsup_{n\to\infty} |p_n(\mu;z)|^{1/n} = 1 \qquad \text{qu.e. on } \partial\Omega.$$

The notations "exterior" and "interior" asymptotic behavior keep their intuitive meaning so long as $\mathrm{Int}(\mathrm{Pc}(S(\mu))) = \emptyset$, since then we have $\partial\Omega = S(\mu)$. We shall show in the next theorem that exterior and interior asymptotic behavior, as well as some other related relations, are equivalent.

A comparison of (1.1) and (1.3) with the asymptotic bounds given in Theorem 1.1.4 shows that regular asymptotic behavior means that the lower bound (1.1.7) is assumed asymptotically. This, of course, will always be the case if the weight measure μ is sufficiently dense everywhere on its support $S(\mu)$, and since this is often the case, the name "regular" for this type of asymptotic behavior is justified. For example, the normalized Jacobi polynomials, which can be seen as the prototypes of orthonormal polynomials associated with a weight measure with compact support, possess the regular asymptotic behavior (1.1) and (1.3). Of course, they also possess the stronger types of regular asymptotic behavior sketched in (I.1) and (I.2) of the introduction (see, for instance, [Sz3], [Ge], or [Fr]).

We remark that $\mathrm{cap}(S(\mu)) = 0$ has not been excluded in (1.1) and (1.3), and neither will it be excluded in the next theorem, although this case is somewhat special: we have $g_\Omega(z;\infty) \equiv \infty$, and therefore the limit (1.1) follows from the lower and upper bounds (1.1.6) and (1.1.7) of Theorem 1.1.4 for any weight measure μ with $\mathrm{cap}(S(\mu)) = 0$.

Our definition of regular (nth-root) asymptotic behavior is based on the following theorem.

Theorem 3.1.1. *The following four assertions are equivalent.*

(i) *The limit*

$$(1.4) \qquad \lim_{n\to\infty} \gamma_n(\mu)^{1/n} = \frac{1}{\mathrm{cap}(S(\mu))}$$

holds true, where $\gamma_n(\mu)$ is the leading coefficient of the orthonormal polynomial $p_n(\mu;z)$.

(ii) *The limit*

$$(1.5) \qquad \lim_{n\to\infty} |p_n(\mu;z)|^{1/n} = e^{g_\Omega(z;\infty)}$$

holds true locally uniformly in $\overline{\mathbb{C}} \setminus \mathrm{Co}(S(\mu))$.

(iii) *The limit*

$$(1.6) \qquad \limsup_{n\to\infty} |p_n(\mu;z)|^{1/n} = 1$$

holds true qu.e. on $\partial\Omega$.

(iv) *The limit* (1.5) *holds true locally uniformly in* $\overline{\mathbb{C}} \setminus \mathrm{Co}(S(\mu))$, *and in capacity in* $\Omega \cap \mathrm{Co}(S(\mu))$, *that is, for every compact set* $V \subseteq \Omega \cap \mathrm{Co}(S(\mu))$ *and every* $\varepsilon > 0$ *we have*

$$\lim_{n\to\infty} \operatorname{cap}\left(\left\{z \in V \mid \left||p_n(\mu; z)|^{1/n} - e^{g_\Omega(z;\infty)}\right| > \varepsilon\right\}\right) = 0,$$

and furthermore the limit (1.6) *holds true qu.e. on* $\partial\Omega$, *and we have*

$$\limsup_{n\to\infty} |p_n(\mu; z)|^{1/n} \le e^{g_\Omega(z;\infty)} \tag{1.7}$$

locally uniformly in $\mathbb{C}$.

Definition 3.1.2. *If one of the four assertions of* Theorem 3.1.1 *holds true, then the orthonormal polynomials* $p_n(\mu; z)$, $n \in \mathbb{N}$, *associated with the measure* μ *are said to have regular* (n-*th root*) *asymptotic behavior, and we write* $\mu \in$ **Reg**. *We shall also often refer to* $\mu \in$ **Reg** *simply as* μ *is regular.*

Remarks. (1) A weight measure μ is regular, that is, $\mu \in$ **Reg**, if μ is sufficiently dense on $S(\mu)$. This can be intuitively best understood as a consequence of the minimality property of the monic orthogonal polynomials

$$q_n(\mu; z) = \frac{1}{\gamma_n(\mu)} p_n(\mu; z)$$

in $L^2(\mu)$. We have

$$\|q_n(\mu; \cdot)\|_{L^2(\mu)} = \inf_{q(z)=z^n+\cdots} \|q\|_{L^2(\mu)} = \frac{1}{\gamma_n(\mu)}, \quad n \in \mathbb{N}.$$

Monic polynomials cannot be arbitrarily small, but if the measure μ is thin, then their $L^2(\mu)$ norm may be small and therefore $\gamma_n(\mu)$ large. Corollary 1.1.7 shows that in the regular case $\gamma_n(\mu)^{1/n}$ assumes asymptotically the smallest possible value. The preceding relation shows that this will be the case only if μ has a certain density on $S(\mu)$.

(2) A measure μ may belong to **Reg**, although the lower and upper bounds given in Theorem 1.1.4 are different. This is, for instance, shown by Examples 1.5.4 and 3.5.1.

(3) Definition 3.1.2 gives definite information about the asymptotic behavior of $p_n(\mu; z)$ as $n \to \infty$ only for $z \in \overline{\Omega}$. Nothing, except the upper bound in (1.7), is said about the asymptotic behavior in the interior of the polynomial convex hull $\mathrm{Pc}(S(\mu))$.

We postpone the proof of Theorem 3.1.1 until after the statement of Theorem 3.1.4.

Besides regular asymptotic behavior in the sense of Definition 3.1.2, we consider a closely related type of asymptotic behavior: regular asymptotic zero distribution of the orthonormal polynomials $p_n(\mu; z)$ as $n \to \infty$. It can only be defined if $\operatorname{cap}(S(\mu)) > 0$, since its definition involves the equilibrium distribution $\omega_{S(\mu)}$.

Definition 3.1.3. *Let* $\operatorname{cap}(S(\mu)) > 0$. *Then the orthonormal polynomials* $p_n(\mu; z)$, $n \in \mathbb{N}$, *associated with the measure* μ *are said to have* regular asymptotic zero distribution *if*

$$\frac{1}{n}\nu_{p_n(\mu;\cdot)} \xrightarrow{*} \omega_{S(\mu)} \quad \text{as} \quad n \to \infty. \tag{1.8}$$

Recall that ν_{p_n} is the counting measure on the zero set of p_n.

When $S(\mu) = [-1, 1]$, this definition gives the classical arcsine distribution:

$$\lim_{n\to\infty} \frac{1}{n}\#\{\text{zeros of } p_n(\mu;\cdot) \text{ lying in } (\alpha, \beta)\} = \frac{1}{\pi}(\arccos\alpha - \arccos\beta)$$

for all $(\alpha, \beta) \subseteq [-1, 1]$.

Having representation (1.2.3) of the Green function $g_\Omega(z; \infty)$ in mind, we are not so surprised that regular asymptotic zero distribution is very closely related to regular asymptotic behavior (in the sense of Definition 3.1.2); however, as the next theorem shows, the two notions are not fully identical. For further results in this direction see Section 6.7.

Theorem 3.1.4. *Assume* $\operatorname{cap}(S(\mu)) > 0$.

(a) *If* $\operatorname{Int}(\operatorname{Pc}(S(\mu))) = \emptyset$, *then* $\mu \in$ **Reg** *implies regular asymptotic zero distribution of the orthonormal polynomials* $p_n(\mu; z)$, $n \in \mathbb{N}$.

(b) *If* $S(\mu) \subseteq \partial\Omega$, *in particular, if* $\operatorname{Int}(\operatorname{Pc}(S(\mu))) = \emptyset$, *and* $c_\mu > 0$, *then regular asymptotic zero distribution of* $\{p_n(\mu; z) \mid n \in \mathbb{N}\}$ *implies* $\mu \in$ **Reg**.

(c) *Thus, if* $c_\mu > 0$ *and* $\operatorname{Int}(\operatorname{Pc}(S(\mu))) = \emptyset$, *then* $\mu \in$ **Reg** *and regular asymptotic zero distribution are equivalent.*

Remarks. (1) In case of a weight measure μ with $S(\mu) \subseteq \mathbb{R}$, we have $\operatorname{Int}(\operatorname{Pc}(S(\mu))) = \emptyset$, and therefore for such weight measures regular asymptotic zero distribution follows from $\mu \in$ **Reg**. That the reverse is in general not true is shown by Example 3.5.2, hence the assumption $c_\mu > 0$ in assertion (b) cannot be dropped.

(2) For weight measures with $\operatorname{Int}(\operatorname{Pc}(S(\mu))) \neq \emptyset$ it can be questioned whether (1.8) is the natural way to define regular asymptotic zero distribution (see, for instance, the very simple Example 1.5.1, which does not satisfy (1.8)). An alternative definition could, for example, allow on the right-hand side of (1.8) any probability measure ν on $\operatorname{Pc}(S(\mu))$ that generates the same logarithmic potential in Ω as the equilibrium distribution

$\omega_{S(\mu)}$. With such modification the assumption $\mathrm{Int}(\mathrm{Pc}(S(\mu))) = \emptyset$ in assertion (a) would no longer be necessary. We will, however, keep Definition 3.1.2 since it is the usual one in the literature.

Proof of Theorem 3.1.1. We shall use some results from the proof of Theorem 1.1.4 from Sections 1.2 and 1.3. Without loss of generality we can assume that $\mathrm{cap}(S(\mu)) > 0$ since otherwise all four assertions of Theorem 3.1.1 are satisfied for any measure μ, which can easily be verified via Theorem 1.1.4.

(i)⇒(iv): Let $N \subseteq \mathbb{N}$ be an infinite subsequence such that the two limits in (1.3.26) exist, that is,

(1.9)

$$\frac{1}{n}\nu_{p_n(\mu;\cdot)} \xrightarrow{*} \nu \quad \text{and} \quad \frac{1}{n}\log\gamma_n(\mu) \to c \in \mathbb{R} \cup \{\infty\} \quad \text{as} \quad n \to \infty,\ n \in N.$$

In part (b) of the proof of Theorem 1.1.4 it has been shown that (1.9) implies inequality (1.3.42), that is,

$$-p(\nu; z) + c \geq g_\Omega(z; \infty) \quad \text{for all} \quad z \in \Omega. \tag{1.10}$$

Assertion (i) implies $c = -\log\mathrm{cap}(S(\mu))$. Together with (1.10) this yields that the function

$$-p(\nu; z) + c - g_\Omega(z; \infty) \tag{1.11}$$

has a zero at infinity and is nonnegative for all $z \in \Omega$, and because $\mathrm{supp}(\nu) \subseteq \mathrm{Pc}(S(\mu))$, it is harmonic in Ω. By the minimum principle for harmonic functions it then follows that the function (1.11) is zero for all $z \in \Omega$. In the same way as in the proof of Theorem 2.2.1(b) we can show, by using the continuity of (1.11) in the fine topology, that this equality extends to $\overline{\Omega}$ (see Appendix II).

From what we have shown so far, and from the fact that $-p(\nu; z)$ is subharmonic in $\mathbb{C}$, it follows that

$$-p(\nu; z) + c = \begin{cases} g_\Omega(z; \infty) & \text{for } z \in \overline{\Omega}; \\ \leq 0 & \text{for } z \in Int(Pc(S(\mu))). \end{cases} \tag{1.12}$$

From (1.12), the principle of descent (Appendix III), and the general estimates (1.1.7) and (1.1.10) we can derive the limits (1.5) and (1.6) provided n tends to infinity through N. Since the right-hand sides of (1.5) and (1.6) are independent of N, (1.5) follows from this.

With (1.6) one has to be a little cautious, because for general functions p_n the fact that (1.6) is true qu.e. on $\partial\Omega$ whenever n tends to infinity through a sequence N for which the limits in (1.9) exist *does not* imply that (1.6) is true for n tending to infinity arbitrarily. Fortunately, however, here we are speaking of potentials and (1.6) easily follows from what we have done before with a simple application of the principle of descent. Equation(1.6) is a consequence of (1.1.10) and (1.7), so it is enough to prove (1.7).

Suppose on the contrary that (1.7) is not true. Then there is a sequence $N^* \subseteq \mathbf{N}$, a point $z^* \in \mathbf{C}$, and for every $n \in N^*$ a point $z_n \in \mathbf{C}$ such that $z_n \to z^*$ as $n \to \infty, n \in N^*$, but

$$\liminf_{n\to\infty, n\in N^*} |p_n(\mu; z_n)|^{1/n} > e^{g_\Omega(z^*;\infty)}.$$

Let $N \subseteq N^*$ be a subsequence of N^* for which the limits in (1.9) hold true. Then we also have

$$\liminf_{n\to\infty, n\in N} |p_n(\mu; z_n)|^{1/n} > e^{g_\Omega(z^*;\infty)},$$

and so the principle of descent yields

$$-p(\nu; z^*) + c > g_\Omega(z^*; \infty).$$

This, however, contradicts (1.12) and this contradiction proves (1.7) and at the same time (1.6) as well.

The proof of convergence in capacity follows in the same way as the proof of the lower estimate (1.1.8) in part (d) of the proof of Theorem 1.1.4; we only have to replace the right-hand side of (1.3.47) by $g_\Omega(z; \infty)$, which is possible because of (1.12).

(iv)⇒(iii): This implication is trivially true.

(iii)⇒(i): Again let $N \subseteq \mathbf{N}$ be an infinite subsequence such that the two limits in (1.9) exist. From assertion (iii), (1.1.9), and the lower envelope theorem it follows that

$$-p(\nu; z) + c = g_\Omega(z; \infty) \qquad \text{qu.e. on } \partial\Omega. \tag{1.13}$$

Let $C \subseteq \partial\Omega$ be the set of points z at which we have equality in (1.13) and at which $g_\Omega(z; \infty) = 0$. We have $\text{cap}(\partial\Omega \setminus C) = 0$. Since we have assumed $\text{cap}(S(\mu)) > 0$, the equilibrium distribution $\omega = \omega_{S(\mu)}$ exists and is of finite energy. Hence, $\omega(\partial\Omega \setminus C) = 0$ (see Appendix I). Together with representation (1.2.3) (or see (A.10) in the Appendix) of the Green function $g_\Omega(z; \infty)$ this implies that

$$\begin{aligned} 0 = \int g_\Omega(x; \infty) d\omega(x) &= -\int p(\nu; z) d\omega(x) + c = -\int p(\omega; z) d\nu(x) + c \\ &= \int g_\Omega(z; \infty) d\nu(x) + \log \text{cap}(S(\mu)) + c. \end{aligned}$$

Here the last integral is nonnegative, and therefore we have

$$c \leq \log \frac{1}{\text{cap}(S(\mu))}. \tag{1.14}$$

In Corollary 1.1.7 the converse inequality was proved. Hence, we have shown that in (1.14) equality holds true, which proves (1.4) for any subsequence N for which the two limits (1.9) exist. Since the right-hand side of (1.14) is independent of N, (1.4) is generally proved.

(iv)⇒(ii): This implication is trivially true.

(ii)⇒(i): By considering (1.5) near infinity the limit (1.4) can easily be verified. □

Proof of Theorem 3.1.4. (a) Let $\mu \in \mathbf{Reg}$ and $\mathrm{Int}(\mathrm{Pc}(S(\mu))) = \emptyset$. Furthermore, let $N \subseteq \mathbb{N}$ be an infinite subsequence such that the two limits in (1.9) exist. Since $\mu \in \mathbf{Reg}$, assertion (i) of Theorem 3.1.1 holds true. As in the proof of Theorem 3.1.1, (i)⇒(iv), we can deduce (1.12). Because $\mathrm{Int}(\mathrm{Pc}(S(\mu))) = \emptyset$ we have therefore shown that

$$-p(\nu; z) + c \equiv g_\Omega(z; \infty).$$

As a consequence we have $\nu = \omega_{S(\mu)}$, which proves the limit (1.8) for the considered subsequence N. Since the right-hand side of (1.8) is independent of N, the limit (1.8), that is, regular asymptotic zero distribution, is generally proved.

Part (b) has essentially been proved in Section 2.2 in Corollary 2.2.3. In fact, in that proof the only change we have to make is that now the constant c should be any accumulation point of the sequence $\{\gamma_n(\mu)^{1/n}\}_{n=1}^{\infty}$, and the the rest of the argument has to be repeated for the subsequence N of the natural numbers for which

$$\lim_{n\to\infty, n\in N} \gamma_n(\mu)^{1/n} = c$$

(the reason for that change is that now we do not assume $\mathrm{Int}(\mathrm{Pc}(S(\mu))) = \emptyset$, which is needed in Theorem 2.2.1(a)). □

3.2 Characterization of Regular Asymptotic Behavior

Regular asymptotic behavior discussed in the preceding section is equivalent to several statements relating the size of general polynomials to their $L^2(\mu)$ (or $L^p(\mu)$) norm. We discuss such relations in this section with respect to $L^2(\mu)$ and $L^\infty(\mu)$ norms, and in Section 3.4 with respect to $L^p(\mu)$ norms, $0 < p < \infty$. These results have applications to several problems of approximation theory.

Theorem 3.2.1. *The following five statements are equivalent for a measure* μ.

(i) *The sequence* $\{p_n(\mu; \cdot)\}_{n=1}^{\infty}$ *has regular (nth-root) asymptotic behavior, that is,* $\mu \in \mathbf{Reg}$.

(ii) *We have*

$$\limsup_{n\to\infty} |p_n(\mu; z)|^{1/n} \le e^{g_\Omega(z;\infty)} \tag{2.1}$$

locally uniformly in $\mathbb{C}$.

(iii) *For any sequence* $\{P_n\}_{n=1}^{\infty}$ *of nonzero polynomials of degree* $\deg(P_n) \le n$ *we have*

$$\limsup_{n\to\infty} \left(\frac{|P_n(z)|}{\|P_n\|_{L^2(\mu)}} \right)^{1/n} \le e^{g_\Omega(z;\infty)} \tag{2.2}$$

locally uniformly for $z \in \mathbb{C}$.

(iv) *We have*

$$\limsup_{n\to\infty} |p_n(\mu; z)|^{1/n} \le 1 \tag{2.3}$$

for quasi every $z \in \partial\Omega$.

(v) *For any sequence* $\{P_n\}$ *as in* (iii) *we have*

$$\limsup_{n\to\infty} \left(\frac{|P_n(z)|}{\|P_n\|_{L^2(\mu)}} \right)^{1/n} \le 1 \tag{2.4}$$

for quasi every $z \in \partial\Omega$.

If in addition we also have $S(\mu) \subseteq \partial\Omega$, *in particular, if* $\mathrm{Int}(\mathrm{Pc}(S(\mu))) = \emptyset$, *then each of* (i)–(v) *is equivalent to the following statement.*

(vi) *Let* $g_n \ge 0$ *be weight functions defined on* $S(\mu)$ *in such a way that* $\lim_{n\to\infty} g_n^{1/n} = g$ *exists uniformly on* $S(\mu)$, *where* $g > 0$ *and* g *is continuous on* $S(\mu)$. *We set* $d\mu_n := g_n d\mu$. *Then for any sequence* $\{P_n\}$ *of polynomials as in assertion* (iii) *we have*

$$\limsup_{n\to\infty} \left(\frac{g_n(z)^{1/2}|P_n(z)|}{\|P_n\|_{L^2(\mu_n)}} \right)^{1/n} \le 1 \tag{2.5}$$

for quasi every $z \in \partial\Omega = S(\mu)$.

Remark 3.2.2. (1) We mention that the $L^2(\mu)$ norm in (iii), (v), and (vi) can be replaced by any $L^p(\mu)$ norm, $1 \le p < \infty$. For more on this see Section 3.4.

(2) It may be useful to note that the pairwise equivalence of (ii) and (iii) and that of (iv) and (v) hold pointwise, that is, say (2.3) is true at a single point z if and only if (2.4) is true there. To show this is actually very easy and it will come from our proof in what follows.

(3) Note that as a consequence of Theorem 1.1.4 we must automatically have equality in (2.1) for $z \notin \mathrm{Co}(S(\mu))$ and in (2.3) for quasi every $z \in \partial\Omega$.

(4) Condition (v) plays a distinguished role and we shall make a separate definition for it in Section 5.2. There we will call a point z μ-regular if for every sequence of nonzero polynomials $\{P_n\}_{n=1}^{\infty}$ of corresponding degree n we have

$$\limsup_{n\to\infty} \left(\frac{|P_n(z)|}{\|P_n\|_{L^2(\mu)}} \right)^{1/n} \leq 1.$$

This is obviously equivalent to

$$\lim_{n\to\infty} \left(\sup_{\substack{\deg P_n \leq n \\ P_n \not\equiv 0}} \frac{|P_n(z)|}{\|P_n\|_{L^2(\mu)}} \right)^{1/n} = 1.$$

It is also clear that if z is a μ-regular point then $z \in \mathrm{Pc}(S(\mu))$. μ-regularity will be of most interest when $z \in \partial\Omega$, that is, when z is on the outer boundary of $S(\mu)$. Paraphrasing the equivalence of (i) and (v), we can say that $\mu \in \mathbf{Reg}$ if and only if quasi every $z \in \partial\Omega$ is μ-regular (for more details see Chapter 5).

(5) If we assume (i), then (2.3) and (2.4) are actually true for quasi every $z \in \mathrm{Pc}(S(\mu))$. This follows from (2.2) if we note that $g_\Omega(z, \infty) = 0$ for qu.e. $z \in \mathrm{Pc}(S(\mu))$ (see also the corresponding proofs in the sequel).

When the support $S(\mu)$ of μ, more precisely its outer boundary, is regular with respect to the Dirichlet problems in the domain Ω, then the conditions in Theorem 3.2.1 hold uniformly whenever $\mu \in \mathbf{Reg}$. Because of its importance, we state this case as a separate theorem.

Theorem 3.2.3. *Suppose that $S(\mu)$ is a regular set with respect to the Dirichlet problem in Ω. Then the following five statements are equivalent.*

(i) $\mu \in \mathbf{Reg}$.

(ii)
$$\limsup_{n\to\infty} |p_n(\mu; z)|^{1/n} e^{-g_\Omega(z;\infty)} \leq 1$$

uniformly on $\mathbb{C}$.

(iii)
$$\lim_{n\to\infty} \left(\sup_{\substack{\deg P_n \leq n \\ P_n \not\equiv 0}} \frac{|P_n(z)|}{\|P_n\|_{L^2(\mu)}} \right)^{1/n} e^{-g_\Omega(z;\infty)} = 1$$

uniformly on $\mathbb{C}$.

(iv)
$$\lim_{n\to\infty} \|p_n(\mu; \cdot)\|_{\sup, S(\mu)}^{1/n} = 1,$$

where $\|\cdot\|_{\sup,S(\mu)}$ *denotes the supremum norm on* $S(\mu)$.

(v)
$$\lim_{n\to\infty}\left(\sup_{\substack{\deg P_n\le n\\ P_n\not\equiv 0}}\frac{\|P_n\|_{\sup,S(\mu)}}{\|P_n\|_{L^2(\mu)}}\right)^{1/n}=1.$$

If in addition $S(\mu)\subseteq\partial\Omega$, *then any of* (i)–(v) *is equivalent to the following statement.*

(vi) *If* $\{g_n\}$ *is a sequence as in assertion* (vi) *of* Theorem 3.2.1, *then*

$$\lim_{n\to\infty}\left(\sup_{\substack{\deg P_n\le n\\ P_n\not\equiv 0}}\frac{\|g_n^{1/2}P_n\|_{\sup,S(\mu)}}{\|P_n\|_{L^2(\mu_n)}}\right)^{1/n}=1,$$

where $d\mu_n := g_n d\mu$ *is the measure from* Theorem 3.2.1.

Proof of Theorem 3.2.1. First we verify the equivalence of the assertions (i)–(v).

(i)⇒(ii): This was verified in Theorem 3.1.1.

(ii)⇒(iii): Let P_n be any polynomial of degree at most n, without loss of generality we may assume its $L^2(\mu)$ norm to be 1. Thus, if

$$P_n(z)=\sum_{k=0}^{n}c_k p_k(\mu;z)$$

is its Fourier expansion in the orthonormal polynomials $\{p_k(\mu;\cdot)\}$, then $\sum_{k=0}^{n}|c_k|^2=1$, that is, each c_k is at most 1 in absolute value.

Now if (2.1) holds at a point z, then for every $\varepsilon>0$ there is a C_ε such that

$$|p_k(\mu;z)|\le C_\varepsilon(e^{g_\Omega(z;\infty)}+\varepsilon)^k,\quad k=0,1,2,\ldots,$$

and so

$$|P_n(z)|\le\sum_{k=0}^{n}|c_k||p_k(\mu;z)|\le nC_\varepsilon(e^{g_\Omega(z;\infty)}+\varepsilon)^n,$$

which proves (2.2) at z.

Thus, if (2.1) holds at every z then so does (2.2). From this the locally uniform convergence in (2.2) can be derived from the following general argument.

Suppose on the contrary that the convergence does not hold locally uniformly. Then there is a sequence $N^*\subseteq\mathbb{N}$, a point $z^*\in\mathbb{C}$, and for every $n\in N^*$ a point $z_n\in\mathbb{C}$ such that $z_n\to z^*$ as $n\to\infty$, $n\in N^*$, but

$$\liminf_{n\to\infty,n\in N^*}|P_n(z_n)|^{1/n}>e^{g_\Omega(z^*;\infty)}.$$

If we replace all zeros ξ of P_n lying outside the disk $\{z \mid |z| > R\}$ by $R\xi/|\xi|$, then for large R and $|z - z^*| < 1$ the ratio

$$\left(\frac{|P_n(z)|}{\|P_n\|_{L^2(\mu)}}\right)^{1/n}$$

changes by a factor that lies in the interval

$$((R - |z^*| - 1)/(R + |z^*| + 1), (R + |z^*| + 1)/(R - |z^*| - 1)),$$

hence by choosing R large enough we can assume that the zeros of the polynomials P_n lie in a fixed compact set (besides the normalization $\|P_n\|_{L^p(\mu)} = 1$ assumed at the beginning of the proof).

Let $N \subseteq N^*$ be a subsequence of N^* for which the limits

$$\frac{1}{n}\nu_{P_n} \xrightarrow{*} \nu \quad \text{and} \quad \frac{1}{n}\log \gamma_n \to c \in \mathbf{R} \cup \{\infty\} \quad \text{as} \quad n \to \infty,\ n \in N,$$

hold true, where γ_n is the leading coefficient of P_n. Then we also have

$$\liminf_{n\to\infty, n\in N} |P_n(z_n)|^{1/n} > e^{g_\Omega(z^*;\infty)},$$

and so the principle of descent yields

$$-p(\nu; z^*) + c > g_\Omega(z^*; \infty).$$

Since both sides are continuous in the fine topology (see Appendix II), it follows that the inequality

$$-p(\nu; z) + c > g_\Omega(z; \infty)$$

must hold on a set of positive capacity (see Appendix II). But we have previously verified (2.2) pointwise for every z, hence the lower envelope principle gives

$$-p(\nu; z) + c \le g_\Omega(z; \infty)$$

for quasy every z and this contradicts the preceding statement. This contradiction shows that in fact, the convergence in (2.2) holds locally uniformly.

(iii)⇒(v): This follows from the fact that

$$g_\Omega(z; \infty) = 0 \qquad \text{for qu.e. } z \in \partial\Omega.$$

(v)⇒(iv): This is obvious and **(iv)⇒(i)** follows from Theorem 3.1.1 (cf. Theorem 3.1.1(iii)) if we take into account (1.1.10).

With this the equivalence of the first five statements (i)–(v) has been proved. Finally, we show that in the case $S(\mu) \subseteq \partial\Omega$ condition (v) implies (vi). Since **(vi)⇒(v)** is trivial (take $g_n \equiv 1$, $n = 1, 2, \ldots$), this will complete the proof.

In the proof of **(v)⇒(vi)** we need the following lemma about approximation on $\partial\Omega$ by absolute values of polynomials.

Lemma 3.2.4. *The set*

$$\{|P| \mid P \quad \textit{is a polynomial}\}$$

is dense in $C_+(\partial\Omega)$, that is, every nonnegative continuous function on $\partial\Omega$ can be uniformly approximated by absolute values of polynomials.

Proof of Lemma 3.2.4. It is known that the set

$$\{\mathrm{Re}P \mid P \quad \text{is a polynomial}\}$$

is dense in the space $C_R(\partial\Omega)$ of real-valued continuous functions on $\partial\Omega$ (see [Ca]). If $g \in C_R(\partial\Omega)$ is strictly positive, then we can uniformly approximate $\log g$ by real parts of polynomials:

$$\lim_{k\to\infty} \mathrm{Re}P_k(z) = \log g(z)$$

uniformly on $\partial\Omega$. Then we will have uniformly on $\partial\Omega$

$$\begin{aligned}\lim_{k\to\infty}\left|\sum_{s=0}^{M(k)} \frac{(P_k(z))^s}{s!}\right| &= \lim_{k\to\infty} |\exp(P_k(z))| \\ &= \lim_{k\to\infty} \exp(\mathrm{Re}P_k(z)) = \exp(\log g(z)) = g(z)\end{aligned}$$

for some appropriately chosen sequence $\{M(k)\}_{k=1}^{\infty}$, which proves Lemma 3.2.4 for positive functions. Of course, it is then true for nonnegative functions as well. □

Consider now the function g from (vi). Since we have assumed $S(\mu) \subseteq \partial\Omega$, which is equivalent to $S(\mu) = \partial\Omega$, Lemma 3.2.4 yields a polynomial $Q = Q_\varepsilon$ with

$$1 - \varepsilon \le g(z)^{1/2}/|Q(z)| \le 1 + \varepsilon, \quad z \in S(\mu),$$

whatever $\varepsilon > 0$ is given. Then for large n we will have

$$1 - 2\varepsilon \le g_n(z)^{1/2n}/|Q(z)| \le 1 + 2\varepsilon,$$

which implies with $R_n(z) = P_n(z)Q^n(z)$

$$\text{(2.6)} \qquad \frac{g_n(z)^{1/2}|P_n(z)|}{\|P_n\|_{L^2(\mu_n)}} \le \frac{(1+2\varepsilon)^n |R_n(z)|}{(1-2\varepsilon)^n \|R_n\|_{L^2(\mu)}}.$$

If $\deg(Q) = k$ then $\deg(R_n) \le n(1+k)$ and (v) applied to $R_n(z)$ implies

$$\limsup_{n\to\infty} \left(\frac{|R_n(z)|}{\|R_n\|_{L^2(\mu)}} \right)^{1/n(1+k)} \le 1$$

for qu.e. $z \in \partial\Omega$. But then

$$\limsup_{n\to\infty} \left(\frac{R_n(z)}{\|R_n\|_{L^2(\mu)}} \right)^{1/n} \le 1$$

is also satisfied quasi everywhere on $\partial\Omega$, and so, in view of (2.6), the left-hand side of (2.5) is at most $(1+2\varepsilon)/(1-2\varepsilon)$ for qu.e. $z \in \partial\Omega = S(\mu)$. Because of the arbitrariness of ε, (2.5) holds. □

Proof of Theorem 3.2.3. Recall that the assumption of Theorem 3.2.3 is equivalent to the fact that $g_\Omega(\cdot;\infty)$ is continuous throughout $\mathbb{C}$.

In the proof we can be very short because basically we only have to repeat the preceding proof. In fact, (i)⇒(ii) follows from Theorem 3.1.1, (1.7), and the maximum principle for harmonic functions if we use the fact that, because of the continuity of $g_\Omega(\cdot;\infty)$, (1.7) holds true uniformly on compact subsets of $\mathbb{C}$ provided it holds locally uniformly on $\mathbb{C}$. Then (iii) follows from (ii) as in the preceding proof. (iii)⇒(v) is true because the regularity of $S(\mu)$ implies that $g_\Omega(z;\infty)$ vanishes on $\mathrm{Pc}(S(\mu))$. (v)⇒(iv) is again trivial and that (iv) implies (i) follows from Theorem 3.2.1.

Finally, (vi) obviously implies (v) and (v)⇒(vi) follows exactly as in Theorem 3.2.1. □

In Theorem 3.2.1 we have verified that (i) and (vi) are equivalent. The implication (vi)⇒(i) means that if (2.5) holds true quasi everywhere on $\partial\Omega = S(\mu)$ for *every* sequence $\{g_n\}$ and corresponding limit function g satisfying the conditions in (vi), then $\mu \in \mathbf{Reg}$. This does not mean that if (2.5) is true for a *single* sequence $\{g_n\}$ and corresponding limit g then $\mu \in \mathbf{Reg}$. Consider, for example, the following:

Example 3.2.5. Let $\mathrm{supp}(\mu) = [-1,1] \cup [2,3]$,

$$\mu = \mu^* + \sum_{k=1}^{\infty} d_k \delta_{x_k},$$

where μ^* is the Lebesgue measure on $[-1,1]$, $d_k \downarrow 0$ sufficiently fast (to be specified subsequently), and $\{x_k\} \subseteq [2,3]$ is a dense subset of $[2,3]$. With $\gamma > 2$ set

$$g_n(x) = \begin{cases} 1 & \text{if } x \in [-1,1]; \\ (x+\sqrt{x^2-1})^{-\gamma n} & \text{if } x \in [2,3]. \end{cases}$$

Then

$$\lim_{n\to\infty} g_n^{1/n}(x) = 2^{-\gamma}\exp(\gamma p(\omega_{[-1,1]};x)),$$

where $\omega_{[-1,1]}$ is the equilibrium distribution of the interval $[-1,1]$.

By Theorem 3.2.3 we have for every $x \in [-1,1]$

$$\limsup_{n\to\infty}\left(\frac{g_n(x)^{1/2}|P_n(x)|}{\|P_n\|_{L^2(\mu_n)}}\right)^{1/n} \le 1$$

for any sequence $P_n \not\equiv 0$, $\deg P_n \le n$, because μ^* is regular in the sense of Definition 3.1.2. But (2.5) actually holds for every $x \in [2,3]$ as well, because

$$\begin{aligned} g_n(x)^{1/2}|P_n(x)| &\le \|P_n\|_{[-1,1]}(x+\sqrt{x^2-1})^{(1-\gamma/2)n} \\ &\le 3n\|P_n\|_{L^2(\mu^*)}(x+\sqrt{x^2-1})^{(1-\gamma/2)n}, \end{aligned}$$

where in the first inequality we applied the Bernstein–Walsh lemma and in the second one Nikolskii's inequality (see [Ti, 4.9.6]).

Thus, (2.5) is true for every $x \in \operatorname{supp}(\mu)$, but if $\{d_k\}$ decreases sufficiently fast then μ will not be in the class **Reg** (cf. Theorem 4.2.8). □

3.3 Regular Behavior in the Case of Varying Weights

In one of our applications we will need to conclude $\mu \in$ **Reg** from (2.5) provided we know it only for a single sequence $\{g_n\}$ and corresponding limit function g. Despite Example 3.2.5 we shall show that such a conclusion is possible if we know some additional properties of g. In our application the following will be sufficient.

Theorem 3.3.1. *Suppose that*

(3.1) $$g(z) = \exp(\gamma p(\sigma; z)), \quad z \in S(\mu),$$

where $0 < \gamma \le 2$ *and* σ *is a probability measure with compact support in* Ω. *Assume further that the assumptions of* Theorem 3.2.1(vi) *are satisfied, that is,* $S(\mu) \subseteq \partial\Omega$, $g_n^{1/n} \to g$ *uniformly as* $n \to \infty$, *and* $d\mu_n = g_n d\mu$. *Then if*

(3.2) $$\limsup_{n\to\infty} g(z)^{1/2}|p_n(\mu_n; z)|^{1/n} \le 1$$

for quasi every $z \in \partial\Omega = S(\mu)$ *then* $\mu \in$ **Reg**.

Remarks. (1) A constant factor in g does not make any difference, so in (3.1) the exponential term can be multiplied by a constant and the result is still true.

(2) The theorem is not true if $\gamma > 2$ is allowed. In fact, Example 3.2.5 already shows this phenomenon, although there the support of σ was not contained in Ω. To get a proper example, move the support of σ a little, for example, let g be defined by (3.1) (with $\gamma > 2$), where σ is the equilibrium measure of the segment $[-1 + i\varepsilon,\ 1 + i\varepsilon]$ for some small $\varepsilon > 0$.

(3) A comparison of the two assertions (ii) and (iv) in Theorem 3.2.1 with assertions (iii) and (v), respectively, shows that the orthonormal polynomials $p_n(\mu;\cdot)$, $n \in \mathbb{N}$, are optimal test functions for the two inequalities (2.2) and (2.4). Whereas in assertions (iii) and (v) the inequalities have to hold for all sequences $\{P_n\}$ of nonzero polynomials with $\deg(P_n) \leq n$, in the two assertions (ii) and (iv) it is enough that the inequalities hold only for the sequence $\{p_n(\mu;\cdot)\}$. It is an interesting question (and it arises also in applications, see, e.g., Sections 6.1 and 6.2) whether in the case of weighted $L^2(\mu_n)$ norms (cf. assertion (vi) in Theorem 3.2.1) the orthonormal polynomials $\{p_n(\mu_n;\cdot)\}_{n\in\mathbb{N}}$ have a similar testing power in $L^2(\mu_n)$. Theorem 3.1.1 implies that this is in fact the case for weights like the ones in (3.1) with $\gamma \leq 2$; furthermore, the preceding remark shows that for general weights $\{\mu_n\}$ this is not so.

Proof of Theorem 3.3.1. The proof will be carried out indirectly; that is, we assume that (3.2) holds but μ is not regular. Then there is an $\alpha < 1$ and a subsequence $\mathcal{N} \subseteq \mathbb{N}$ such that for the monic polynomials

$$q_n(\mu; z) = p_n(\mu; z)/\gamma_n(\mu)$$

we have

$$\int |q_n(\mu;\cdot)|^2 d\mu \leq \alpha^{2n}(\operatorname{cap}(S(\mu)))^{2n}, \quad n \in \mathcal{N}. \tag{3.3}$$

Let σ' be the balayage of σ onto $S(\mu)$. Then

$$p(\sigma'; z) = p(\sigma; z) + \int g_\Omega(z; \infty) d\sigma(x)$$

for quasi every $z \in S(\mu)$ (see Appendix VII) and hence for the measure

$$\rho = \frac{\gamma}{2}\sigma' + \left(1 - \frac{\gamma}{2}\right)\omega_{S(\mu)},$$

where $\omega_{S(\mu)}$ denotes the equilibrium distribution of $S(\mu)$, we have

$$p(\rho; z) = \frac{\gamma}{2}p(\sigma; z) + c = \log(g^{1/2}(z)) + c \tag{3.4}$$

for quasi every $z \in S(\mu)$, where the constant c is given by

$$c := \frac{\gamma}{2}\int g_\Omega(x; \infty) d\sigma(x) + \left(1 - \frac{\gamma}{2}\right)\log\frac{1}{\operatorname{cap}(S(\mu))}. \tag{3.5}$$

In fact, formula (3.4) is an immediate consequence of the definition of the measure ρ and the relation

$$p(\omega_{S(\mu)}; z) = \log \frac{1}{\operatorname{cap}(S(\mu))} \quad \text{for qu.e. } z \in S(\mu).$$

We shall show subsequently that relation (3.3) implies that there exists an infinite subsequence $\mathcal{N}^* \subseteq \mathbb{N}$ with

$$\lim_{n\to\infty, n\in\mathcal{N}^*} \frac{1}{n} \log(\gamma_n(\mu_n)) = c_0 > c, \tag{3.6}$$

where μ_n is the varying measure $d\mu_n = g_n d\mu$. But first we prove that (3.6) and (3.2) are incompatible, and the so obtained contradiction will prove our theorem.

Let ν be a weak* limit point of the zero distributions of the orthogonal polynomials $\{q_n(\mu_n;\cdot)\}_{n\in\mathcal{N}^*}$, and suppose that the weak* limit is ν along the subsequence $\mathcal{N}^{**} \subseteq \mathcal{N}^*$. It follows from Theorem 2.1.1 that ν is a probability measure with support in $S(\mu)$. By the lower envelope theorem of potential theory (see Appendix III)

$$\limsup_{\substack{n\to\infty\\ n\in\mathcal{N}^{**}}} |q_n(\mu_n; z)|^{1/n} = \exp(-p(\nu; z))$$

for qu.e. $z \in S(\mu)$, and hence, in view of (3.6),

$$\limsup_{\substack{n\to\infty\\ n\in\mathcal{N}^{**}}} |p_n(\mu_n; z)|^{1/n} = \exp(-p(\nu; z) + c_0) \tag{3.7}$$

for qu.e. $z \in S(\mu)$. Together with the representation (3.4) of g we deduce from (3.2) and (3.7) the inequality

$$-p(\nu; z) + c_0 \le -p(\rho; z) + c \tag{3.8}$$

for quasi every $z \in S(\mu) = \partial\Omega$. Since ρ is of finite logarithmic energy, this inequality extends to all $\mathbb{C}$ by the principle of domination (see Appendix III). Considering (3.8) around infinity we can conclude $c_0 \le c$, which contradicts (3.6), and this contradiction proves our theorem.

Thus, it remains to be shown that (3.3) implies (3.6). In order to prove this we need some preparations.

Let $\varepsilon < \frac{1}{2}\operatorname{dist}(S(\mu), \operatorname{supp}(\sigma))$, and

$$K = K_\varepsilon = \{z | \operatorname{dist}(z, S(\mu)) \le \varepsilon\}.$$

We shall specify $\varepsilon > 0$ at a later stage of the proof. It is obvious that K is a compact subset of $\mathbb{C}$, $K \cap \operatorname{supp}(\sigma) = \emptyset$, K contains $S(\mu)$ in its interior, and Ω_K, the unbounded component of $\mathbb{C} \setminus K$, contains $\operatorname{supp}(\sigma)$ for sufficiently small $\varepsilon > 0$, say, $\varepsilon \le \varepsilon_0 < \frac{1}{2}\operatorname{dist}(S(\mu), \operatorname{supp}(\sigma))$; furthermore, $\partial\Omega_K$ is regular with respect to the Dirichlet problem in Ω_K. For the time being the only requirement on $\varepsilon > 0$ is that all these properties are satisfied, that is, $\varepsilon \leq \varepsilon_0$.

Let σ'' be the balayage of σ onto K. We know that $\mathrm{supp}(\sigma'') \subseteq \partial\Omega_K$ and the regularity of $\partial\Omega_K$ implies

$$p(\sigma''; z) = p(\sigma; z) + \int g_{\Omega_K}(x; \infty) d\sigma(x), \quad z \in K \tag{3.9}$$

(see Appendix VII). Our first objective is to compare σ'' and the equilibrium distribution ω_K of K.

Let $h \geq 0$ be an arbitrary continuous function on $\partial\Omega_K$. By the regularity of $\partial\Omega_K$ there is a function, which we continue to denote by h, that is continuous on $\overline{\Omega}_K$ and harmonic on Ω_K (including ∞). By Appendix VII we have then

$$\int_{\partial\Omega_k} h d\sigma'' = \int_{\Omega_k} h d\sigma = \int_{\mathrm{supp}(\sigma)} h d\sigma. \tag{3.10}$$

In a similar fashion, since the balayage of δ_∞ onto K is ω_K, we get

$$\int_{\partial\Omega_K} h d\omega_K = h(\infty). \tag{3.11}$$

But h is nonnegative, so by Harnack's inequality (see Appendix III) there are two positive constants c_1 and c_2 such that independently of the choice of $h \geq 0$

$$c_1 h(\infty) \leq h(z) \leq c_2 h(\infty), \quad z \in \mathrm{supp}(\sigma).$$

Combining this with (3.10) and (3.11), we can see that for the signed measures

$$\sigma_1 = \sigma'' - c_1\omega_K, \qquad \sigma_2 = c_2\omega_K - \sigma''$$

with support in $\partial\Omega_K$ the integrals

$$\int_{\partial\Omega_K} h d\sigma_j, \quad j = 1, 2,$$

are always nonnegative when $h \in C(\partial\Omega_K)$ is nonnegative. But then it follows from the Riesz representation theorem for positive functionals that both signed measures σ_1 and σ_2 are nonnegative, that is,

$$c_1\omega_K \leq \sigma'' \leq c_2\omega_K. \tag{3.12}$$

Now let $1 < \lambda < 1 + c_1/c_2$ be of the form $\lambda = k/(k-1)$, $k \in \mathbb{N}$, with a fixed k. It follows from (3.12) that the signed measure

$$\tilde{\sigma} = \lambda\sigma'' - (\lambda - 1)\omega_K$$

is actually a probability measure that satisfies (cf. (3.9))

$$p(\tilde{\sigma}; z) = \lambda p(\sigma; z) + \tilde{d}, \quad z \in K,$$

(3.13)

$$\tilde{d} = \lambda \int g_{\Omega_K}(x; \infty) d\sigma(x) - (\lambda - 1)\log(1/\mathrm{cap}(K)).$$

From our discussion it is clear that the same constants c_1, c_2, λ can serve for all $K = K_\varepsilon$ with $\varepsilon \leq \varepsilon_0$, that is, k can be chosen independently of $0 < \varepsilon \leq \varepsilon_0$.

Choose now a sequence of discrete measures

$$\frac{1}{n}\sum_{i=1}^{n} \delta_{x_{i,n}}$$

with support in $\partial\Omega_K$ that converges in the weak* topology to $\tilde{\sigma}$, and set

$$R_n(z) = \prod_{i=1}^{n}(z - x_{i,n}).$$

Since $S(\mu)$ and $\partial\Omega_K$ are disjoint, we get from (3.13) that

$$\lim_{n\to\infty} |R_n(z)|^{1/n} = \exp(-\lambda p(\sigma; z) - \tilde{d})$$

uniformly on $S(\mu)$.

On the other hand, if $\{y_{i,n}\}_{1\leq i\leq n,\ n\geq 0} \subseteq \partial K$ are chosen in such a way that the discrete measures

$$\frac{1}{n}\sum_{i=1}^{n} \delta_{y_{i,n}}$$

converge to ω_K, then for the polynomials

$$R_n^*(z) = \prod_{i=1}^{n}(z - y_{i,n})$$

we have uniformly on $S(\mu)$

$$\lim_{n\to\infty} |R_n^*(z)|^{1/n} = \mathrm{cap}(K)$$

because $p(\omega_K; z) \equiv \log(1/\mathrm{cap}(K))$ on $S(\mu)$.

Thus, for the polynomials

$$Q_n(z) = R_{[n\gamma/2]}(z) R^*_{[n(1-\gamma/2)]}(z)(z-a)^{n-[n\gamma/2]-[n(1-\gamma/2)]}$$

of degree n, where $a \notin S(\mu)$ is any fixed number introduced to attain $\deg(Q_n) = n$, we have uniformly on $S(\mu)$

$$\lim_{n\to\infty} |Q_n(z)|^{1/n} = \exp\left(-\frac{\lambda\gamma}{2} p(\sigma; z) - d^*\right), \tag{3.14}$$

where

$$(3.15)\quad d^* = \frac{\lambda\gamma}{2}\int g_{\Omega_K}(x;\infty)d\sigma(x) + \left(\left(1-\frac{\gamma}{2}\right) - \frac{\gamma}{2}(\lambda-1)\right)\log\frac{1}{\mathrm{cap}(K)}.$$

After these preparations we are ready to prove that (3.3) implies (3.6). Multiplying both sides of (3.3) by

$$(Q_n(z))^{2(k-1)}g_{kn}(z)$$

and making use of the facts

$$(g_{kn}(z)/g^{kn}(z))^{1/kn} \to 1,$$

$$g^{kn}(z) = \left(\exp\left(\frac{\lambda\gamma}{2}p(\sigma;z)\right)\right)^{2(k-1)n},$$

$$2(k-1)\left(\left(1-\frac{\gamma}{2}\right)-\frac{\gamma}{2}(\lambda-1)\right)+2 = 2k\left(1-\frac{\gamma}{2}\right)$$

(recall that $\lambda = k/(k-1)$) as $n\to\infty$, we get for the monic polynomials

$$P_{kn}(z) = q_n(\mu;z)\cdot Q_n^{k-1}(z)$$

of degree kn from (3.14)–(3.15) and from (3.3) the estimate

$$(3.16)\qquad \limsup_{\substack{n\to\infty\\ n\in\mathcal{N}}}\left(\int|P_{kn}|^2 g_{kn}d\mu\right)^{1/2kn} \le \alpha^{1/k}e^{-d},$$

where

$$(3.17)\qquad d = d_K = \frac{\gamma}{2}\int g_{\Omega_K}(x;\infty)d\sigma(x) + \left(1-\frac{\gamma}{2}\right)\log\frac{1}{\mathrm{cap}(K)} - \frac{1}{k}\log\frac{\mathrm{cap}(S(\mu))}{\mathrm{cap}(K)}.$$

If we replace on the left-hand side of (3.16) the polynomials P_{kn} by the corresponding monic orthogonal polynomial $q_{kn}(\mu_{kn};\cdot)$, then the left-hand side does not increase, hence

$$(3.18)\qquad \limsup_{\substack{n\to\infty\\ n\in\mathcal{N}}}(\gamma_{kn}(\mu_{kn}))^{-1/kn} \le \alpha^{1/k}e^{-d}.$$

Recall now that $d = d_K$ (cf. (3.17)) depends on $K = K_\varepsilon$. If $\varepsilon\to 0$ then $\mathrm{cap}(K_\varepsilon)\to\mathrm{cap}(S(\mu))$ and $g_{\Omega_K}(x;\infty)\to g_\Omega(x;\infty)$ uniformly on the support of σ, and these imply that then $d = d_{K_\varepsilon}$ tends to

$$c = \frac{\gamma}{2}\int g_\Omega(x;\infty)d\sigma(x) + \left(1-\frac{\gamma}{2}\right)\log\frac{1}{\mathrm{cap}(S(\mu))}$$

(see (3.5)). Thus, if $\alpha^{1/k} < \beta < 1$ is fixed then we can choose $0 < \varepsilon < \varepsilon_0$ in such a way that together with (3.18) we also have

$$\limsup_{\substack{n\to\infty \\ n\in N}} (\gamma_{kn}(\mu_{kn}))^{-1/kn} \le \beta e^{-c},$$

and this proves (3.6) with $c_0 = c + \log \frac{1}{\beta}$. □

With respect to the applications in Section 6.1 it is desirable to complete Theorem 3.3.1 by further results about orthonormal polynomials with respect to varying weights. In the next theorem the lower asymptotic bounds (1.1.7) and (1.1.10) of Theorem 1.1.4 are extended to these more general polynomials, and the analogue of Theorem 3.1.1 is proved.

Theorem 3.3.2. *Let $g(z) = \exp(\gamma p(\sigma; z))$ be defined as in* (3.1)*; that is, $0 < \gamma \le 2$ and σ is a probability measure with compact support in Ω. Suppose further that the assumptions of assertion* (vi) *of* Theorem 3.2.1 *are satisfied, that is, $S(\mu) \subseteq \partial\Omega$, $g_n^{1/n} \to g$ uniformly on $S(\mu)$ as $n \to \infty$, and $d\mu_n = g_n d\mu$. Then for any infinite subsequence $N \subseteq \mathbb{N}$ we have*

$$\limsup_{n\to\infty, n\in N} |g_n(z)^{1/2} p_n(\mu_n; z)|^{1/n} \ge 1 \tag{3.19}$$

for quasi every $z \in \partial\Omega$, and

$$\liminf_{n\to\infty} |p_n(\mu_n; z)|^{1/n} \ge \exp\left(\frac{\gamma}{2}(g(\sigma, \Omega; z) - p(\sigma; z)) + \frac{2-\gamma}{2} g_\Omega(z; \infty)\right) \tag{3.20}$$

locally uniformly for $z \in \mathbb{C} \setminus \mathrm{Co}(S(\mu))$, where

$$g(\sigma, \Omega; z) := \int g_\Omega(z; x) d\sigma(x) \tag{3.21}$$

denotes the Green potential of the measure σ in the domain Ω.

Theorem 3.3.3. *With the same assumptions as in* Theorem 3.3.2, *the following four assertions are equivalent.*

(i) $\mu \in \mathbf{Reg}$.

(ii) $$\lim_{n\to\infty} \gamma_n(\mu_n)^{1/n} = \left(\frac{1}{\mathrm{cap}(S(\mu))}\right)^{1-\frac{\gamma}{2}} \exp\left(\frac{\gamma}{2}\int g_\Omega(x; \infty) d\sigma(x)\right).$$

(iii) $$\lim_{n\to\infty} |p_n(\mu_n; z)|^{1/n} = \exp\left(\frac{\gamma}{2}(g(\sigma, \Omega; z) - p(\sigma; z)) + \frac{2-\gamma}{2} g_\Omega(z; \infty)\right)$$

locally uniformly for $z \in \mathbb{C} \setminus \mathrm{Co}(S(\mu))$.

(iv)
$$\limsup_{n\to\infty} |g_n(z)^{1/2} p_n(\mu_n; z)|^{1/n} = 1$$

for quasi every $z \in \partial\Omega$.

Proof of Theorem 3.3.2. The proof is practically a repetition of the proof of the estimates (1.1.7) and (1.1.10) in Theorem 1.1.4. We will therefore closely follow the proof given in Section 1.3.

The estimate (3.19) will be proved first. In this proof we can assume that the potential $p(\sigma;\cdot)$ of σ is continuous throughout $\mathbb{C}$. In fact, if this is not the case then we can replace σ by its balayage onto a regular compact set lying in $\mathbb{C}\setminus \mathrm{supp}(\sigma)$ and containing $\mathrm{Pc}(S(\mu))$ in its interior.

As in (1.3.26), let $N \subseteq \mathbb{N}$ be an infinite subsequence such that the two limits

$$\text{(3.22)} \qquad \frac{1}{n}\nu_{p_n(\mu_n;\cdot)} \xrightarrow{*} \nu \quad \text{and} \quad \frac{1}{n}\log \gamma_n(\mu_n) \to c \in \mathbb{R}\cup\{\infty\}$$

exist as $n \to \infty$, $n \in N$. We have $\mathrm{supp}(\nu) \subseteq \mathrm{Pc}(S(\mu))$ as a consequence of Theorem 2.1.1. With the same arguments as applied for the derivation of (1.3.27) and (1.3.28), and with the fact that $g_n^{1/n} \to g$ uniformly on $S(\mu)$, we derive from (3.22) that

$$\text{(3.23)} \qquad \limsup_{n\to\infty, n\in N} \frac{1}{n}\log |g_n(z)^{1/2} p_n(\mu_n; z)|^{1/n} = p\left(\frac{\gamma}{2}\sigma - \nu; z\right) + c$$

for quasi every $z \in \partial\Omega$, and

$$\text{(3.24)} \qquad \lim_{n\to\infty, n\in N} \frac{1}{n}\log |p_n(\mu_n; z)|^{1/n} = -p(\nu; z) + c$$

locally uniformly for $z \in \mathbb{C}\setminus \mathrm{Co}(S(\mu))$. The lower estimate (3.19) is proved if we show that

$$\text{(3.25)} \qquad p\left(\frac{\gamma}{2}\sigma - \nu; z\right) + c \geq 0$$

for every $z \in \partial\Omega$, which is the analogue of (1.3.30) in the proof of Theorem 1.1.4. This inequality can be proved by contradiction in exactly the same way as has been done in the case of (1.3.30). The general strategy is to replace the measure ν everywhere by $\nu - \frac{\gamma}{2}\sigma$, and the weight measure μ by μ_n. The details are as follows.

Let us assume that, on the contrary, (3.25) does not hold true for all $z \in \partial\Omega$. Then

$$\text{(3.26)} \qquad \partial\Omega \cap A_\delta \neq \emptyset \quad \text{for} \quad A_\delta := \left\{ z \in \mathbb{C} \mid -p\left(\nu - \frac{\gamma}{2}\sigma; z\right) + c < -\delta \right\}$$

for some $\delta > 0$. From the upper semicontinuity of $-p(\nu - \frac{\gamma}{2}\sigma; z)$ (recall that $p(\sigma;\cdot)$ has been assumed to be continuous) it follows that A_δ is open.

Every component A_δ^* of A_δ has the same boundary ∂A_δ^* in the fine topology and in the ordinary planar topology, and this combined with that fact that potentials are continuous in the fine topology implies that on ∂A_δ^* we must have

$$-p\left(\nu - \frac{\gamma}{2}\sigma; z\right) + c \equiv -\delta.$$

Thus, if $\eta > \delta$ and A_η^* is a component of A_η lying in A_δ^*, then we have $\overline{A_\eta^*} \subseteq A_\delta^*$. By the minimum principle for superharmonic functions this is possible only if $A_\delta^* \cap S(\nu) \neq \emptyset$. Hence every component of the set A_δ has a nonempty intersection with $S(\nu)$. Therefore there exists a simply connected domain B with

$$B \cap S(\nu) \neq \emptyset, \qquad B \cap \Omega \neq \emptyset, \qquad \text{and} \qquad \overline{B} \subseteq A_\delta. \tag{3.27}$$

Since B is connected, the set

$$S := \mathrm{Pc}(S(\mu)) \setminus B \tag{3.28}$$

is polynomially convex. Let now $V \subseteq B$ be a compact set with

$$\mathrm{Int}(V) \cap S(\nu) \neq \emptyset. \tag{3.29}$$

Since S is polynomially convex and S and V are disjoint, part (a) of Lemma 1.3.2 can be applied. Let $a < 1$ and $m \in \mathbb{N}$ be the two numbers introduced in part (a) of Lemma 1.3.2. Because of (3.29) and the first limit of (3.22), for sufficiently large $n \in N$ there exist at least m zeros $x_{1,n}, \dots, x_{m,n}$ of the polynomial $p_n(\mu_n; z)$ on V. From Lemma 1.3.2 we know that there also exist m points $y_{1,n}, \dots, y_{m,n} \in \mathbb{C}$ so that the rational function

$$r_m(z) = \prod_{j=1}^{m} \frac{z - y_{j,n}}{z - x_{j,n}}$$

satisfies the inequality

$$\|r_m\|_S \leq a. \tag{3.30}$$

Consider now the modified monic polynomials

$$\tilde{q}_n(z) := r_m(z) q_n(\mu_n; z) = z^n + \cdots .$$

Since the polynomials $\tilde{q}_n$ and $q_n(\mu; z)$, $n \in N$, differ only in m zeros, these two sequences of polynomials have the same asymptotic zero distribution, that is, we have

$$\frac{1}{n}\nu_{\tilde{q}_n} \xrightarrow{*} \nu \quad \text{as} \quad n \to \infty,\ n \in N, \tag{3.31}$$

where ν is the same probability measure as in (3.22). From (3.31), (3.26), the principle of descent of potential theory, and from the limit relation $g_n^{1/n} \to g$, which holds uniformly on $S(\mu)$, we derive that

$$\text{(3.32)} \qquad \limsup_{n\to\infty,\ n\in N} \frac{1}{n}\log|g_n^{1/2}\tilde{q}_n(z)| \le -p\left(\nu - \frac{\gamma}{2}\sigma; z\right) < -\delta - c$$

locally uniformly for $z \in A_\delta \cap S(\mu)$, and so the inequality between the left- and right-hand sides holds true uniformly for $z \in B \cap S(\mu)$. From (3.32) and the second limit in (3.22) it follows that there exists a zero sequence $\{\varepsilon_n > 0\}$ such that the supremum norm of $g_n^{1/2}\tilde{q}_n$ on $B \cap S(\mu)$ satisfies

$$\text{(3.33)} \qquad \|g_n^{1/2}\tilde{q}_n\|_{B\cap S(\mu)} \le \frac{1}{\gamma_n(\mu_n)}e^{-n\delta/2} \le \frac{\varepsilon_n}{\gamma_n(\mu_n)}$$

for all $n \in N$ sufficiently large. For the $L^2(\mu_n)$ norm of $\tilde{q}_n$ we get from (3.28), (3.30) and (3.33)

(3.34)

$$\begin{aligned} \|\tilde{q}_n\|^2_{L^2(\mu_n)} &= \int_{B\cap S(\mu)} |\tilde{q}_n|^2 g_n d\mu + \int_{S\cap S(\mu)} |\tilde{q}_n|^2 g_n d\mu \\ &\le \mu(B)\|g_n^{1/2}\tilde{q}_n\|^2_{B\cap S(\mu)} + \|r_m\|^2_S \int_{S\cap S(\mu)} |q_n(\mu;.)|^2 g_n d\mu \\ &\le \frac{\mu(B)\varepsilon_n^2}{\gamma_n(\mu_n)^2} + \frac{a^2}{\gamma_n(\mu_n)^2} < \frac{1}{\gamma_n(\mu_n)^2} \end{aligned}$$

for all sufficiently large $n \in N$. The estimate (3.34) contradicts the minimality (1.3.10) of the monic orthogonal polynomials $q_n(\mu_n; z)$. Hence, (3.25) and thereby also (3.19) is proved.

Next we prove (3.20). We have

$$\text{(3.35)} \qquad \frac{\gamma}{2}g(\sigma, \Omega; z) + \frac{2-\gamma}{2}g_\Omega(z;\infty) = 0$$

for all $z \in \partial\Omega$ which are regular points with respect to the solution of Dirichlet problems in Ω, and hence this inequality holds quasi everywhere on $\partial\Omega$. We claim that

$$\text{(3.36)} \qquad p\left(\frac{\gamma}{2}\sigma - \nu; z\right) + c \ge \frac{\gamma}{2}g(\sigma,\Omega; z) + \frac{2-\gamma}{2}g_\Omega(z;\infty)$$

for all $z \in \Omega$.

Suppose that on the contrary there is an $\varepsilon > 0$ such that the set

$$A_\varepsilon = \left\{ z \in \Omega \mid p\left(\frac{\gamma}{2}\sigma - \nu; z\right) + c - \frac{\gamma}{2}g(\sigma,\Omega;z) - \frac{2-\gamma}{2}g_\Omega(z;\infty) < -\varepsilon \right\}$$

is not empty. We shall make use of the fact that the expression

$$p(\sigma; z) - g(\sigma, \Omega; z) \tag{3.37}$$

is harmonic in Ω. In fact, this follows from (1.3.16) and (1.3.17) if we consider that the measure ω_ν in (1.3.17) is supported on the boundary of Ω. By the minimum principle for harmonic functions (see, e.g., Theorem III.28 of [Ts]), for every component A_ε^* of A_ε we must have

$$\operatorname{cap}(\partial A_\varepsilon^* \cap \partial\Omega) > 0.$$

Thus, there is a point $z_0 \in \partial A_\varepsilon^* \cap \partial\Omega$ for which (3.35) holds with z replaced by z_0. Since z_0 is also a boundary point of A_ε^* in the fine topology, and the left-hand side of the following inequality is continuous everywhere in this topology, we have

$$p\left(\frac{\gamma}{2}\sigma - \nu; z_0\right) + c - \frac{\gamma}{2} g(\sigma, \Omega; z_0) - \frac{2-\gamma}{2} g_\Omega(z_0; \infty) \le -\varepsilon,$$

which contradicts the choice of z_0. This contradiction proves (3.36).

We get (3.20) from (3.24) and (3.36) if the limit on the left is taken through the subsequence N for which the limits in (3.22) exist. Since the right-hand side is independent of the choice of N, (3.20) is fully proved.

□

Proof of Theorem 3.3.3. We prove the implications (i) ⇔ (iv) and (iii) ⇒ (iv) ⇒ (ii) ⇒ (iii).

(i)⇔(iv): (iv)⇒(i) immediately follows from Theorem 3.3.1. The direction (i)⇒(iv) is obtained from Theorem 3.2.1, (i)⇒(vi) and from (3.19).

(iii)⇒(iv): From (3.23) it follows that assertion (iv) is equivalent to

$$p\left(\frac{\gamma}{2}\sigma - \nu; z\right) + c = 0 \tag{3.38}$$

for quasi every $z \in \partial\Omega$ and for any subsequence $N \subseteq \mathbb{N}$ for which the two limits in (3.22) exist (see also the proof of Theorem 3.1.1 concerning the sufficiency of considering the limits in question for subsequences).

On the other hand, assertion (iii) is equivalent to

$$-p(\nu; z) + c = \frac{\gamma}{2}(g(\sigma, \Omega; z) - p(\sigma; z)) + \frac{2-\gamma}{2} g_\Omega(z; \infty) \tag{3.39}$$

for all $z \in \mathbb{C} \setminus \operatorname{Co}(S(\mu))$ and for any subsequence $N \subseteq \mathbb{N}$ for which the two limits in (3.22) exist. Since $\operatorname{supp}(\nu) \subseteq \operatorname{Pc}(S(\mu))$, we get from the harmonicity of the function in (3.37) that (3.39) holds also for all $z \in \Omega$, and since both sides of (3.39) are logarithmic potentials plus constants (cf. formulas (1.3.16)–(1.3.17)), the equality in (3.39) extends to all $z \in \partial\Omega$, for these functions are continuous in Cartan's fine topology.

Using (3.35) we see immediately that (3.39) implies (3.38) and this proves (iii)⇒(iv).

(iv)⇒(ii): Choose again an $N \subseteq \mathbb{N}$ with property (3.22). If we consider (3.36) around infinity we get

$$c \geq \frac{\gamma}{2} \int g_\Omega(x;\infty) d\sigma(x) + \frac{2-\gamma}{2} \log \frac{1}{\operatorname{cap}(S(\mu))}. \tag{3.40}$$

From (iv) we can conclude (3.38) and hence

$$p\left(\frac{\gamma}{2}\sigma - \nu; z\right) + c = 0$$

holds true for quasi every $z \in \partial\Omega$. Hence the same equality holds true $\omega_{S(\mu)} =: \omega-$almost everywhere and we can write (cf. also the proof of (1.14) in Section 3.1)

$$\begin{aligned} 0 &= \int p(\frac{\gamma}{2}\sigma - \nu; z) d\omega(z) + c = \int p(\omega; z) d(\frac{\gamma}{2}\sigma - \nu)(z) + c \\ &= -\frac{\gamma}{2} \int g_\Omega(z;\infty) d\sigma(z) + (\frac{\gamma}{2} - 1) \log \frac{1}{\operatorname{cap}(S(\mu))} + \int g_\Omega(z;\infty) d\nu(z) + c \\ &\geq -\frac{\gamma}{2} \int g_\Omega(z;\infty) d\sigma(z) - \frac{2-\gamma}{2} \log \frac{1}{\operatorname{cap}(S(\mu))} + c, \end{aligned}$$

which gives the opposite of (3.40). Thus, we have in (3.22)

$$c = \frac{\gamma}{2} \int g_\Omega(x;\infty) d\sigma(x) + \frac{2-\gamma}{2} \log \frac{1}{\operatorname{cap}(S(\mu))},$$

which proves (ii) if the limit in (ii) is taken along N. But then (ii) is generally proved by standard arguments.

(ii)⇒(iii): Let N and c be as in (3.22). If we assume (ii) then the two sides in (3.36) coincide at infinity, hence we must have equality in (3.36) for all $z \in \Omega$, and this proves (iii). □

3.4 Characterization of Regular Asymptotic Behavior in $L^p(\mu)$

Whereas in Section 3.2 regular asymptotic behavior was characterized by relating the size of general polynomials to their $L^2(\mu)$ norms, we now investigate the same problem with respect to $L^p(\mu)$ norms for $1 \leq p < \infty$, or $L^p(\mu)$ quasinorms for $0 < p < 1$.

The minimality of monic orthogonal polynomials $q_n(\mu; z)$ in $L^2(\mu)$ (see (1.3.10)), which is equivalent to their orthogonality, motivates us to consider also minimal monic polynomials

$$q_n(\mu, p; z) = z^n + \cdots, \quad n \in \mathbb{N}, \tag{4.1}$$

in $L^p(\mu)$ for $0 < p \leq \infty$, that is,

$$\|q_n(\mu, p; z)\|_{L^p(\mu)} = \inf_{q(z)=z^n+\cdots} \|q\|_{L^p(\mu)}. \tag{4.2}$$

The existence of $q_n(\mu, p; z)$ follows for every $n \in \mathbb{N}$ from a simple compactness argument, and we get from Lemma 1.3.2 that all zeros of $q_n(\mu, p; z)$ lie in the convex hull of $S(\mu)$ (see the proof of Theorem 2.1.1). In general, these polynomials are not unique if $0 < p < 1$. We note that $\|\cdot\|_{L^p(\mu)}$ is a norm only for $1 \leq p \leq \infty$, whereas for $0 < p < 1$ the triangle inequality does not hold, and $\|\cdot\|_{L^p(\mu)}$ is therefore only a quasinorm.

For $p = 2$ the polynomials (4.1) are the monic orthogonal polynomials with respect to μ, and for $p = \infty$ they are the Chebyshev polynomials on $S(\mu)$ since for continuous functions $\|\cdot\|_{L^\infty(\mu)}$ is identical with the supremum norm $\|\cdot\|_{\sup, S(\mu)}$ on $S(\mu)$.

Theorem 3.4.1. *If*

$$\lim_{n\to\infty} \left(\inf_{q(z)=z^n+\cdots} \|q\|_{L^p(\mu)} \right)^{1/n} = \operatorname{cap}(S(\mu)) \tag{4.3}$$

holds for one p *with* $0 < p < \infty$, *then* (4.3) *holds true for all* $0 < p \leq \infty$.

Remarks. (1) The case $p = \infty$ is special as then the minimal polynomials $q_n(\mu, \infty; z)$, $n \in \mathbb{N}$, depend only on the support $S(\mu)$ and not on the measure μ itself, and (4.3) holds for any measure μ with compact support (see [Ts, Theorem III. 26]).

(2) By the inequalities

$$\mu(\mathbb{C})^{-1/p} \|f\|_{L^p(\mu)} \leq \mu(\mathbb{C})^{-1/q} \|f\|_{L^q(\mu)} \leq \|f\|_{\sup, S(\mu)}, \tag{4.4}$$

which hold for $0 < p \leq q \leq \infty$ and any bounded measurable function f, we have

(4.5)

$$\limsup_{n\to\infty} \left(\inf_{q_n(z)=z^n+\cdots} \|q_n\|_{L^p(\mu)} \right)^{1/n} \leq \limsup_{n\to\infty} \left(\inf_{q_n(z)=z^n+\cdots} \|q_n\|_{L^q(\mu)} \right)^{1/n}$$
$$\leq \operatorname{cap}(S(\mu))$$

for all $0 < p \leq q \leq \infty$ and any weight measure μ. The inequalities in (4.5) show that in (4.3) only the proper limit and an lower estimate by $\operatorname{cap}(S(\mu))$ are critical.

Equality in (4.3) means that the weight measure μ has a certain sufficient density in a neighborhood of $\partial\Omega$. From Theorem 3.4.1 together with the definition of regularity of a weight measure μ in (1.4) and the identity $\|q_n(\mu, 2; \cdot)\|_{L^2(\mu)} = 1/\gamma_n(\mu)$ (see (1.3.10)) we derive the next corollary, which shows that the definition of regularity (Definition 3.1.2) is not specific for orthogonal polynomials; it can be equivalently based on the size of monic minimal polynomials $q_n(\mu, p; z)$, $n \in \mathbb{N}$, in $L^p(\mu)$, $0 < p < \infty$.

Corollary 3.4.2. *Let* $0 < p < \infty$. *Then* $\mu \in \mathbf{Reg}$ *if and only if* (4.3) *holds.*

Proof of Theorem 3.4.1. If $\text{cap}(S(\mu)) = 0$, then (4.5) proves our claim. Hence, let $\text{cap}(S(\mu)) > 0$, and then, by making use of a linear transformation, we may assume without loss of generality that $\text{cap}(S(\mu)) = 1$.

In view of (4.5) it is enough to show that if for some $0 < p < \infty$ we have

$$\liminf_{n\to\infty} \|q_n(\mu, p; \cdot)\|_{L^p(\mu)}^{1/n} < 1, \tag{4.6}$$

then for all r with $p < r$

$$\liminf_{n\to\infty} \left(\inf_{q_n(z)=z^n+\dots} \|q_n\|_{L^r(\mu)}^{1/n} \right) < 1. \tag{4.7}$$

By (4.6) there is an $\alpha < 1$ and an infinite sequence N_1 such that for all $n \in N_1$

$$\|q_n(\mu, p; \cdot)\|_{L^p(\mu)}^{1/n} < \alpha. \tag{4.8}$$

If s is any fixed natural number and $[n/s]$ denotes the largest integer not greater than n/s, then the relation

$$\|q_{s([n/s]+1)}(\mu, p; \cdot)\|_{L^p(\mu)}^{1/n} \le C_s \|q_n(\mu, p; \cdot)\|_{L^p(\mu)}^{1/n},$$

which follows, with some constant C_s depending only on $S(\mu)$ and the number s, from the definition of the extremal polynomials $q_n(\mu, p; \cdot)$, shows that we can replace all large $n \in N_1$ by $s([n/s]+1)$. Hence, we can assume without loss of generality that *for every natural number s all but finitely many terms in N_2 are divisible by s.*

We can also select a subsequence $N_2 \subseteq N_1$ such that the limits

$$\lim_{n\to\infty, n\in N_2} \nu_{q_n(\mu,p;\cdot)} = \nu, \qquad \lim_{n\to\infty, n\in N_2} \frac{1}{n} \log \frac{1}{\|q_n(\mu, p; \cdot)\|_{L^p(\mu)}} = c \tag{4.9}$$

exist. It is easy to check that the proof of Theorem 2.1.1, which gives information on the location of minimal $L^2(\mu)$ polynomials, carries over to the $L^p(\mu)$-extremal polynomials $q_n(\mu, p; \cdot)$. Therefore, we know that ν is a probability measure with support in $\text{Pc}(S(\mu))$.

As a further simplification, we may assume $c < \infty$, for in the opposite case (4.7) trivially holds by choosing $q_n(z) = q_n(\mu, p; z)$ for $n = 1, 2, \dots$ (taking into account that the norms $\|q_n(\mu, p; \cdot)\|_{\text{Pc}(S(\mu))}^{1/n}$ are uniformly bounded in n because all the zeros of $q_n(\mu, p; \cdot)$ are contained in the convex hull of $S(\mu)$).

Now we need some basic facts from the theory of weighted potentials discussed in Appendix IX. Set $\Sigma = \mathrm{Pc}(S(\mu))$, $Q(z) = p(\nu; z)$, and $w(z) = \exp(-Q(z)) = \exp(-\lambda p(\nu; z))$ for $z \in \Sigma$. Then w^λ is admissible for all $\lambda > 0$ in the sense of Appendix IX, so we can consider the problem of finding the minimal energy

$$\inf\left\{\int\int \log\frac{1}{|t-z|}d\omega(t)d\omega(z) + 2\lambda\int Q(t)d\omega(t)\right\},$$

where the infimum is taken for all probability measures ω supported on Σ. Let ω_{w^λ} be the unique minimizing measure. Then we know that for some constant F_{w^λ}

$$p(\omega_{w^\lambda}; z) + \lambda Q(z) = p(\omega_{w^\lambda}; z) + \lambda p(\nu; z) \geq F_{w^\lambda} \tag{4.10}$$

for quasi every $z \in \mathrm{Pc}(S(\mu))$, and we know that the constant F_{w^λ} and the corresponding weighted Chebyshev constant, which is defined by

$$t^{w^\lambda} = \lim_{n\to\infty}\left\{\inf_{P_n(z)=z^n+\cdots} \|w^{\lambda n}P_n\|_\Sigma^{1/n}\right\},$$

are connected by

$$t^{w^\lambda} = \exp(-F_{w^\lambda}) \tag{4.11}$$

(see (A.30) in Appendix IX). Furthermore, if S_{w^λ} is the support of ω_{w^λ}, then we know that S_{w^λ} decreases as λ increases.

We shall need some further preparation before we can turn to the proof of (4.7). First we show that

$$t^{w^\lambda} \to \mathrm{cap}(S(\mu)) = 1 \quad \text{as} \quad \lambda \to 0+0. \tag{4.12}$$

It is clear that $t^{w^\lambda} \leq t^1 C^\lambda$, where t^1 denotes the Chebyshev constant corresponding to the weight identically 1 on Σ, and C is an upper bound of w on Σ. By Szegő's theorem (see Appendix IX, especially (A.30) for $w \equiv 1$) $t^1 = \mathrm{cap}(\Sigma)$, hence from $\Sigma = \mathrm{Pc}(S(\mu))$ we can deduce

$$\limsup_{\lambda\to 0} t^{w^\lambda} \leq \mathrm{cap}(S(\mu)).$$

On the other hand, if $\gamma < 1$ then there is an $\epsilon > 0$ such that for the set

$$E_\epsilon = \{z \in \mathrm{Pc}(S(\mu)) \,|\, w(z) > \epsilon\}$$

we have

$$\mathrm{cap}(E_\epsilon) > \gamma\,\mathrm{cap}(S(\mu)),$$

and so for any monic polynomial P_n of degree n

$$\|w^{n\lambda}P_n\|_{\mathrm{Pc}(S(\mu))}^{1/n} \geq \epsilon^\lambda \|P_n\|_{E_\epsilon}^{1/n} \geq \epsilon^\lambda \mathrm{cap}(E_\epsilon) > \epsilon^\lambda \gamma \mathrm{cap}(S(\mu)),$$

where we have used the fact that

$$\|P_n\|_K \geq \mathrm{cap}(K)$$

(see (A.29) in Appendix IX for $w \equiv 1$) for every compact K. This, together with what we have shown previously proves (4.12).

Paraphrasing (4.12) we can write (cf. (4.11))

$$\lim_{\lambda\to 0} F_{w^\lambda} = 0. \tag{4.13}$$

In (4.17) we shall prove a stronger version of this limit relation, but first we show that (4.13) implies

$$\mathrm{cap}(\mathcal{S}_{w^\lambda}) \to 1 \quad \text{as} \quad \lambda \to 0. \tag{4.14}$$

In fact, for any monic polynomial P_n of degree n we have (cf. (A.29) in Appendix IX)

$$\exp(-F_{w^\lambda}) \leq \|w^{n\lambda}P_n\|_{\mathcal{S}_{w^\lambda}}^{1/n} \leq C^\lambda \|P_n\|_{\mathcal{S}_{w^\lambda}}^{1/n},$$

where C is an upper bound for w on $\mathrm{Pc}(S(\mu))$. Hence (4.14) follows from (4.13) and the fact that there are monic polynomials P_n for which

$$\lim_{n\to\infty} \|P_n\|_{\mathcal{S}_{w^\lambda}}^{1/n} = \mathrm{cap}(\mathcal{S}_{w^\lambda})$$

(see (A.30) in Appendix IX).

Now let

$$\mathcal{S}_0 = \bigcup_{\lambda>0} \mathcal{S}_{w^\lambda}.$$

Then $\mathrm{cap}(\mathcal{S}_0) = 1 = \mathrm{cap}(S(\mu))$ and $\mathcal{S}_0 \subseteq \mathrm{Pc}(S(\mu))$, hence

$$g_{\mathbf{C}\setminus\mathcal{S}_0}(z;\infty) = g_{\mathbf{C}\setminus\mathrm{Pc}(S(\mu))}(z;\infty),$$

furthermore, from the decreasing character of the sets $\mathcal{S}_{w^\lambda}$ we get

$$g_{\mathbf{C}\setminus\mathcal{S}_{w^\lambda}}(z;\infty) \searrow g_{\mathbf{C}\setminus\mathcal{S}_0}(z;\infty) = g_{\mathbf{C}\setminus\mathrm{Pc}(S(\mu))}(z;\infty)$$

monotonically as $\lambda \searrow 0$ (see the proof of Theorem A.1 in Appendix IV). Hence by the monotone convergence theorem

$$\int g_{\mathbf{C}\setminus\mathcal{S}_{w^\lambda}}(z;\infty)d\nu(z) \searrow \int g_{\mathbf{C}\setminus\mathrm{Pc}(S(\mu))}(z;\infty)d\nu(z).$$

Next we verify that here the right-hand side is zero, that is,

$$\int g_{\mathbf{C}\setminus \mathrm{Pc}(S(\mu))}(z;\infty)d\nu(z) = 0, \tag{4.15}$$

which, together with the previous relations, will prove

$$\lim_{\lambda\to 0}\int g_{\mathbf{C}\setminus S_{w^\lambda}}(z;\infty)d\nu(z) = 0. \tag{4.16}$$

Since $\mathrm{supp}(\nu) \subseteq \mathrm{Pc}(S(\mu))$, and $g_{\mathbf{C}\setminus \mathrm{Pc}(S(\mu))}(z;\infty) = 0$ for quasi every $z \in \mathrm{Pc}(S(\mu))$, the relation (4.15) is false only if the set

$$\mathrm{Pc}(S(\mu))_I = \{z \in \mathrm{Pc}(S(\mu)) \,|\, g_{\mathbf{C}\setminus \mathrm{Pc}(S(\mu))}(z;\infty) > 0\}$$

of the so-called irregular points of $\mathrm{Pc}(S(\mu))$ (see Appendix VI) has positive ν measure.

Now we show that this is not the case. Indeed, as the proof of (1.3.30) can be repeated word for word for the polynomials $q_n(\mu, p; \cdot)$ having minimal $L^p(\mu)$ norm, we get that

$$p(\nu; z) \le c, \quad z \in \partial\Omega$$

(for the definition of c see (4.9)). In fact, the key ingredient of that proof is Lemma 1.3.2, which is independent of p. Since we have assumed that the constant c is finite, it follows that the restriction of ν to $\partial\Omega$ has finite logarithmic energy. But the set of irregular points is of zero capacity and lies on $\partial\Omega$, hence $\nu(\mathrm{Pc}(S(\mu))_I) = 0$ must hold, as we have claimed.

With (4.16) we are prepared to prove that

$$\lim_{\lambda\to 0} F_{w^\lambda}/\lambda = 0. \tag{4.17}$$

From formula (A.28) in Appendix IX we know that

$$F_{w^\lambda} = \log\frac{1}{\mathrm{cap}(S_{w^\lambda})} - \lambda\int p(\nu; z)d\omega_{S_{w^\lambda}}(z) \ge \lambda\int p(\omega_{S_{w^\lambda}}; z)d\nu(z),$$

where $\omega_{S_{w^\lambda}}$ denotes the equilibrium measure of the compact set S_{w^λ}, and where we have also used the fact that this latter set is part of $\mathrm{Pc}(S(\mu))$, and hence it has capacity at most 1. This inequality together with (4.14), (4.16), and the formula

$$p(\omega_{S_{w^\lambda}}; z) = g_{\mathbf{C}\setminus S_{w^\lambda}}(z;\infty) + \log\frac{1}{\mathrm{cap}(S_{w^\lambda})}$$

proves

$$\liminf_{\lambda\to 0} F_{w^\lambda}/\lambda \ge 0.$$

On the other hand, using (4.10) and the fact that the set of exceptional points of $\mathrm{Pc}(S(\mu))$ for which the inequality (4.10) does not hold is of capacity zero, and hence it has $\omega_{S(\mu)}$ measure zero, we can write

$$F_{w^\lambda} \le \int p(\omega_{S_{w^\lambda}} + \lambda\nu; z)d\omega_{S(\mu)}(z) = \int p(\omega_{S(\mu)}; z)d(\omega_{S_{w^\lambda}} + \lambda\nu) \le 0$$

because by (A.8)(α)

$$p(\omega_{S(\mu)}; z) \le \log \frac{1}{\mathrm{cap}(S(\mu))} = 0$$

for all $z \in \mathbf{C}$. These equations prove (4.17).

After these preparations we are ready to prove that (4.6) implies (4.7). By (4.11) and (4.17), if $\epsilon > 0$ is given, then there is a $\lambda_\epsilon > 0$ such that for all $0 < \lambda < \lambda_\epsilon$ there is a monic polynomial R_{n_λ} of degree n_λ such that

$$(4.18) \qquad w^\lambda(z)|R_{n_\lambda}(z)|^{1/n_\lambda} \le (1+\epsilon)^\lambda \quad \text{for all} \quad z \in \mathrm{Pc}(S(\mu)).$$

Let $r > p$ be arbitrary. Choose an $\epsilon > 0$ so that

$$(4.19) \qquad (1+2\epsilon)^{r-p}\alpha^p < 1$$

is satisfied (for the definition of α see (4.8)) and for this ϵ choose $0 < \lambda < \lambda_\epsilon$ of the form $\lambda = (r-p)/rk$, where k is an integer. From (4.18) and the principle of descent we get

$$(4.20) \qquad \left(\limsup_{m\to\infty, m\in N_2} |q_m(\mu, p; z_m)|^{1/m}\right)^\lambda |R_{n_\lambda}(z)|^{1/n_\lambda} \le (1+\epsilon)^\lambda$$

for every $z \in \mathrm{Pc}(S(\mu))$ and every sequence $\{z_m\}$ converging to z. We claim that this implies

$$(4.21) \qquad \limsup_{m\to\infty, m\in N_2} \left\| (|q_m(\mu, p; \cdot)|^{1/m})^\lambda |R_{n_\lambda}|^{1/n_\lambda} \right\|_{\mathrm{Pc}(S(\mu))} \le (1+\epsilon)^\lambda.$$

In fact, in the opposite case there is a $\gamma > (1+\epsilon)^\lambda$, a subsequence $N_3 \subseteq N_2$, and for each $m \in N_3$ a point $z_m \in \mathrm{Pc}(S(\mu))$ such that

$$(4.22) \qquad \left(|q_m(\mu, p; z_m)|^{1/m}\right)^\lambda |R_{n_\lambda}(z_m)|^{1/n_\lambda} \ge \gamma.$$

Without loss of generality we can assume that the sequence $\{z_m\}_{m\in N_3}$ converges to some point z of $\mathrm{Pc}(S(\mu))$. But the sequence $\{\|q_m(\mu, p; \cdot)\|^{1/m}_{\mathrm{Pc}(S(\mu))}\}$ is bounded and the function $|R_{n_\lambda}(z)|^{1/n_\lambda}$ is continuous on $\mathrm{Pc}(S(\mu))$, hence (4.22) contradicts (4.20), and this contradiction proves (4.21).

We have assumed that all but finitely many terms in N_2 are divisible by n_λ, hence these terms can be written in the form $m = n_\lambda l$ with an integer l. From (4.21) and $\lambda = (r-p)/rk$ we get for each large m

$$\left\| |q_m(\mu,p;\cdot)|^{r-p} |R_{n_\lambda}|^{rkl} \right\|_{\mathrm{Pc}(S(\mu))} \le (1+2\epsilon)^{(r-p)m}. \tag{4.23}$$

Consider now for $m = ln_\lambda \in N_2$ the monic polynomials $q_m(\mu,p;z)\times R_{n_\lambda}(z)^{lk}$ of degree $m + n_\lambda lk = m(1+k)$. For their $L^r(\mu)$ norm we have

$$\begin{aligned}
\|q_m(\mu,p;\cdot)R_{n_\lambda}^{lk}\|_{L^r(\mu)}^{1/m(1+k)} &= \left(\int |q_m(\mu,p;\cdot)R_{n_\lambda}^{lk}|^r d\mu\right)^{1/rm(1+k)} \\
&\le \left(\int \left\| |q_m(\mu,p;\cdot)|^{r-p}|R_{n_\lambda}|^{rkl}\right\|_{\mathrm{Pc}(S(\mu))} |q_m(\mu,p;\cdot)|^p d\mu\right)^{1/rm(1+k)} \\
&\le \left((1+2\epsilon)^{(r-p)m}\alpha^{mp}\right)^{1/rm(1+k)} = \left((1+2\epsilon)^{r-p}\alpha^p\right)^{1/r(1+k)} < 1,
\end{aligned}$$

where we have used (4.23), (4.8), and (4.19). The last relation verifies (4.7), and the proof is complete. □

Like the definition of regularity of a weight measure μ, also its characterization given in Theorem 3.2.1 could be extended to L^p norms or quasinorms. However, the proofs tend to be rather technical, so instead we only consider the case when the support $S(\mu)$ is a regular set with respect to the Dirichlet problem, that is, we consider the analogue of Theorem 3.2.3 but now with L^p norms or quasinorms instead of L^2 norms.

Theorem 3.4.3. *Suppose that $S(\mu)$ is a regular set with respect to the Dirichlet problem in Ω and $0 < p < \infty$. Then the following five statements are equivalent.*

(i) $\mu \in \mathbf{Reg}$.

(ii)
$$\limsup_{n\to\infty} \left(\frac{|q_n(\mu,p;z)|}{\|q_n(\mu,p;\cdot)\|_{L^p(\mu)}}\right)^{1/n} e^{-g_\Omega(z;\infty)} \le 1$$

uniformly on $\mathbb{C}$.

(iii)
$$\lim_{n\to\infty} \left(\sup_{\substack{\deg P_n \le n \\ P_n \not\equiv 0}} \frac{|P_n(z)|}{\|P_n\|_{L^p(\mu)}}\right)^{1/n} e^{-g_\Omega(z;\infty)} = 1$$

uniformly on $\mathbb{C}$.

(iv)
$$\lim_{n\to\infty} \left(\frac{\|q_n(\mu,p;\cdot)\|_{\sup,S(\mu)}}{\|q_n(\mu,p;\cdot)\|_{L^p(\mu)}}\right)^{1/n} = 1,$$

where $\|\cdot\|_{\sup,S(\mu)}$ *denotes the supremum norm on* $S(\mu)$.

(v)
$$\lim_{n\to\infty}\left(\sup_{\substack{\deg P_n\le n\\ P_n\not\equiv 0}}\frac{\|P_n\|_{\sup,S(\mu)}}{\|P_n\|_{L^p(\mu)}}\right)^{1/n}=1.$$

If in addition $S(\mu)\subseteq\partial\Omega$, *then any of* (i)–(v) *is equivalent to the following statement.*

(vi) *If* $\{g_n\}$ *is a sequence as in assertion* (vi) *of* Theorem 3.2.1, *then*

$$\lim_{n\to\infty}\left(\sup_{\substack{\deg P_n\le n\\ P_n\not\equiv 0}}\frac{\|g_n^{1/p}P_n\|_{\sup,S(\mu)}}{\|P_n\|_{L^p(\mu_n)}}\right)^{1/n}=1,$$

where $d\mu_n := g_n d\mu$ *are the measures from* Theorem 3.2.1.

Proof of Theorem 3.4.3. We mention first of all that if for a sequence $\{P_n\}$ of polynomials we have

$$\lim_{n\to\infty}\left(\frac{\|P_n\|_{\sup,S(\mu)}}{\|P_n\|_{L^2(\mu)}}\right)^{1/n}=1,$$

then the same is true with $L^p(\mu)$ norms:

$$\lim_{n\to\infty}\left(\frac{\|P_n\|_{\sup,S(\mu)}}{\|P_n\|_{L^p(\mu)}}\right)^{1/n}=1.$$

For $p\ge 2$ this follows from Hölder's inequality, whereas for $p<2$ we can apply the estimate

$$\int |P_n|^2 d\mu \le \|P_n\|_{\sup,S(\mu)}^{2-p}\int |P_n|^p d\mu.$$

Hence, in this case all the $L^p(\mu)$ norms are asymptotically equivalent in the nth-rooth sense.

But then any of the implications (i)⇒(j) where (j) is any of (ii)–(v) follows from the corresponding implications in Theorem 3.2.3 (use also Theorem 3.2.3, (i)⇒(v)). As for (i)⇒(vi), the proof of this is also identical with the proof of the corresponding implication in Theorem 3.2.3 if we use (i)⇒(v) and replace 2 by p throughout in that proof. Now since each of the statements (ii)–(vi) implies (iv), it remains only to prove that (iv) implies (i).

But this is a consequence of Theorem 3.4.1. Indeed, for any monic polynomial of degree n, hence in particular for $q_n(\mu,p;\cdot)$, we have

$$\|q_n(\mu,p;\cdot)\|_{\sup,S(\mu)}^{1/n}\ge \operatorname{cap}(S(\mu)),$$

(see (A.29) in Appendix IX with the special choice $w \equiv 1$, or [Ts, Section III.5]), and so (iv) implies

$$\lim_{n\to\infty} \|q_n(\mu, p; \cdot)\|_{L^p(\mu)}^{1/n} = \operatorname{cap}(S(\mu)) \tag{4.24}$$

(recall that the inequality

$$\limsup_{n\to\infty} \|q_n(\mu, p; \cdot)\|_{L^p(\mu)}^{1/n} \le \operatorname{cap}(S(\mu))$$

automatically holds since for the Chebyshev polynomials $T_n(z) = z^n + \cdots$ associated with $S(\mu)$ we have

$$\lim_{n\to\infty} \|T_n\|_{\sup, S(\mu)}^{1/n} = \operatorname{cap}(S(\mu))$$

(see [Ts, Theorem III.26] or (A.30) in Appendix IX with the special choice $w \equiv 1$)). Now having (4.24) at our disposal, all we have to do to conclude the implication (iv)$\Rightarrow$(i) is to apply Theorem 3.4.1. □

3.5 Examples

Three examples will be given, which supplement our results from Sections 3.1 and 3.2.

Example 3.5.1. We construct a weight measure μ with $S(\mu) = [-2, 2]$, $c_\mu = 0$, and $\mu \in \mathbf{Reg}$.

This example shows that $\mu \in \mathbf{Reg}$ can happen even if the lower and upper bounds in Theorem 1.1.4 are different.

Actually such an example trivially follows from the regularity criteria to be discussed in the next chapter but we need the bases of the present construction in the more sophisticated Example 3.5.2.

Let $t_m(x) := 2\cos(m \arccos(x/2))$, $-2 \le x \le 2$, $m \in \mathbb{N}$, be the monic Chebyshev polynomials of the first kind on $[-2, 2]$, and set $S_n := Z(t_m)$, the set of zeros of t_m for $m = 3^n$, $n \in \mathbb{N}$. Each set S_n consists of 3^n distinct points that are contained in $[-2, 2]$, and $S_n \subseteq S_{n+1}$. Set

$$\mu_n := 3^{-n} \sum_{x \in S_n} \delta_x \quad \text{for} \quad n \in \mathbb{N}, \tag{5.1}$$

and define the weight measure μ by

$$\mu := \sum_{n=1}^{\infty} a_n \mu_n \quad (a_n > 0, \ \sum a_n < \infty), \tag{5.2}$$

where in this example we set

$$a_n := (1 - b)b^{n-1} \quad \text{for some} \quad 0 < b < 1 \quad \text{and} \quad n = 1, 2, \ldots. \tag{5.3}$$

We have $\|\mu\| = \sum a_n = 1$ since the μ_n are probability measures. Since $S_1 \cup S_2 \cup \cdots$ is a countable carrier of μ, we have $c_\mu = 0$.

From the Gauss–Jacobi quadrature formula it follows that

(5.4)

$$\int p d\mu_n = \frac{1}{\pi} \int_{-2}^{2} \frac{p(x)dx}{\sqrt{4-x^2}} \quad \text{for polynomials} \quad p \quad \text{with} \quad \deg(p) \leq 2 \cdot 3^n - 1,$$

and therefore

$$p_m(\mu_n; z) \equiv \frac{1}{\sqrt{2}} t_m(z) \qquad \left(\gamma_m(\mu_n) = \frac{1}{\sqrt{2}}\right) \quad \text{for} \quad m < 3^n. \tag{5.5}$$

From Lemma 1.5.3 together with (5.2) we know that

$$\gamma_m(\mu) \leq \sqrt{\frac{1}{a_n}} \gamma_m(\mu_n) \leq \frac{b^{(1-n)/2}}{\sqrt{2(1-b)}} \quad \text{for} \quad m < 3^n. \tag{5.6}$$

If we use estimate (5.6) for $3^{n-1} \leq m < 3^n$, $n = 1, 2, \ldots$, then we easily derive

$$\limsup_{m \to \infty} \gamma_m(\mu)^{1/m} \leq 1. \tag{5.7}$$

Together with the lower bound of Corollary 1.1.7 this implies that $\gamma_m(\mu)^{1/m} \to 1$ as $m \to \infty$, which shows $\mu \in \mathbf{Reg}$ (see Definition 3.1.2).

□

The next example shows that, in the case $c_\mu = 0$, regular asymptotic distribution of the zeros of the orthonormal polynomials $p_n(\mu; z)$, $n \in \mathbb{N}$, does not necessarily imply $\mu \in \mathbf{Reg}$. The example further shows that the assumption $c_\mu > 0$ is really necessary in Theorem 2.2.1.(a) and Corollary 2.2.3.

Example 3.5.2. We construct a weight measure μ with $S(\mu) = [-2, 2]$, $c_\mu = 0$, and $\mu \notin \mathbf{Reg}$, but in spite of $\mu \notin \mathbf{Reg}$, the orthonormal polynomials $p_n(\mu; z)$, $n \in \mathbb{N}$, have regular asymptotic zero distribution. Furthermore, we will have

$$1 < \liminf_{m \to \infty} \gamma_m(\mu)^{1/m} < \limsup_{m \to \infty} \gamma_m(\mu)^{1/m}. \tag{5.8}$$

Another measure μ with similar properties but for which $\gamma_n(\mu)^{1/n}$ tends to infinity can be obtained from the construction that we used in the proof of Theorem 2.1.4.

The measure μ is defined by (5.1) and (5.2) as in Example 3.5.1, but now the coefficients $a_n > 0$, $n \in \mathbb{N}$, tend faster to zero and they are defined by

$$a_n := (1 - b^3) b^{3^n} \quad \text{for some} \quad 0 < b < 1 \quad \text{and} \quad n = 1, 2, \ldots. \tag{5.9}$$

We first show that $\mu \notin \mathbf{Reg}$. Let $\tau = \omega_{[-2,2]}$ be the equilibrium distribution on $[-2,2]$, that is,

$$d\tau(x) := \frac{dx}{\pi\sqrt{4-x^2}}, \quad -2 \le x \le 2, \tag{5.10}$$

and set

$$\mu^{(n)} := \sum_{j=n}^{\infty} a_j \mu_j. \tag{5.11}$$

From (5.9) it follows that

$$(1-b^3)b^{3^n} \le \|\mu^{(n)}\| = \sum_{j=n}^{\infty} a_j \le b^{3^n} \quad \text{for} \quad n = 1, 2, \ldots. \tag{5.12}$$

The measure τ has already been used on the right-hand side of (5.4). The monic Chebyshev polynomials t_m are not only orthogonal with respect to μ_n for $m < 3^n$ (see (5.5)), but also with respect to the measure τ, and we have

$$\|t_m\|_{L^2(\tau)} = \|t_m\|_{L^2(\mu_n)} = \sqrt{2} \quad \text{for} \quad m < 3^n. \tag{5.13}$$

On $[-2,2]$ all t_m are bounded by 2. In order to derive a lower estimate for the leading coefficients $\gamma_m(\mu)$ we use the test polynomials

$$q_m(z) := t_{m-3^{n-1}}(z)\, t_{3^{n-1}}(z) \quad \text{for} \quad 3^{n-1} \le m < 3^n \tag{5.14}$$

in the minimality property (1.3.10) of the monic orthogonal polynomial $q_m(\mu; z)$. The test polynomial q_m has zeros at every point of the set S_{n-1} if $3^{n-1} \le m < 3^n$. From (1.3.10) and (5.13) it follows that

$$\frac{1}{\gamma_m(\mu)^2} \le \int |q_m|^2 d\mu = \int |q_m|^2 d\mu^{(n)} \le 4 \int |t_{3^{n-1}}|^2 d\mu^{(n)} = 8\|\mu^{(n)}\| \tag{5.15}$$

for $3^{n-1} \le m < 3^n$, which, with (5.12), implies that

$$\gamma_m(\mu)^2 \ge \frac{1}{8\|\mu^{(n)}\|} \ge \frac{1}{8}\left(\frac{1}{b}\right)^{3^n} \quad \text{for} \quad 3^{n-1} \le m < 3^n, \tag{5.16}$$

and, furthermore,

$$\liminf_{m\to\infty} \gamma_m(\mu)^{1/m} \ge \sqrt{\frac{1}{b}} > 1. \tag{5.17}$$

The estimate (5.17) shows that $\mu \notin \mathbf{Reg}$.

In order to show that (5.8) holds true we have also to establish an upper estimate for the coefficients $\gamma_m(\mu)$. From Lemma 1.5.3 together with (5.4) and (5.12) it follows that

(5.18)
$$\gamma_m(\mu) \le \gamma_m(\mu^{(n)}) = \frac{1}{\sqrt{\|\mu^{(n)}\|}}\gamma_m(\tau) \le (2(1-b^3)b^{3^n})^{-1/2} \quad \text{for} \quad m < 3^n.$$

From (5.16) and (5.18) we derive

$$(5.19) \quad 8^{-1/2} \le b^{3^n/2}\gamma_m(\mu) \le (2(1-b^3))^{-1/2} \quad \text{for} \quad 3^{n-1} \le m < 3^n.$$

If $N := \{3^n \mid n \in \mathbb{N}\}$, then from (5.19) it follows that

$$(5.20) \qquad \lim_{m\to\infty,\ m\in N} \gamma_m(\mu)^{1/m} = \left(\frac{1}{b}\right)^{3/2}$$

and

$$(5.21) \qquad \lim_{m\to\infty,\ m\in N} \gamma_{m-1}(\mu)^{1/(m-1)} = \left(\frac{1}{b}\right)^{1/2}.$$

Since these two limits are different, inequality (5.8) is proved.

Last we show that the orthonormal polynomials $p_m(\mu; z)$, $m \in \mathbb{N}$, have regular asymptotic zero distribution. For this purpose we consider the monic orthogonal polynomials $q_m(\mu; z)$ instead of the orthonormal ones. We have

$$8\|\mu^{(n)}\| \ge \|q_m(\mu;\cdot)\|^2_{L^2(\mu)} \ge \|q_m(\mu;\cdot)\|^2_{L^2(\mu^{(n)})} = \|\mu^{(n)}\|\ \|q_m(\mu;\cdot)\|^2_{L^2(\tau)}$$

for $3^{n-1} \le m < 3^n$. Indeed, the first inequality was proved in (5.15) , the second one follows from $\mu \ge \mu^{(n)}$, and the equality follows from (5.4) and (5.11). From this we derive

$$(5.22) \qquad \|q_m(\mu;\cdot)\|_{L^2(\tau)} \le \sqrt{8}.$$

From Theorem 1.1.8 it follows with (5.22) that

$$(5.23) \quad \limsup_{m\to\infty} |q_m(\mu; z)|^{1/m} \le \limsup_{m\to\infty} \left|\frac{q_m(\mu; z)}{\|q_m(\mu;\cdot)\|_{L^2(\tau)}}\right|^{1/m} \le e^{g_\tau(z;\infty)}$$

locally uniformly for $z \in \mathbb{C}$. Since τ is identical with the equilibrium distribution $\omega_{[-2,2]}$ of the interval $[-2, 2]$, we know from Lemma 1.2.7 that $g_\tau(z;\infty) \equiv g_\Omega(z;\infty)$ with $\Omega = \overline{\mathbb{C}} \setminus [-2,2]$. Hence, from (5.23) it follows that

$$\limsup_{m\to\infty} |q_m(\mu; z)|^{1/m} \le 1 = \exp(p(\omega_{[-2,2]}; z) - \max_{z\in[-2,2]} p(\omega_{[-2,2]}; z))$$

uniformly on $[-2,2]$, and so

$$\frac{1}{m}\nu_{q_m(\mu;\cdot)} \xrightarrow{*} \omega_{[-2,2]}, \quad m \to \infty, \tag{5.24}$$

follows from Lemma 2.1.6. This proves that the orthonormal polynomials $p_m(\mu; z)$, $m \in \mathbb{N}$, have regular asymptotic zero distribution. □

Example 3.5.3. In connection with Theorem 3.2.3 we show by an example that it is possible for a regular measure μ to have orthonormal polynomials with

$$\limsup_{n\to\infty} \|p_n(\mu;\cdot)\|^{1/n}_{\sup,S(\mu)} > 1.$$

Of course, then $S(\mu)$ is not regular with respect to the Dirichlet problem in Ω.

Let

$$S = \{0\} \cup \bigcup_{n\in N} (2^{-n}, 2^{-n} + \exp(-n^3)),$$

for some subsequence $N = \{n_1 = 1, n_2, n_3, \ldots\}$ of the natural numbers, and let

$$S^* = \{0\} \cup \bigcup_{n\in\mathbb{N}} (2^{-n}, 2^{-n} + \exp(-n^3)).$$

By Wiener's test [Ts, Theorem III. 62] 0 is not a regular point of S^*, so $g_{\mathbb{C}\setminus S^*}(0;\infty) =: \alpha > 0$. With $N_m = \{n_1, n_2, n_3, \ldots, n_m\}$ and

$$S_m = \bigcup_{n\in N_m} (2^{-n}, 2^{-n} + \exp(-n^3))$$

we obviously have

$$\alpha = g_{\mathbb{C}\setminus S^*}(0;\infty) \le g_{\mathbb{C}\setminus S_m}(0;\infty)$$

whatever the sequence N be.

Suppose that the numbers $\{n_1, n_2, \ldots, n_m\}$ have already been chosen. Let $P_n^{(m)}$ be the orthonormal polynomials with respect to the restriction of linear Lebesgue measure to S_m. Obviously, this is a regular measure, so we have

$$\|P_n^{(m)}\|_{L^2(S_m)} = 1 \quad \text{and} \quad \lim_{n\to\infty} |P_n^{(m)}(0)|^{1/n} = \exp(g_{\mathbb{C}\setminus S_m}(0;\infty)) \ge e^{\alpha}.$$

Choose M_m in such a way that

$$|P_{M_m}^{(m)}(0)|^{1/M_m} \ge e^{\alpha/2},$$

and then a ρ_m so that with

$$S_m^* = S_m \cup \bigcup_{n \ge \rho_m} (2^{-n}, 2^{-n} + \exp(-n^3))$$

we have $\|P_{M_m}^{(m)}\|_{L^2(S_m^*)} < 2$. Finally, let us choose $n_{m+1} > \rho_m$ arbitrarily, and this completes the inductive choice of the numbers $\{n_k\}$.

If μ is the restriction of the Lebesgue measure to S, then by the construction of the polynomials $P_{M_m}^{(m)}$ we have

$$\limsup_{m\to\infty} \left(\frac{|P_{M_m}^{(m)}(0)|}{\|P_{M_m}^{(m)}\|_{L^2(\mu)}} \right)^{1/M_m} \ge e^{\alpha/2} > 1,$$

hence we must have for the orthonormal polynomials $p(\mu; \cdot)$ the relation

$$\limsup_{n\to\infty} \|p_n(\mu; \cdot)\|_{\sup, S(\mu)}^{1/n} > 1$$

(see the proof of Theorem 3.2.3, (iv)⇒(v)). On the other hand, μ is regular: For every $z \in S = S(\mu)$, $z \ne 0$, and any sequence $\{P_n\}$ condition (2.4) is satisfied because it is satisfied with the restriction of μ to the interval of S containing z (the restriction of Lebesgue measure to any interval is regular), so we can apply Theorem 3.2.1. □

3.6 Regular Behavior and Monic Polynomials

In many applications it is not regular nth-root asymptotic behavior but regular zero distribution that plays a central role. We saw in Section 3.1 that these two notions are strongly related (cf. Theorem 3.1.4) but they are not fully equivalent (see Examples 2.2.5 and 3.5.2). In this section we briefly discuss characterization of regular zero distribution in the sense of Section 3.1. Incidentally we get a variant of Theorem 3.1.1 for the monic orthogonal polynomials $\{q_n(\mu; \cdot)\}$ as well. However, we would like to point out that, although our next result is very similar to Theorem 3.1.1, it is much less deep than the latter one.

Theorem 3.6.1. *Let* $\mathrm{cap}(S(\mu)) > 0$. *The following three assertions are equivalent.*

(i)
$$\lim_{n\to\infty} |q_n(\mu; z)|^{1/n} = e^{g_\Omega(z;\infty)} \mathrm{cap}(S(\mu))$$

locally uniformly in $\overline{\mathbb{C}} \setminus \mathrm{Co}(S(\mu))$.

(ii)
$$\limsup_{n\to\infty} |q_n(\mu; z)|^{1/n} = \mathrm{cap}(S(\mu))$$

for quasi every $z \in \partial\Omega$.

(iii)
$$\lim_{n\to\infty} \hat{\nu}_{q_n(\mu;\cdot)} = \omega_{S(\mu)}$$

in the weak topology, where* $\hat{\nu}$ *denotes the balayage of the measure* ν *onto* $\overline{\Omega}$.

Furthermore, each of (i)–(iii) *follows from* $\mu \in$ **Reg**.

If $\text{Int}(\text{Pc}(S(\mu))) = \emptyset$, *that is,* $S(\mu)$ *has empty interior and connected complement, then* (iii) *takes the form*

(iii)′
$$\lim_{n\to\infty} \nu_{q_n(\mu;\cdot)} = \omega_{S(\mu)},$$

which is nothing else than the regular asymptotic zero distribution of $\{p_n(\mu;\cdot)\}$.

Proof. That (i)–(iii) follow from $\mu \in$ **Reg** is an immediate consequence of Theorem 3.1.1, hence we only have to prove the pairwise equivalence of (i)–(iii).

(i)⇒(iii): Let σ be a weak* limit point of $\{\hat{\nu}_{q_n(\mu;z)}\}_{n\in\mathbb{N}}$. By Theorem 2.1.1 the measure σ is supported on $\partial\Omega$. We have to show that $\sigma = \omega_{S(\mu)}$. Since for $z \in \Omega$ we have

$$\frac{1}{n}\log|q_n(\mu;z)| = -p(\hat{\nu}_{q_n(\mu;\cdot)};z), \tag{6.1}$$

and (i) is equivalent to

$$\lim_{n\to\infty}\frac{1}{n}\log|q_n(\mu;z)| = -p(\omega_{S(\mu)};z)$$

for $z \in \mathbb{C}\setminus \text{Co}(S(\mu))$, it follows that

$$p(\sigma;z) = p(\omega_{S(\mu)};z) \tag{6.2}$$

for $z \in \mathbb{C}\setminus\text{Co}(S(\mu))$. Since here both sides are harmonic in Ω, the equality extends to Ω, and then, by the fine-topology argument of the proof of Theorem 2.2.1(b), to all $z \in \overline{\Omega}$. From (6.2) for all $z \in \partial\Omega$ we get

$$\begin{aligned} I(\sigma) :&= \int p(\sigma;z)d\sigma(z) = \int p(\omega_{S(\mu)};z)d\sigma(z) \\ &= \int p(\sigma;z)d\omega_{S(\mu)}(z) = \int p(\omega_{S(\mu)};z)d\omega_{S(\mu)}(z) = I(\omega_{S(\mu)}) < \infty, \end{aligned}$$

that is, together with $\omega_{S(\mu)}$ the measure σ also has finite logarithmic energy. Hence we can apply the principle of domination twice to extend (6.2) (for all $z \in \partial\Omega$) to all $z \in \mathbb{C}$, and so $\sigma = \omega_{S(\mu)}$ follows from the one-to-one relation between measures and potentials (see Appendix III).

(iii)⇒(ii): If we use

$$p(\omega_{S(\mu)}; z) = \log \frac{1}{\operatorname{cap}(S(\mu))} \quad \text{for qu.e. } z \in \partial\Omega$$

and (6.1) for qu.e. $z \in \partial\Omega$, then (ii) is an immediate consequence of (iii) and the lower envelope theorem (see Appendix III).

(ii)⇒(i): Let σ be a weak* limit point of the zero distributions $\{\nu_{q_n(\mu;\cdot)}\}$. By Theorem 2.1.1 we know that σ is supported on $\operatorname{Pc}(S(\mu))$ and every zero of $q_n(\mu;\cdot)$ is contained in $\operatorname{Co}(S(\mu))$. Hence, we only have to show that (ii) implies $p(\sigma; z) = p(\omega_{S(\mu)}; z)$ for all $z \in \Omega$. But (ii) and the lower envelope theorem yield

$$p(\sigma; z) \geq p(\omega_{S(\mu)}; z) \tag{6.3}$$

for qu.e. $z \in \partial\Omega = \operatorname{supp}(\omega_{S(\mu)}; z)$, hence we get from the principle of domination that the same holds true for all $z \in \overline{\mathbb{C}}$ (note that $\omega_{S(\mu)}$ has finite logarithmic energy and so the set of $z \in \partial\Omega$ for which (6.3) is false has zero $\omega_{S(\mu)}$ measure). But at infinity we have equality in (6.3), and since both sides are harmonic in Ω, the equality sign in (6.3) must hold for all $z \in \Omega$ by the minimum principle for harmonic functions, as we have claimed. □

4

Regularity Criteria

For applications of our results concerning regular (nth-root) behavior of orthonormal polynomials we need easy-to-use criteria for the inclusion $\mu \in \mathbf{Reg}$. In Section 4.1 we review the existing regularity criteria and compare their strength. In Section 4.2 we give several new criteria. Their common feature is that they directly relate the size of μ on disks to $\mu \in \mathbf{Reg}$. These new criteria turn out to be much stronger in applications than those in Section 4.1 and they are very easy to use. The proofs that they imply the regularity of μ are given in Section 4.3. In Section 4.2 we also discuss the sharpness of our criteria – they turn out to be "almost optimal." Their sharpness is shown in Section 4.5 after some potential-theoretic preparations in Section 4.4. Finally, in Section 4.6 we give a review of all criteria discussed by us and establish their relative strength.

4.1 Existing Regularity Criteria and Their Generalizations

We review the existing criteria that ensure regular (nth-root) asymptotic behavior of the orthonormal polynomials $\{p_n(\mu;\cdot)\}$, in short, the regularity of μ ($\mu \in \mathbf{Reg}$).

It follows from Corollary 1.1.7 and Definition 3.1.2 that every measure with support of zero capacity is regular. Therefore, in what follows until the end of this chapter we shall always assume $\operatorname{cap}(S(\mu)) > 0$. This also implies the existence of $\omega_{S(\mu)}$, the equilibrium distribution of $S(\mu)$.

The definition of regularity, that is,

$$\lim_{n\to\infty} \gamma_n(\mu)^{1/n} = \frac{1}{\operatorname{cap}(S(\mu))}$$

(cf. Theorem 3.1.1), and also the equivalent formulations of it presented in Section 3.1 refer directly to the orthogonal polynomials in question, and therefore, except for a few concrete cases, it is very difficult to check its validity. Besides this, the story is usually the opposite, namely, we want to derive different properties of the orthogonal polynomials (say, zero asymptotics, asymptotics away from the support, etc.) from the knowledge of the regularity of μ (see the results in the preceding chapter). Therefore, we are looking for criteria that directly connect μ to its regularity.

The first such criterion was given in an influential paper by P. Erdős and P. Turán, [ErTu]. They claimed that if μ is a measure on $[-1,1]$ and $\mu'(x) > 0$ almost everywhere on $[-1,1]$, then μ is regular (of course, $\mu'(x)$ denotes the Radon–Nikodym derivative of μ with respect to Lebesgue measure, which, in the present case, coincides a.e. with the usual derivative of the function $\mu([-1,x])$). Seeing that $\omega_{[-1,1]}$ and the Lebesgue measure on $[-1,1]$ are mutually absolutely continuous with respect to each other, we have the following criterion as an obvious generalization:

Erdős–Turán criterion:

$$(1.1) \qquad \frac{d\mu}{d\omega_{S(\mu)}} > 0 \qquad \omega_{S(\mu)}\text{-almost everywhere.}$$

Inequality (1.1) should be understood in the following sense: Let $\mu = \mu_a + \mu_s$ be the decomposition of μ into its absolutely continuous and singular parts with respect to $\omega_{S(\mu)}$. Then the Radon–Nikodym derivative $d\mu_a/d\omega_{S(\mu)}$ exists $\omega_{S(\mu)}$–almost everywhere and (1.1) means the $\omega_{S(\mu)}$–a.e. positivity of this derivative. Thus, the Erdős–Turán criterion actually refers only to the absolutely continuous part of μ (with respect to $\omega_{S(\mu)}$), but, of course, the regularity of μ_a implies that of μ.

Theorem 4.1.1. *The Erdős–Turán criterion implies* $\mu \in$ **Reg**.

Remark. Condition (1.1) can be equivalently expressed in at least two other ways:

(i) Inequality (1.1) is equivalent to

$$\Gamma(\mu) \subseteq \Gamma(\omega_{S(\mu)}),$$

where $\Gamma(\cdot)$ denotes the set of carriers defined in (1.1.1).

(ii) Inequality (1.1) is equivalent to the assertion that the measures μ_a and $\omega_{S(\mu)}$ are equivalent in the measure-theoretic sense, that is, $\mu_a(B) = 0$ if and only if $\omega_{S(\mu)}(B) = 0$ for all Borel sets B, which is also equivalent to the equality

$$\Gamma(\mu_a) = \Gamma(\omega_{S(\mu)}),$$

that is, to the mutual absolute continuity of μ_a and $\omega_{S(\mu)}$ with respect to each other.

Corollary 4.1.2. *If $S(\mu) = A_1 \cup \cdots \cup A_n \cup E$ with $A_1, \ldots, A_n$ smooth and compact Jordan arcs in $\mathbb{C}$, $A_1 \cup \cdots \cup A_n$ does not separate $\overline{\mathbb{C}}$, and $E \subseteq \mathbb{C}$ is a Borel set with* $\operatorname{cap}(E) = 0$, *then*

$$\frac{d\mu}{ds} > 0 \quad \textit{a.e. on } A_j \textit{ for } j = 1, \ldots, n,$$

implies $\mu \in$ **Reg**. *Here ds is the line element of the arc length on A_j.*

Corollary 4.1.3. *If $S(\mu) = [-1, 1]$, then*

$$\frac{d\mu(x)}{dx} > 0 \quad \textit{a.e. on } [-1, 1] \tag{1.2}$$

implies $\mu \in$ **Reg**.

Remark. As has already been mentioned, Corollary 4.1.3 contains the original result of Erdős and Turán [ErTu]. It may be interesting to note that more than 40 years later E. A. Rahmanov ([Ra1] and [Ra3]) succeeded in showing that (1.2) is not only sufficient for regular nth-root asymptotic behavior, but also for the stronger ratio asymptotic behavior (I.2) of the introduction. However, in Rahmanov's result the fact that $S(\mu)$ is an interval is essential. The weaker assumption of Corollary 4.1.2, even in its most special form of two disjoint real intervals A_1 and A_2 and $E = \emptyset$, is no longer sufficient for regular ratio asymptotics (see [Ra2]).

A weaker condition than the Erdős–Turán one was introduced by J. L. Ullman in [Ul1]. It claimed for a measure μ with $S(\mu) = [-1, 1]$ that $c_\mu = \frac{1}{2}$ be satisfied. Thus, we have the following criterion.

Ullman's criterion: $c_\mu = \operatorname{cap}(S(\mu))$.
Recall that c_μ is the minimal carrier capacity of μ (see (1.1.2)).

Theorem 4.1.4. *Ullman's criterion implies the regularity of μ.*

This is obvious from Corollary 1.1.7. It is not difficult to show that the Erdős–Turán criterion implies Ullman's criterion. In fact, by dilation we can assume $\operatorname{diam}(S(\mu)) < 1$. Now (1.1) implies that if C is a carrier of μ and $K_1 \subseteq K_2 \subseteq \cdots \subseteq C$ are compact subsets with $\mu(K_n) \to \|\mu\|$ as $n \to \infty$, then we have also $\omega_{S(\mu)}(K_n) \to 1$, and so for the logarithmic energy

$$I(\nu_n) = \int\int \log \frac{1}{|z-t|} d\nu_n(z) d\nu_n(t)$$

of the unit measures

$$\nu_n = \omega_{S(\mu)}|_{K_n} / \omega_{S(\mu)}(K_n)$$

we have

$$\text{(1.3)} \qquad \limsup_{n\to\infty} I(\nu_n) \le \limsup_{n\to\infty} I(\omega_{S(\mu)})/(\omega_{S(\mu)}(K_n))^2 = I(\omega_{S(\mu)}).$$

Since $\log(1/\text{cap}(K_n))$ is at most $I(\nu_n)$ and $I(\omega_{S(\mu)})$ equals $\log(1/\text{cap}(S(\mu)))$, we can conclude

$$\text{(1.4)} \qquad \liminf_{n\to\infty} \text{cap}(K_n) \ge \text{cap}(S(\mu)),$$

by which $\text{cap}(C) \ge \text{cap}(S(\mu))$, and so $c_\mu \ge \text{cap}(S(\mu))$, is established. This proves that in the presence of (1.1) Ullman's criterion $c_\mu = \text{cap}(S(\mu))$ is satisfied.

Thus, Theorem 4.1.4 proves also Theorem 4.1.1. Corollary 4.1.2 follows from Theorem 4.1.1 because $\omega_{S(\mu)}$ and the arc measure on $A_1 \cup \cdots \cup A_n$ are mutually absolutely continuous (cf. [Wi2, p. 141]) and $\text{cap}(E) = 0$ implies $\omega_{S(\mu)}(E) = 0$.

In the case when $\text{Int}(\text{Pc}(S(\mu))) \ne \emptyset$, the Erdős–Turán criterion can be trivially weaker in strength than Ullman's criterion, see, for example, Example 1.5.1, where the measure μ is the planar Lebesgue measure on the open disc $\{z \mid |z| < 1\}$ and $\mu \in \mathbf{Reg}$. The equilibrium distribution $\omega_{S(\mu)}$ is in this case supported on the circle $\{z \mid |z| = 1\}$, and the open disc $\{z \mid |z| < 1\}$ is a carrier of μ. Hence, the derivative (1.1) is zero $\omega_{S(\mu)}$–a.e., and the Erdős–Turán criterion is not satisfied, despite the fact that $\mu \in \mathbf{Reg}$. It is much less trivial that Ullman's criterion can be significantly stronger in strength than the Erdős–Turán criterion even if $\text{Int}(\text{Pc}(S(\mu))) = \emptyset$, say, if $S(\mu) \subseteq \mathbf{R}$. In Lemma 1.3., part (b) of [Ul1] it was shown that for every $\varepsilon > 0$ there exist weight measures μ with $S(\mu) = [-1,1]$, $c_\mu = \text{cap}([-1,1])$, and a carrier $C \in \Gamma(\mu)$ with the linear Lebesgue measure $\lambda(C) < \varepsilon$. Since the equilibrium distribution $\omega_{[-1,1]}$ of the interval $[-1,1]$ is absolutely continuous with respect to λ, it follows that $\omega_{[-1,1]}(C) < 1$.

In the next example the relation between the minimal-carrier capacity c_μ and the linear Lebesgue measure λ of minimal carriers is carried to an extreme. We show that there exist weight measures on $[0,1]$ with minimal carriers of linear Lebesgue measure zero and $c_\mu = \text{cap}([0,1])$. Thus, in this example "linear Lebesgue measure" and the "minimal-carrier capacity" are maximally different.

Example 4.1.5. We construct a measure μ such that $S(\mu) = [0,1]$, $c_\mu = \text{cap}([0,1]) = \frac{1}{4}$, and there exist carriers $C \in \Gamma(\mu)$ with linear Lebesgue measure $\lambda(C) = 0$.

The basic idea of the construction is to paste linearly transformed Cantor sets together. The measure μ is then defined as a weighted sum of the equilibrium distribution on these transformed Cantor sets.

Let K be the classical Cantor set on $[0,1]$. Then $\lambda(K) = 0$ and $\text{cap}(K) > 0$. A set $A \subseteq [0,1]$ is constructed by filling up the gaps in the Cantor set K with linearly transformed Cantor sets. We describe the construction more formally as follows.

For any set $S \subseteq \mathbb{R}$ the linearly transformed set $aS + b$ is defined by

$$aS + b := \{ax + b \mid x \in S\} \qquad \text{with } a, b \in \mathbb{R},\ a > 0. \tag{1.5}$$

The set A is defined as

$$\begin{aligned} A := K \bigcup \left(\frac{1}{3}K + \frac{1}{3}\right) \bigcup \left[\left(\frac{1}{9}K + \frac{1}{9}\right)\right. \\ \cup \left(\frac{1}{9}K + \frac{4}{9}\right) \cup \left.\left(\frac{1}{9}K + \frac{7}{9}\right)\right] \bigcup \cdots, \end{aligned} \tag{1.6}$$

or in a compact formulation as

$$A := K \cup \bigcup_{j=0}^{\infty} \left[\bigcup_{k=0}^{3^j-1} K_{jk} \right], \tag{1.7}$$

where

$$K_{jk} := \frac{1}{3^{j+1}} K + \frac{3k+1}{3^{j+1}} \tag{1.8}$$

for $j, k \in \mathbb{N}$ and $0 \leq k < 3^j$.

Let μ_{jk} denote the equilibrium distribution of the transformed Cantor set (1.8). The weight measure μ is then defined as

$$\mu := \frac{1}{2}\omega_K + \sum_{j=0}^{\infty} \frac{1}{3^{2j+1}} \sum_{k=0}^{3^j-1} \mu_{jk}. \tag{1.9}$$

It is easy to see that μ is a probability measure, $A \in \Gamma(\mu)$ is a minimal carrier, and $\lambda(A) = 0$. It remains to be shown that

$$c_\mu = \text{cap}(A) = \text{cap}([0,1]) = \tfrac{1}{4}. \tag{1.10}$$

The Cantor set K is regular with respect to the Dirichlet problem in $\mathbb{C} \setminus K$ (see Theorem III.63 of [Ts]). From Lemma 1.2.7 and Definition 1.1.1 it therefore follows that

$$g_\mu(z;\infty) \leq g_{\mu_{jk}}(z;\infty) = g_{\overline{\mathbb{C}} \setminus K_{jk}}(z;\infty) \tag{1.11}$$

for all $z \in \mathbb{C}$ and $j, k \in \mathbb{N}$ $(0 \leq k < 3^j)$. From the regularity of the Cantor sets K_{jk} it then follows that

$$g_\mu(z;\infty) = 0 \quad \text{for all} \quad z \in A. \tag{1.12}$$

We next show that for any $x \in [0,1]$ and the sets

$$A_n := A \cap \{z \mid 2^{-n} \geq |x - z| \geq 2^{-(n+1)}\}, \quad n = 1, 2, \dots, \tag{1.13}$$

we have

$$\sum_{n=1}^{\infty} \frac{n}{\log(1/\mathrm{cap}(A_n))} = \infty. \tag{1.14}$$

Indeed, because of the regular structure of the set A, for every subinterval $I \subseteq [0,1]$ of length $|I| \geq 3^{-(j-1)}$, $j \in \mathbb{N}$, there exists at least one transformed Cantor set (1.8) with $K_{jk} \subseteq I$ $(0 \leq k < 3^j)$. Let $j \in \mathbb{N}$ be the first number with $3^{-(j-1)} \leq 2^{-(n+1)}$. Then there exists a $K_{jk} \subseteq A_n$ with $0 \leq k < 3^j$, and therefore

$$\mathrm{cap}(A_n) \geq \frac{1}{3^{j+1}}\mathrm{cap}(K) \geq \frac{1}{27} 2^{-(n+1)} \mathrm{cap}(K). \tag{1.15}$$

This proves (1.14).

From (1.14) it follows that all $x \in [0,1]$ are regular points of the Borel set A, which implies that

$$g_{\overline{\mathbb{C}} \setminus A}(z; \infty) = 0 \quad \text{for all } z \in [0,1]. \tag{1.16}$$

As a reference for this last result see [La], Chapter V, Section 1/6, especially the second to last paragraph. Since [La] contains only a very sketchy treatment of irregular points of Borel sets for logarithmic potentials, we also refer to [Ts], where also only closed sets are considered, but the proofs of Theorem III.61 and Theorem III.62 can be generalized without using earlier results so that the derivation of (1.16) from (1.14) is covered.

From (1.14) and (1.16) it follows that

$$g_\mu(z; \infty) \equiv g_{\overline{\mathbb{C}} \setminus A}(z; \infty) \equiv g_{\overline{\mathbb{C}} \setminus [0,1]}(z; \infty),$$

which proves (1.10). □

Obviously, in Example 1.5.1 the collapse of the Erdős–Turán criterion is somewhat artificial. However, the example is instructive for the following strengthening of the Erdős–Turán criterion, which is, up to a small modification, identical with a regularity criterion introduced by Widom in [Wi1]. The original formulation of the criterion is given as Corollary 4.1.7.

Widom's criterion: *For every carrier C of the measure μ there exists a sequence of compact sets $V_n \subseteq S(\mu)$, $\mathrm{cap}(V_n) > 0$, $n \in \mathbb{N}$, with*

(i) $\mathrm{cap}(V_n) \to \mathrm{cap}(S(\mu))$ *as* $n \to \infty$

and

(ii) $$\omega_{V_n}(C) \to 1 \quad as \quad n \to \infty.$$

Theorem 4.1.6. *Widom's criterion implies $\mu \in$ **Reg**.*

Remarks. (1) In Theorem 4.1.6 it is sufficient to check the assumption for only one minimal carrier $C \in \Gamma_0(\mu)$. However, it is essential that the carrier be really minimal, which is in general as difficult to decide as to decide whether Ullman's criterion is satisfied or not.

(2) If all sets V_n are taken to be identical to $S(\mu)$, then condition (i) is trivially satisfied and condition (ii) becomes equivalent to condition (1.1) of the Erdős–Turán criterion. Hence, Widom's criterion is in general stronger than the Erdős–Turán criterion.

As a corollary to Theorem 4.1.6 we state the original formulation of Widom's criterion from [Wi1].

Corollary 4.1.7. *If there exists a family of compact sets $V_t \subseteq S(\mu)$, $0 \leq t < t_0$, with*

(i) $$\lim_{t \to t_0} \operatorname{cap}(V_t) = \operatorname{cap}(S(\mu)),$$

and for any carrier $C \in \Gamma(\mu)$ we have

(ii) $$\limsup_{t \to t_0} \omega_{V_t}(C) = 1,$$

*then $\mu \in$ **Reg**.*

Remarks. (1) Similarly as in Theorem 4.1.6 we can choose all V_t $(0 \leq t < t_0)$ to be identical to $S(\mu)$, and condition (ii) of Corollary 4.1.7 is then equivalent to condition (1.1) of the Erdős–Turán criterion. Hence, Widom's criterion in its original formulation is also stronger than the Erdős–Turán criterion.

(2) Since in Theorem 4.1.6 for every carrier $C \in \Gamma(\mu)$ we may select a different sequence $\{V_n\}$, the new formulation of the criterion is easier to apply than Widom's criterion in its original form, where the family $\{V_t\}$ of compact sets has to be the same for all carriers $C \in \Gamma(\mu)$. The proof of the corollary will show that the criterion of Theorem 4.1.6 is at least as strong as the one in Corollary 4.1.7.

Proof of Corollary 4.1.7 and Theorem 4.1.6. Corollary 4.1.7 immediately follows from Theorem 4.1.6. In fact, let V_t, $0 \leq t \leq t_0$, be a family of compact sets satisfying the assumptions of Corollary 4.1.7. Let further

$C \in \Gamma(\mu)$ be an arbitrary carrier. From (ii) of the corollary it then follows that there exists a sequence $t_n \in [0, t_0)$ with $t_n \to t_0$ such that for the sequence of sets $V_n := V_{t_n}$ the two limits (i) and (ii) in Theorem 4.1.6 hold true. Hence, by Theorem 4.1.6 we have $\mu \in \mathbf{Reg}$.

As for Theorem 4.1.6 itself, we only have to prove that Widom's criterion implies Ullman's criterion and then apply Theorem 4.1.4. But this is easy: If C is a carrier of μ and $\{V_n\}$ are as in (i) and (ii) in Widom's criterion, then for each n there is a compact set $K_n \subseteq C$ such that $\omega_{V_n}(K_n) \to 1$ as $n \to \infty$. Now setting

$$\nu_n = \omega_{V_n}|_{K_n} / \omega_{V_n}(K_n)$$

and taking into account that for the logarithmic energy of ω_{V_n} we have

$$\lim_{n\to\infty} I(\omega_{V_n}) = \lim_{n\to\infty} \log(1/\mathrm{cap}(V_n)) = \log(1/\mathrm{cap}(S(\mu))),$$

we can conclude exactly as in (1.3)–(1.4) that

$$\mathrm{cap}(C) \geq \liminf_{n\to\infty} \mathrm{cap}(K_n) \geq \mathrm{cap}(S(\mu)),$$

proving $c_\mu = \mathrm{cap}(S(\mu))$. □

Thus, we have verified that Widom's criterion implies Ullman's criterion. Now we show that they are actually equivalent. In fact, if we assume Ullman's criterion, that is, $c_\mu = \mathrm{cap}(S(\mu))$, and C is a carrier of μ then there are compact sets $V_n \subseteq C$ such that $\mathrm{cap}(V_n) \to \mathrm{cap}(S(\mu))$ as $n \to \infty$. For this sequence $\{V_n\}$ the two conditions (i) and (ii) in Widom's criterion are satisfied.

As an illustration we show how to apply Widom's criterion to Example 1.5.1. In Example 1.5.1 the measure μ is the planar Lebesgue measure of the open disc$\{z \mid |z| < 1\}$. Set $V_t := \{z \mid |z| \leq t\}$, $0 < t < 1$. The equilibrium distribution ω_t of the set V_t is uniformly distributed on the circle $\{z \mid |z| = t\}$. Let $C \in \Gamma(\mu)$ be arbitrary. Then for almost all $t \in (0, 1)$ we have

$$\lambda(C \cap \{z \mid |z| = t\}) = 2\pi t,$$

where λ denotes the linear Lebesgue measure on $\{z \mid |z| = t\}$. Hence, there exists a sequence $\{t_n\}$ with $t_n \to 1$ such that the corresponding sets $V_n := V_{t_n}$ satisfy the two conditions (i) and (ii) of Theorem 4.1.5. Hence, by Widom's criterion, we have $\mu \in \mathbf{Reg}$.

Of course, Widom's criterion is of a different quality than Ullman's or the Erdős–Turán criterion. Its application depends on the skill of the investigator in selecting the sequences $\{V_n\}$. For every carrier $C \in \Gamma(\mu)$ a different sequence may be selected. This different quality gives a practical value to the equivalence between Ullman's and Widom's criteria.

As a summary of our findings we state the following theorem.

Theorem 4.1.8. *Let* $\operatorname{cap}(S(\mu)) > 0$. *The following dependencies hold:*

Erdős–Turán criterion
$\Rightarrow$ *Widom's criterion* $\Leftrightarrow$ *Ullman's criterion*
$\Rightarrow \mu \in \mathbf{Reg}$.

All the these criteria are rather weak. The problem with them is that the two stronger criteria, namely, Ullman's and Widom's criteria, simply do not take into account the size of μ, they only refer to the carriers of μ. Furthermore, they work only when in Theorem 1.1.4 and Corollary 1.1.7 the upper and lower bounds coincide, that is, only in the "trivial" case. For example, they never can be used to check the regularity of a discrete measure. In the next section we present some incomparably weaker (and hence in applications stronger) criteria for regularity of measures. Their common feature will be that they relate the size of μ on disks to $\mu \in \mathbf{Reg}$.

4.2 New Criteria and Their Sharpness

Let $\Delta_r(z) = \{z' \mid |z' - z| \le r\}$. We start with

Criterion Λ:

$$\operatorname{cap}\left(\left\{z \,\middle|\, \limsup_{r\to 0+} \frac{\log 1/\mu(\Delta_r(z))}{\log 1/r} < \infty\right\}\right) = \operatorname{cap}(S(\mu)). \tag{2.1}$$

Theorem 4.2.1. *Criterion* Λ *implies* $\mu \in \mathbf{Reg}$.

We remark that Λ is weaker than Ullman's criterion (which is equivalent to Widom's criterion) because the set

$$\left\{z \,\middle|\, \limsup_{r\to 0+} \frac{\log 1/\mu(\Delta_r(z))}{\log 1/r} \le 2\right\} \tag{2.2}$$

is always a carrier for μ (cf. Theorem 4.6.1).

Theorem 4.2.1 is about as far as anyone can go even if $S(\mu) = [0,1]$; namely, we show that in (2.1) the function $\log 1/r$ cannot be replaced by a much bigger one.

Theorem 4.2.2. *There exists a measure* μ *such that* $S(\mu) = [0,1]$, *the set*

$$\left\{x \,\middle|\, \lim_{r\to 0} \frac{\log 1/\mu([x-r, x+r])}{(\log 1/r)^{1+\eta}} = 0 \quad \textit{for every} \quad \eta > 0\right\} \tag{2.3}$$

is of capacity $\frac{1}{4}$ *but* $\mu \notin \mathbf{Reg}$.

Condition Λ says that the measure μ is sufficiently dense in the points of a set of full capacity. Note that from a measure-theoretic point of view, in general, this requires denseness only in a relatively sparse set (see Example 4.1.5 or Corollary 4.4.4). But actually, we do not need this; all we need is a kind of "denseness in the average." Therefore we introduce the following "moving variant" of Λ.

Criterion Λ^*: *There exists a constant $L > 0$ such that*

$$\lim_{r\to 0} \mathrm{cap}(\{z \mid \mu(\Delta_r(z)) \geq r^L\}) = \mathrm{cap}(S(\mu)). \tag{2.4}$$

Theorem 4.2.3. *Assume that $S(\mu) = \mathrm{supp}(\mu)$ is regular with respect to the solutions of Dirichlet problems in $\Omega = \mathbb{C} \setminus \mathrm{Pc}(S(\mu))$. Then criterion Λ^* implies the regularity of μ.*

We shall show in Section 4.6 that Λ^* can be significantly weaker than Λ even for $S(\mu) = [0,1]$; namely, we will exhibit a μ with $S(\mu) = [0,1]$ for which (2.4) is true for every $L > 1$ but the set

$$\left\{ x \;\middle|\; \limsup_{r\to 0+0} \frac{\log 1/\mu([x-r, x+r])}{\log 1/r} < \infty \right\}$$

is of zero capacity.

Theorem 4.2.3 is the best possible, as is shown by the following theorem.

Theorem 4.2.4. *If $\gamma\colon (0,1) \to (0,1)$ is any function tending faster to zero than any x^L, $L > 0$, as $x \to 0+0$, then there exists a measure μ such that $S(\mu) = [0,1]$ and*

$$\lim_{r\to 0} \mathrm{cap}(\{x \mid \mu([x-r, x+r]) \geq \gamma(r)\}) = \tfrac{1}{4},$$

but $\mu \notin$ **Reg**.

Conditions Λ and Λ^* require that at a relatively large set of the points x the μ measure of the disk of radius r around x be at most polynomially small in r. We have already remarked that this relatively large set must be large only in capacity and not, say, in Lebesgue measure (when, e.g., $S(\mu) = [0,1]$). Now we show that the μ measure of those disks can be almost exponentially small in r provided we assume it to hold at more points. We only consider the classical and most important case $S(\mu) = [0,1]$.

Criterion λ: *$S(\mu) = [0,1]$ and for almost every $x \in [0,1]$*

$$\liminf_{r\to 0} r \log \mu([x-r, x+r]) \geq 0. \tag{2.5}$$

Inequality (2.5) is a much weaker – "almost exponentially weaker" – density assumption than the one defining the set (2.1). The price we pay for it is that it must hold at more points (recall that $\text{meas}(E) = 1$ for a set $E \subseteq [0,1]$ implies $\text{cap}(E) = \frac{1}{4}$).

Theorem 4.2.5. *Criterion λ implies $\mu \in$* **Reg**.

This is the best possible in the sense of the following theorem.

Theorem 4.2.6. *If γ: $(0,1) \to (0,1)$ is any function with $\gamma(r)/r \to 0$ as $r \to 0+0$, then there exists a measure μ such that $S(\mu) = [-1,1]$ and for all $x \in [0,1]$*

$$\lim_{r\to 0+0} \gamma(r) \log \mu([x-r, x+r]) = 0,$$

but $\mu \notin$ **Reg**.

The "moving variant" of criterion λ – in the sense of Theorem 4.2.3 – is

Criterion λ*: *$S(\mu) = [0,1]$ and for every $\eta > 0$*

$$\lim_{n\to\infty} \text{meas}\left(\left\{x \,\middle|\, \mu\left(\left[x - \frac{1}{n}, x + \frac{1}{n}\right]\right) \geq e^{-\eta n}\right\}\right) = 1.$$

Here "meas" denotes linear Lebesgue measure on $\mathbf{R}$.

Theorem 4.2.7. *Criterion λ^* implies the regularity of μ.*

Since condition λ^* is obviously weaker than λ, we do not have to demonstrate the unimprovability of Theorem 4.2.7 in the sense of Theorem 4.2.6 because it follows from that of the latter one.

In Section 4.6 we shall show that condition λ^* implies condition Λ^* (and so Theorem 4.2.7 is a consequence of Theorem 4.2.3). However, to verify this requires much more sophisticated arguments than directly proving Theorem 4.2.7, so in Section 4.3 we shall present a direct proof for Theorem 4.2.7. This is also warranted by the simplicity of criteria λ and λ^*.

Finally, we present a necessary condition for $\mu \in$ **Reg** in the case when $S(\mu) = [0,1]$.

Theorem 4.2.8. *Let μ be regular and $S(\mu) = [0,1]$. Then for every $\eta > 0$*

$$\lim_{n\to\infty} \operatorname{cap}\left(\left\{x \,\middle|\, \mu\left(\left[x - \frac{1}{n}, x + \frac{1}{n}\right]\right) \geq e^{-\eta n}\right\}\right) = \frac{1}{4}.$$

This should be compared to criteria λ^* and Λ^*. To our knowledge Theorem 4.2.8 is the only nontrivial necessity result for regularity of measures.

Let us shortly illustrate the aforementioned criteria by the examples of Section 1.5. In Section 4.1 we showed how one can apply Widom's criterion to prove that the planar measure on the unit disk from Example 1.5.1 is regular. Since for this μ we have for every $0 < r < 1$, $|z| = 1$, the estimate $\mu(\Delta_r(z)) \geq \frac{\pi}{4}r^2$, criterion Λ is trivially satisfied and $\mu \in \mathbf{Reg}$ follows from Theorem 4.2.1. In a similar fashion, if μ is the measure from Example 1.5.4, then for $|z| = 1$

$$\mu(\Delta_{5/n}(z)) \geq \frac{1}{(n+1)n^2}$$

immediately follows from the definition of μ in (1.5.10) and (1.5.11); therefore again $\mu \in \mathbf{Reg}$ by Theorem 4.2.1.

Although our criteria are easy to work with, neither of them gives necessary and sufficient conditions for regularity (with the possible exception of Λ^*). Thus, giving a necessary and sufficient condition for $\mu \in \mathbf{Reg}$ that is *not* a mere reformulation of the definition of regularity is an open problem. Theorems 4.2.2, 4.2.4, and 4.2.6 at least show that our criteria that connect regularity to the size of μ are not very far from the truth.

4.3 Proof of the Regularity Criteria

In this section we prove Theorems 4.2.1, 4.2.3, 4.2.5, and 4.2.7.

Proof of Theorem 4.2.1 Let $\varepsilon > 0$. Then there exist constants $K, \alpha > 0$ such that the set

$$S_{K,\alpha} = \left\{x \,\middle|\, \frac{\log 1/\mu(\Delta_r(x))}{\log 1/r} \leq K \quad \text{for} \quad 0 < r \leq \alpha\right\}$$

has capacity

$$\operatorname{cap}(S_{K,\alpha}) > \operatorname{cap}(S(\mu)) - \varepsilon.$$

By [An1,An2] there is a compact set $E \subseteq S_{K,\alpha}$ such that

$$\operatorname{cap}(E) > \operatorname{cap}(S(\mu)) - \varepsilon$$

and E is regular with respect to the solution of Dirichlet problems in the unbounded component of $\mathbb{C} \setminus E$, which is equivalent to the continuity of $g_{\mathbb{C}\setminus \operatorname{Pc}(E)}(z;\infty)$ throughout $\mathbb{C}$. Thus, for some $\delta > 0$

(3.1) $$g_{\mathbb{C}\setminus \operatorname{Pc}(E)}(z) \leq \log(1+\varepsilon)^{1/K}$$

if $d(z;E) < \delta$.

Now let $P_n(z) = z^n + \cdots$ be an arbitrary monic polynomial. The norm of P_n on E is at least as large as $(\mathrm{cap}(E))^n$ (see [Ts, Theorem III.15]); therefore there exists a $z_n \in E$ such that

$$M := |P_n(z_n)| = \|P_n\|_E \geq (\mathrm{cap}(S(\mu)) - \varepsilon)^n. \tag{3.2}$$

Inequality (3.1) implies via the Bernstein–Walsh lemma [Wa, p. 77]

$$|P_n(z)| \leq M(1+\varepsilon)^{n/K}$$

for $d(z; E) < \delta$. But then for $d(z; E) < \delta/2$ we have

$$|P_n'(z)| \leq \frac{2}{\delta} M(1+\varepsilon)^{n/K},$$

and so for $r_n = \delta(1+\varepsilon)^{-n/K}/4$ and $z \in \Delta_{r_n}(z_n)$ we get

$$|P_n(z)| = \left| P_n(z_n) + \int_{z_n}^{z} P_n'(\xi)d\xi \right| \geq M - r_n \frac{2}{\delta} M(1+\varepsilon)^{n/k} = \frac{M}{2}.$$

For large n we will have $r_n < \alpha$, and since $z_n \in S_{K,\alpha}$, we can write for such ns

$$\begin{aligned} \int |P_n|^2 d\mu &\geq \int_{\Delta_{r_n}(z_n)} |P_n|^2 d\mu \geq \left(\frac{M}{2}\right)^2 \mu(\Delta_{r_n}(z_n)) \\ &\geq \frac{(\mathrm{cap}(S(\mu)) - \varepsilon)^{2n}}{4} r_n^K \geq \frac{\delta^K}{4^{K+1}} \left(\frac{\mathrm{cap}(S(\mu)) - \varepsilon}{1+\varepsilon} \right)^{2n} \end{aligned}$$

Since here the infimum of the left-hand side for P_n with $\deg(P_n) = n$ is $\gamma_n(\mu)^{-2}$, it follows that

$$\limsup_{n\to\infty} \gamma_n(\mu)^{1/n} \leq \frac{1+\varepsilon}{\mathrm{cap}(S(\mu)) - \varepsilon},$$

and for $\varepsilon \to 0$ we get

$$\limsup_{n\to\infty} \gamma_n(\mu)^{1/n} \leq \frac{1}{\mathrm{cap}(S(\mu))}.$$

With Corollary 1.1.7 and Definition 3.1.2 this proves the regularity of μ.

□

Proof of Theorem 4.2.3. We break the proof into several steps.

Step 1. If K, K_j, $j = 1, 2, \ldots$ are compact sets and $K_j \subseteq K$ with

$$\lim_{j\to\infty} \mathrm{cap}(K_j) = \mathrm{cap}(K) > 0, \tag{3.3}$$

then $\omega_{K_j} \to \omega_K$ as $j \to \infty$ in the weak* topology.

Of course, this is a well-known statement but for completeness we present its proof.

Without loss of generality we may assume that $\omega_{K_j} \to \sigma$ for some σ in the weak* sense (subtract a suitable subsequence if necessary), and we have to show that $\sigma = \omega_K$. We know that for quasi every $x \in K$

$$p(\omega_K; x) = \log \frac{1}{\mathrm{cap}(K)},$$

and the left-hand side is never larger than the right-hand one (see [Ts, Theorem III.12]). Applying the same identity to K_j in place of K we get from $K_j \subseteq K$ that

$$p(\omega_{K_j}; x) \le p(\omega_K; x) + \log \frac{\mathrm{cap}(K)}{\mathrm{cap}(K_j)} \tag{3.4}$$

for qu.e. $x \in K_j$. Since the exceptional set where (3.4) is not true is of zero capacity and ω_{K_j} has finite logarithmic energy, it follows that (3.4) is true ω_{K_j}–almost everywhere. But then by the principle of domination inequality (3.4) holds for all $x \in \mathbb{C}$. As $j \to \infty$ we obtain from the principle of descent and (3.3) that

$$p(\sigma; x) \le p(\omega_K; x), \quad x \in \mathbb{C}. \tag{3.5}$$

However, $p(\omega_K; \cdot) - p(\sigma; \cdot)$ vanishes at ∞, so by the minimum principle for harmonic functions we have equality in (3.5) for $x \in \mathbb{C} \setminus \mathrm{Pc}(K)$. Using fine-topology argument (cf. the Appendix and the proof of (1.1.10) in Section 1.3) the equality in (3.5) extends to all $x \in \partial\mathrm{Pc}(K)$, which means that for $x \in \mathrm{supp}(\omega_K)$

$$p(\sigma; x) \ge p(\omega_K; x). \tag{3.6}$$

Applying once more the principle of domination we deduce (3.6) for all $x \in \mathbb{C}$. This and (3.5) yield $p(\omega_K; \cdot) \equiv p(\sigma; \cdot)$, proving $\omega_K = \sigma$.

Step 2. With the assumptions of Step 1, for regular K (here and in what follows by "regular K" we mean the regularity of K with respect to Dirichlet problems in $\mathbb{C} \setminus \mathrm{Pc}(K)$) we have

$$\liminf_{j\to\infty} \min_{x \in K} p(\omega_{K_j}; x) \ge \min_{x \in K} p(\omega_K; x) = \log \frac{1}{\mathrm{cap}(K)}. \tag{3.7}$$

In fact, we know from Step 1 that ω_{K_j} tends to ω_K in the weak* topology. The principle of descent says that if $x_j \in K$, $x_j \to x^*$ as $j \to \infty$, then

$$\liminf_{j\to\infty} p(\omega_{K_j}; x_j) \ge p(\omega_K; x^*).$$

If we apply this to convergent subsequences of a sequence of points $\{y_j\}_{j=1}^{\infty} \subseteq K$, where y_j is chosen to satisfy

$$p(\omega_{K_j}; y_j) < \min_{x \in K} p(\omega_{K_j}; x) + 1/j,$$

we get (3.7).

Step 3. If K is regular, then for every $\varepsilon > 0$ there exists a $\delta > 0$ such that for every compact subset V of K for which $\operatorname{cap}(V) > \operatorname{cap}(K) - \delta$ and for every polynomial P_n of degree at most n we have for the supremum norms $\|P_n\|_K$ and $\|P_n\|_V$ the inequality

$$\|P_n\|_K \leq (1+\varepsilon)^n \|P_n\|_V. \tag{3.8}$$

To prove this, first of all we deduce from Step 2 that for every $\eta > 0$ there is a $\delta > 0$ such that $\operatorname{cap}(V) > \operatorname{cap}(K) - \delta$ implies

$$p(\omega_V; x) \geq \log \frac{1}{\operatorname{cap}(K)} - \eta, \quad x \in K.$$

But then for $x \in K$ we get from the Bernstein–Walsh lemma that

$$\begin{aligned} |P_n(x)| &\leq \|P_n\|_V \exp\left(n\left(-p(\omega_V; x) + \log \frac{1}{\operatorname{cap}(V)}\right)\right) \\ &\leq \|P_n\|_V \exp\left(n\left(\log \frac{\operatorname{cap}(K)}{\operatorname{cap}(V)} + \eta\right)\right) \\ &\leq \|P_n\|_V \exp(n(\log(1 - \delta/\operatorname{cap}(K))^{-1} + \eta)), \end{aligned}$$

and if here δ and η are sufficiently small, (3.8) holds.

Step 4. If K is regular, then for every $\varepsilon > 0$ there is a $\delta > 0$ and an $n_\varepsilon \in \mathbb{N}$ such that if $V \subseteq K$ is compact, $\operatorname{cap}(V) > \operatorname{cap}(K) - \delta$, and $P_n(z) = z^n + \cdots$ is a monic polynomial of degree $n \geq n_\varepsilon$, then we can find a $z_0 = z_0(V, P_n) \in V$ such that for $|z - z_0| \leq (1+\varepsilon)^{-n}$ we have

$$|P_n(z)| \geq (\operatorname{cap}(K))^n (1-\varepsilon)^n. \tag{3.9}$$

By Step 3 if $\varepsilon_1 > 0$ then there is a $0 < \delta_1 < \varepsilon/2$ such that $\operatorname{cap}(V) > \operatorname{cap}(K)(1 - \delta_1)$ implies

$$\|P_n\|_K \leq (1+\varepsilon_1)^n \|P_n\|_V.$$

The regularity of K yields a $\delta_2 > 0$ such that if $d(z; K) < \delta_2$ then $g_{\mathbb{C}\setminus \operatorname{Pc}(K)}(z; \infty) < 1 + \varepsilon_1$, hence by the Bernstein–Walsh lemma

$$|P_n(z)| \leq (1+\varepsilon_1)^n \|P_n\|_K \leq (1+\varepsilon_1)^{2n} \|P_n\|_V.$$

Thus, for $d(z; K) < \delta_2/2$

$$|P_n'(z)| \leq \frac{2}{\delta_2}(1+\varepsilon_1)^{2n} \|P_n\|_V.$$

Now if $z_0 \in V$ is chosen to satisfy $|P_n(z_0)| = \|P_n\|_V$ and $|z - z_0| < \delta_2(1+\varepsilon_1)^{-2n}/4$, then

$$\begin{aligned} |P_n(z)| &\geq \tfrac{1}{2}|P_n(z_0)| = \tfrac{1}{2}\|P_n\|_V \geq \tfrac{1}{2}(\operatorname{cap}(V))^n \\ &\geq \tfrac{1}{2}(\operatorname{cap}(K)(1-\delta_1))^n \geq \tfrac{1}{2}(\operatorname{cap}(K))^n \left(1 - \frac{\varepsilon}{2}\right)^n, \end{aligned}$$

where in the second inequality we have used the fact that for monic polynomial $P_n(z) = z^n + \cdots$

$$\|P_n\|_V \geq (\text{cap}(V))^n$$

(see [Ts, Theorem III.15]). Thus, if ε_1 is chosen so that $(1+\varepsilon_1)^2 < 1+\varepsilon$, then for large n, say, $n \geq n_\varepsilon$, (3.9) will hold.

Step 5. Finally, we prove that condition Λ^* implies $\mu \in \mathbf{Reg}$.

Let $L > 0$ be such that

$$\lim_{r\to 0} \text{cap}(\{z \mid \mu(\Delta_r(z)) \leq r^L\}) = \text{cap}(S(\mu)), \tag{3.10}$$

and for $K = S(\mu)$ and $\varepsilon > 0$ let δ, n_ε be as in Step 4.

From (3.10) we can deduce that for every sufficiently large n, say, $n \geq n(\delta)$, there exists a compact set $V = V_n$ of capacity bigger than $\text{cap}(S(\mu)) - \delta$ such that for $x \in V_n$ and $r_n = (1+\varepsilon)^{-n}$

$$\mu(\Delta_{r_n}(x)) \geq r_n^L. \tag{3.11}$$

Thus, if $P_n(z) = z^n + \cdots$ is a monic polynomial of degree $n \geq \max(n(\delta), n_\varepsilon)$ and $z_0 = z_0(V_n, P_n) \in V_n$ is the point ensured by Step 4, then (3.11) applied to z_0 yields (see (3.9))

$$\int |P_n|^2 d\mu \geq \int_{\Delta_{r_n}(z_0)} |P_n|^2 d\mu \geq \left(\min_{z\in\Delta_{r_n}(z_0)} |P_n(z)|\right)^2 \mu(\Delta_{r_n}(z_0))$$
$$\geq (\text{cap}(K))^{2n} \left(\frac{(1-\varepsilon)^2}{(1+\varepsilon)^L}\right)^n,$$

proving

$$\limsup_{n\to\infty} \gamma_n(\mu)^{1/n} \leq \frac{(1+\varepsilon)^L}{(1-\varepsilon)^2} \frac{1}{\text{cap}(S(\mu))}.$$

Now for $\varepsilon \to 0$ the regularity of μ follows from this, Corollary 1.1.7, and Definition 3.1.2. □

Proof of Theorems 4.2.5 and 4.2.7. Since criterion λ implies criterion λ^*, we only have to prove Theorem 4.2.7.

Choose $\varepsilon > 0$ arbitrarily small and let $P_n(z) = z^n + \cdots$ be a monic polynomial. Remez's inequality says that for $\alpha > 0$

$$\|P_n\|_{[0,1]} \leq T_n\left(\frac{2}{\text{meas}(\{x \in [0,1] \mid |P_n(x)| \leq \alpha\})} - 1\right)\alpha,$$

where

$$T_n(x) = \tfrac{1}{2}((x + \sqrt{x^2-1})^n + (x - \sqrt{x^2-1})^n)$$

are the Chebyshev polynomials (see [Fr, Chapter III, Lemma 7.3]). Seeing that

$$T_n\left(\frac{2}{1/(1+\varepsilon)}-1\right)=T_n(1+2\varepsilon)\le(1+4\sqrt{\varepsilon})^n$$

and that $\|P_n\|_{[0,1]}\ge 4^{-n}$, we have that the set

$$A_n=\{x\in[0,1]\mid |P_n(x)|\le 4^{-n}(1+4\sqrt{\varepsilon})^{-n}\}$$

has Lebesgue measure

$$\mathrm{meas}(A_n)\le(1+\varepsilon)^{-1}\le 1-\varepsilon/2.$$

The set $[0,1]\backslash A_n$ consists of at most $n+1$ intervals, and if we disregard those that have length at most $\varepsilon/8n$, then the remaining intervals $I_1,\dots,I_k$, each of length $>\varepsilon/8n$, form a set B_n of total measure $\ge\varepsilon/4$. If I_j' is the interval that we obtain from I_j by shrinking it around its center by a factor of 2, then the so-obtained set $B_n'=\cup_{j=1}^k I_j'$ has measure $\ge\varepsilon/8$.

Now condition λ^* implies that there is an $m_0\in\mathbb{N}$ such that for $m\ge m_0$ the set

$$C(m)=\{x\mid \mu([x-1/m,x+1/m])\ge e^{-\varepsilon^2 m}\}$$

has measure at least $1-\varepsilon/16$. Thus, for large n and $m=[32n/\varepsilon]+1$ ($>m_0$) the sets B_n' and $C(m)$ intersect each other and thus we can find a point $x_n\in B_n'$ for which

$$\mu\left(\left[x_n-\frac{1}{m},x_n+\frac{1}{m}\right]\right)\ge\exp(-\varepsilon^2 m)\ge e^{-64\varepsilon n}. \tag{3.12}$$

But then $x_n\in I_j'$ for some j and $|I_j|>\varepsilon/8n$, hence

$$\left[x_n-\frac{1}{m},x_n+\frac{1}{m}\right]\subseteq\left[x_n-\frac{\varepsilon}{32n},x_n+\frac{\varepsilon}{32n}\right]\subseteq I_j\subseteq[0,1]\setminus A_n,$$

which means that for all $t\in\left[x_n-\frac{1}{m},x_n+\frac{1}{m}\right]$ we have

$$|P_n(t)|\ge 4^{-n}(1+4\sqrt{\varepsilon})^{-n}.$$

This together with (3.12) shows that

$$\int|P_n|^2d\mu\ge\int_{x_n-1/m}^{x_n+1/m}|P_n|^2d\mu\ge 4^{-2n}(1+4\sqrt{\varepsilon})^{-2n}e^{-64\varepsilon n}$$

and so

$$\limsup_{n\to\infty}\gamma_n(\mu)^{1/n}\le 4(1+4\sqrt{\varepsilon})e^{32\varepsilon},$$

and the arbitrariness of $\varepsilon>0$ proves the regularity of μ in the usual way (see Corollary 1.1.7). □

4.4 Preliminaries for Proving the Sharpness of the Criteria

In this section we prove some results concerning the equilibrium distribution and capacity of sets consisting of finitely many intervals. Our proofs in the next section will heavily rely on these lemmas. We have separated the content of Section 4.4 because the results to be proved are interesting in themselves, for example, Corollary 4.4.4 provides the "best" method for constructing sets with small linear measure but large capacity.

Lemma 4.4.1. *Let $H \subseteq [0,1]$ be a set consisting of k intervals $I_j = [a_j, b_j]$, $b_j < a_{j+1}$, $j = 1, \dots, k$. Then there are numbers $y_j \in (b_j, a_{j+1})$, $j = 1, \dots, k-1$ such that ω_H, the equilibrium measure of H, is given by*

$$d\omega_H(x) = \frac{\prod_{j=1}^{k-1} |x - y_j|}{\pi |\prod_{j=1}^{k} (x - a_j)(b_j - x)|^{1/2}} dx, \quad x \in H. \tag{4.1}$$

Proof. Consider the function

$$Q(z) = \int \frac{d\omega_H(t)}{z - t} \tag{4.2}$$

on the Riemann sphere $\overline{\mathbb{C}}$ cut along H. On the cut the real part of Q vanishes because there $\operatorname{Re} Q$ is the derivative of the equilibrium potential of H and the latter is constant on H. For the same reason there is a point $y_j \in (b_j, a_{j+1})$ where Q vanishes. Since $Q(\overline{z}) = \overline{Q(z)}$, the function $Q^2(z)$ is continuous and hence analytic on $\cup_{j=1}^{k}(a_j, b_j)$, so Q^2 is analytic on $\overline{\mathbb{C}} \setminus \{a_j, b_j\}_{j=1}^{k}$. Clearly, Q^2 can have at most single poles at the points a_j, b_j, and hence Q^2 is a rational function on $\overline{\mathbb{C}}$. Q^2 has at least $2k$ zeros (two at each y_j, $j = 1, \dots, k-1$, and at ∞) and at most $2k$ poles, hence we must have

$$Q^2(z) = \text{const.} \times \prod_{j=1}^{k-1} (z - y_j)^2 / \prod_{j=1}^{k} (z - a_j)(z - b_j).$$

But $Q^2(z) \sim 1/z^2$ as $z \to \infty$, so therefore the constant is 1. In summary,

$$Q(z) = \left(\prod_{j=1}^{k-1} (z - y_j)^2 / \prod_{j=1}^{k} (z - a_j)(z - b_j) \right)^{1/2},$$

where that branch of the square root is taken which is positive for positive z. Cauchy's formula gives

$$Q(z) = \frac{1}{2\pi i} \oint_H \frac{Q(\xi)}{\xi - z} d\xi = \frac{1}{\pi i} \int_H \frac{Q(t)}{t - z} dt, \tag{4.3}$$

where the last integral is taken on the upper part of the cut.

Since on the upper part of the cut

$$-\frac{1}{i}Q(t) = \prod_{j=1}^{k-1} |t - y_j| / \left| \prod_{j=1}^{k} (t - a_j)(b_j - t) \right|^{1/2},$$

and the Cauchy transform uniquely determines the measure, (4.1) follows from (4.2) and (4.3). □

Lemma 4.4.2. *Let $H \subseteq [0,1]$ be as in the previous lemma and let $v_H = \omega_H'$, the density of the equilibrium measure (with respect to Lebesgue measure). Then for $x \in (a_j, b_j)$, $j = 1, \ldots, k$,*

$$v_H(x) \le \frac{1}{\pi}((x - a_j)(b_j - x))^{-1/2} \tag{4.4}$$

and

$$|v_H'(x)| \le ((x - a_j)(b_j - x))^{-3/2}. \tag{4.5}$$

Proof. Inequality (4.4) follows immediately from the representation (4.1). Similarly, to prove (4.5) we only have to remark that, for example, for $x \in (a_j, b_j)$

$$\left| d\left(\prod_{l=1}^{j-1} (x - y_l)/((x - a_l)(x - b_l))^{1/2} \right) /dx \right|$$

$$= \left| \left(\prod_{l=1}^{j-1} (x - y_l)/((x - a_l)(x - b_l))^{1/2} \right) ((x - y_{j-1})^{-1} \right.$$

$$- \tfrac{1}{2}((x - a_{j-1})^{-1} + (x - b_{j-1})^{-1}) + (x - y_{j-2})^{-1} - \cdots$$

$$\left. + (x - y_1)^{-1} - \tfrac{1}{2}((x - a_1)^{-1} + (x - b_1)^{-1}) \right| \le (x - y_{j-1})^{-1}$$

$$\le (b_j - x)((x - a_j)(b_j - x))^{-1} \le \tfrac{1}{2}(b_j - x)((x - a_j)(b_j - x))^{-3/2}.$$

A similar estimate for the derivative of

$$\prod_{l=j}^{k-1} (x - y_l)/((a_{l+1} - x)(b_{l+1} - x))^{1/2}$$

and the estimate

$$\left| \frac{d(((x - a_j)(b_j - x))^{-1/2})}{dx} \right| \le \frac{1}{2}((x - a_j)(b_j - x))^{-3/2}$$

yield (4.5). □

Lemma 4.4.3. *Let* $H = \cup_{j=1}^{k}[a_j, b_j]$ *be as in* Lemma 4.4.1,

$$b = \min_{1 \le j \le k} (b_j - a_j),$$

and

$$E(n, \alpha) = \bigcup_{i=1}^{n-1} \left[\frac{i}{n} - e^{-\alpha n}, \frac{i}{n} + e^{-\alpha n} \right]. \tag{4.6}$$

If

$$\alpha < 4^{-10} b^4 \tag{4.7}$$

and

$$n > \alpha^{-3} \tag{4.8}$$

are satisfied then we have

$$\operatorname{cap}(H \cap E(n, \alpha)) \ge \operatorname{cap}(H) \exp(-\alpha^{1/5}). \tag{4.9}$$

Remarks. (1) Inequality (4.9) claims that the capacity of $H \cap E(n, \alpha)$ is about as large as that of H although $E(n, \alpha)$ consists of almost exponentially small n intervals.

(2) Typical applications of Lemma 4.4.3 will use $n = 2^m$, $b = 2^{-o(m)}$, $\alpha = b^5$.

We would like to emphasize that the estimate (4.9) is uniform in H in the sense that the thresholds (4.7)–(4.8) depend only on b and do not depend on the actual size and position of the intervals in H.

Setting $H = [0, 1]$ in Lemma 4.4.3 we get the following corollary.

Corollary 4.4.4. *For* $\alpha < 4^{-10}$ *and* $\alpha > n^{-1/3}$ *we have*

$$\operatorname{cap}(E(n, \alpha)) \ge \tfrac{1}{4} \exp(-\alpha^{1/5}).$$

Note that $E(n, \alpha)$ consists of intervals of total length $< 2ne^{-\alpha n}$; that is, when $\alpha_n \to 0$ as $n \to \infty$, the capacity of $E(n, \alpha_n)$ tends to the capacity of $[0, 1]$ but $E(n, \alpha_n)$ has "almost exponentially" small Lebesgue measure: $m(E(n, \alpha_n)) < \exp(-(\alpha_n/2)n)$. This is sharp in the sense that $\operatorname{cap}(E(n, \alpha_n)) \le e^{-\alpha_n}$ (see, e.g., (4.27)) and so for $\alpha_n \ge \log 8$ we have $\operatorname{cap}(E(n, \alpha_n)) \le \frac{1}{2}\operatorname{cap}([0, 1])$.

Proof of Lemma 4.4.3. We keep the notations from the preceding lemma, and fix α, n with properties (4.7) and (4.8).

Let

$$H^{(\delta)} = \bigcup_{j=1}^{k} [a_j + \delta, b_j - \delta],$$

$$S^{(n)} = \left\{ i \,\middle|\, \frac{i}{n} \in H^{\delta} \right\},$$

and for $i \in S^{(n)}$ let $x_{i,n}$ be a point in the interval

$$J_{i,n} = \left[\frac{2i-1}{2n}, \frac{2i+1}{2n} \right]$$

where $v = v_H$ attains its minimum on $J_{i,n}$. If $\nu_{i,n}$ denotes the equilibrium measure of the interval $[\frac{i}{n} - e^{-\alpha n}, \frac{i}{n} + e^{-\alpha n}]$ then we shall prove the claim in the lemma with the aid of the measure

$$\nu = \frac{1}{n} \sum_{i \in S^{(n)}} v(x_{i,n}) \nu_{i,n}.$$

We shall show that ν has sufficiently small logarithmic energy and norm $\|\nu\|$ close to 1, and thereby normalizing ν so that it becomes a probability measure, we will be able to prove the statement in Lemma 4.4.3 concerning the capacity of $H \cap E(n, \alpha)$.

The δ in the definition of $H^{(\delta)}$ will be chosen subsequently (we will set

$$\delta = \frac{1}{9 \cdot 16} b^{7/3} \alpha^{2/3})$$

and that will complete the definition of the preceding sets and measure. Note that in any case we have $\mathrm{supp}(\nu) \subseteq H \cap E(n, \alpha)$ whenever $n \geq 2/\delta$.

According to Lemma 4.4.2 we have for $v(x) := v_H(x) = \omega_H'(x)$

$$(4.10)\quad v(x) \leq \frac{1}{\pi} ((x - a_j)(b_j - x))^{-1/2}, \quad |v'(x)| \leq ((x - a_j)(b_j - x))^{-3/2}$$

for

$$x \in (a_j, b_j) =: I_j, \quad j = 1, 2, \ldots, k,$$

which imply for $\delta < b/4$ and $n \geq 2/\delta$

$$\int_{H^{(2\delta)}} v(x) dx \geq 1 - 3k(\delta/b)^{1/2},$$

where k, the number of intervals in H, is at most $1/b$. Thus

(4.11)

$$\begin{aligned} \|\nu\| &= \frac{1}{n} \sum_{i \in S^{(n)}} v(x_{i,n}) \geq \sum_{i \in S^{(n)}} \int_{(2i-1)/2n}^{(2i+1)/2n} \left(v(x) - \frac{2^{1/2}}{(b\delta/2)^{3/2} n} \right) dx \\ &\geq \int_{H^{(2\delta)}} v(x) dx - 4(b\delta)^{-3/2} n^{-1} \\ &\geq 1 - 3\delta^{1/2} b^{-3/2} - 4(b\delta)^{-3/2} n^{-1} =: \gamma_n^{-1}, \end{aligned}$$

where, in the first inequality, we have used that by the second estimate in (4.10)

$$v(x_{i,n}) \geq v(x) - \frac{2^{1/2}}{(b\delta/2)^{3/2}n}$$

when $x \in \left[\frac{2i-1}{2n}, \frac{2i+1}{2n}\right]$, $i \in S^{(n)}$.

We shall also need that 1 is an upper bound for the left-hand side of (4.11):

$$\frac{1}{n}\sum_{i\in S^{(n)}} v(x_{i,n}) \leq \sum_{i\in S^{(n)}} \int_{J_{i,n}} v(x)dx \leq \int_H v(x)dx = 1 \tag{4.12}$$

where, in the first inequality, we used that $x_{i,n}$ was chosen to be an (absolute) minimum point for v on $J_{i,n}$.

In order to estimate the logarithmic energy of ν we mention first of all that because

$$\operatorname{cap}\left(\left[\frac{i}{n} - e^{-\alpha n}, \frac{i}{n} + e^{-\alpha n}\right]\right) = e^{-\alpha n}/2,$$

we have

$$(\nu_{i,n}, \nu_{i,n}) := \int\int \log\frac{1}{|x-t|} d\nu_{i,n}(x)d\nu_{i,n}(t) = \alpha n + \log 2. \tag{4.13}$$

Next, for $|i-j| > 1$ we get from

$$|x-t| \geq \left|\frac{i}{n} - \frac{j}{n}\right| - \frac{1}{n}, \quad x \in \operatorname{supp}(\nu_{i,n}),\ t \in \operatorname{supp}(\nu_{j,n}) \tag{4.14}$$

that

$$(\nu_{i,n}, \nu_{j,n}) := \int\int \log\frac{1}{|x-t|} d\nu_{i,n}(x)d\nu_{j,n}(x) \leq \log\frac{n}{|i-j|-1}.$$

When $|i-j| = 1$ then we apply (4.13) and the Cauchy-type inequality

$$(\nu_{i,n}, \nu_{j,n}) \leq (\nu_{i,n}, \nu_{i,n})^{1/2}(\nu_{j,n}, \nu_{j,n})^{1/2},$$

which follows from the fact that for $\operatorname{supp}(\nu), \operatorname{supp}(\mu) \subseteq [0,1]$ the inner product (ν, μ) is positive definite: $(\nu, \nu) \geq 0$ (see [La, Chapter I, Theorem 1.16]). From it we deduce

$$(\nu_{i,n}, \nu_{j,n}) \leq \alpha n + \log 2.$$

All these show that the logarithmic energy of ν can be estimated as

$$(\nu, \nu) \leq \left(\alpha + \frac{\log 2}{n}\right)\left(\frac{1}{n}\sum_{i\in S^{(n)}} v^2(x_{i,n}) + \frac{1}{n}\sum_{\substack{i,j\in S^{(n)},\\ |i-j|=1}} v(x_{i,n})v(x_{j,n})\right)$$

$$+ \frac{1}{n^2}\sum_{\substack{i,j\in S^{(n)},\\ |i-j|\geq 2}} \log\frac{n}{|i-j|-1} v(x_{i,n})v(x_{j,n}) = \Sigma_0 + \Sigma_1.$$

According to (4.10)

$$v(x_{l,n}) \leq (\delta b)^{-1/2}, \quad l \in S^{(n)}, \tag{4.15}$$

hence (4.12) yields

$$\Sigma_0 \leq 4\alpha(\delta b)^{-1/2} \tag{4.16}$$

because $\alpha > 1/n$. For Σ_1 we can write

$$\begin{aligned}\Sigma_1 &= \frac{1}{n^2} \sum_{\substack{i,j \in S^{(n)} \\ |i-j| \geq 2}} \log \frac{1}{|i/n - j/n| + 1/n} \cdot v(x_{i,n}) v(x_{j,n}) \\ &\quad + \frac{1}{n^2} \sum_{\substack{i,j \in S^n \\ |i-j| \geq 2}} \log \frac{|i-j|+1}{|i-j|-1} v(x_{i,n}) v(x_{j,n}) \\ &= \Sigma_{11} + \Sigma_{12}.\end{aligned} \tag{4.17}$$

Here, because of the choice of $x_{i,n}$ and $x_{j,n}$,

$$\begin{aligned}\Sigma_{11} &\leq \sum_{i,j \in S^{(n)}} \int_{J_{i,n}} \int_{J_{j,n}} \log \frac{1}{|x-t|} v(x) v(t) dx dt \\ &\leq \int \int \log \frac{1}{|x-t|} v(x) v(t) dx dt = \log \frac{1}{\operatorname{cap}(H)}.\end{aligned} \tag{4.18}$$

In Σ_{12} we use again (4.15) and (4.12) in the form

$$\begin{aligned}\Sigma_{12} &\leq (\delta b)^{-1/2} \frac{1}{n^2} \sum_{r=2}^{n-1} \log \frac{r+1}{r-1} \sum_{i \in S^{(n)}} v(x_{i,n}) \\ &\leq (\delta b)^{-1/2} \frac{1}{n} \log \prod_{r=2}^{n-1} \frac{r+1}{r-1} = (\delta b)^{-1/2} \frac{\log n}{n}.\end{aligned} \tag{4.19}$$

We obtain from (4.11) and (4.16)–(4.19) the upper estimate for the logarithmic energy $(\tilde{\nu}, \tilde{\nu})$ of the probability measure $\tilde{\nu} = \nu / \|\nu\|$:

$$\int \int \log \frac{1}{|x-t|} d\tilde{\nu}(x) d\tilde{\nu}(t) \leq \left(\log \frac{1}{\operatorname{cap}(H)} \right) \gamma_n^2 + \beta_n, \tag{4.20}$$

where

$$\gamma_n = (1 - 3\delta^{1/2} b^{-3/2} - 4(b\delta)^{-3/2} n^{-1})^{-1} \tag{4.21}$$

was defined in (4.11) and

$$\beta_n = \left(4\alpha(\delta b)^{-1/2} + (\delta b)^{-1/2} \frac{\log n}{n} \right) \gamma_n^2. \tag{4.22}$$

Setting now

$$\delta = \frac{1}{9 \cdot 16} b^{7/3} \alpha^{2/3}$$

and taking into account $\alpha < 4^{-10} b^4$, $n > \alpha^{-3}$ we get for the terms in γ_n and β_n

$$3\delta^{1/2} b^{-3/2} \le \tfrac{1}{4} (\alpha/b)^{1/3},$$

$$4(b\delta)^{-3/2} n^{-1} \le \tfrac{1}{4} \alpha^{1/3},$$

$$4\alpha(\delta b)^{-1/2} \le \tfrac{3}{4} (\alpha/b)^{1/3},$$

$$(\delta b)^{-1/2} \frac{\log n}{n} \le \frac{1}{4^2} \alpha^{1/3}.$$

Obviously,

$$\log \frac{1}{\mathrm{cap}(H)} \le \log \frac{4}{b},$$

by which

$$\left(\log \frac{1}{\mathrm{cap}(H)} \right) \gamma_n^2 \le \log \frac{1}{\mathrm{cap}(H)} + (\gamma_n^2 - 1) \log \frac{4}{b}.$$

Since $\alpha < 4^{-10} b^4$ also implies

$$(\alpha/b)^{1/3} \le 4^{-5/6} \alpha^{1/4},$$

the above estimates together with (4.20)–(4.22) easily yield that the logarithmic energy of $\tilde{\nu}$ is at most

$$\log \frac{1}{\mathrm{cap}(H)} + \alpha^{1/5}.$$

This and $\mathrm{supp}(\tilde{\nu}) \subseteq H \cap E(n, \alpha)$ proves (4.9). □

We shall also need the following sharper form of Corollary 4.4.4.

Lemma 4.4.5. *Let $n^{-1/30} < \alpha < 4^{-10}$, $0 \le \varepsilon \le 1$, and $E_\varepsilon^*(n, \alpha)$ be a Borel set such that for at least $(1 - \varepsilon)n$ of the js in the set $\{j\}_1^n$ we have*

$$\mathrm{cap} \left(E_\varepsilon^*(n, \alpha) \bigcap \left[\frac{2j-1}{2n}, \frac{2j+1}{2n} \right] \right) > e^{-\alpha n}. \tag{4.23}$$

Then for $\varepsilon \le \alpha/2$

$$\mathrm{cap}(E_\varepsilon^*(n, \alpha)) \ge \tfrac{1}{4} \exp(-\alpha^{1/5} - 2\varepsilon/\alpha).$$

Proof. Let $J \subseteq \{j\}_1^n$, be the set of those js that satisfy (4.23), and for each $j \in J$ let K_j be a compact subset of

$$E_\varepsilon^*(n, \alpha) \bigcap \left[\frac{2j-1}{2n}, \frac{2j+1}{2n} \right]$$

of capacity

$$\operatorname{cap}(K_j) > e^{-\alpha n}.$$

If $j \notin J$ then we set

$$K_j = \left[\frac{j}{n} - e^{-\alpha n}, \frac{j}{n} + e^{-\alpha n}\right] \tag{4.24}$$

and consider the sets

$$E_\varepsilon^* = \bigcup_{j \in J} K_j,$$

$$E^* = \bigcup_{j=1}^{n} K_j.$$

Since $E_\varepsilon^* \subseteq E_\varepsilon^*(n, \alpha)$, it is enough to prove

$$\operatorname{cap}(E_\varepsilon^*) \geq \tfrac{1}{4} \exp(-\alpha^{1/5} - 2\varepsilon/\alpha), \tag{4.25}$$

which will be done by estimating the capacity of E^* first and then by showing that $E^* \setminus E_\varepsilon^*$ has small capacity (depending on ε).

As for E^*, we can repeat the proof of Lemma 4.4.3 for $H = [0, 1]$ with ω_{K_j} replacing $\nu_{j,n}$. We have formulated the proof of Lemma 4.4.3 in such a way that all the estimates hold in the present case (actually with a lot of simplifications because $H = [0, 1]$ and $v(x) = v_H(x) = (\pi(x(1-x))^{1/2})^{-1}$). Thus, we can conclude

$$\operatorname{cap}(E^*) > \tfrac{1}{4} \exp(-\alpha^{1/5}). \tag{4.26}$$

$E^* \setminus E_\varepsilon^*$ consists of at most εn intervals of the form (4.24), hence the inequality

$$\frac{1}{\log(1/\operatorname{cap}(E))} \leq \sum_s \frac{1}{\log(1/\operatorname{cap}(E_s))}, \quad E = \bigcup_s E_s, \tag{4.27}$$

which is valid whenever $\operatorname{diam}(E) \leq 1$ (see [Ts, Chapter III, Theorem III.17]), shows that

$$\operatorname{cap}(E^* \setminus E_\varepsilon^*) \leq e^{-\alpha/\varepsilon}. \tag{4.28}$$

Applying once again (4.27) with the choice $E = E^*, E_1 = E_\varepsilon^*$, $E_2 = E^* \setminus E_\varepsilon^*$, we get from (4.26) and (4.28) that

$$\frac{1}{\log 4 + \alpha^{1/5}} \leq \frac{1}{\log(1/\operatorname{cap}(E_\varepsilon^*))} + \frac{\varepsilon}{\alpha},$$

which proves (4.25). □

4.5 Proof of the Sharpness of the Regularity Criteria

In this section we verify Theorems 4.2.2, 4.2.4, 4.2.6, and 4.2.8. We present the proof of the necessity result Theorem 4.2.8 here because it is verified via a construction like the other proofs of this section. In these proofs we will heavily rely on the results from Section 4.4.

Proof of Theorem 4.2.2. Consider the sets

$$E(n,\alpha) = \bigcup_{i=1}^{n-1} \left[\frac{i}{n} - e^{-\alpha n}, \frac{i}{n} + e^{-\alpha n}\right]$$

from Lemma 4.4.3, and set $\varepsilon_0 = \frac{1}{10}$, $M_0 = 4$, $M_{k+1} = 2^{M_k}$, $\varepsilon_{k+1} = \exp(-5M_k\varepsilon_k)$, $k = 0, 1, \ldots$, and $\delta_k = \exp(-kM_{k-1})$. One can easily show that $\varepsilon_k \to 0$ and

$$M_k\varepsilon_k = \exp(M_{k-1}(\log 2 - 5\varepsilon_{k-1})) \to \infty$$

as $k \to \infty$, hence if $\eta > 0$ then we have for large k

$$\log 1/\delta_k < (M_{k-1}\varepsilon_{k-1})^{1+\eta}/k.$$

Let $\mu = \sum_{k=1}^{\infty} \mu_k$, where μ_k assigns mass δ_k to each point of the set

$$S_{M_k} := \left\{\frac{j}{M_k}\right\}_{j=1}^{M_k-1}$$

Obviously, μ is a finite measure with $\operatorname{supp}(\mu) = [0,1]$.

If

$$\exp(-M_k\varepsilon_k) \le r < \exp(-M_{k-1}\varepsilon_{k-1})$$

and $x \in E(M_k, \varepsilon_k)$, then at least one point of S_{M_k} belongs to $[x-r, x+r]$ and so for large k

$$\log 1/\mu([x-r, x+r]) \le \log 1/\delta_k < (M_{k-1}\varepsilon_{k-1})^{1+\eta}/k \le (\log 1/r)^{1+\eta}/k.$$

This shows that if x belongs to almost all the sets $E(M_k; \varepsilon_k)$ (i.e., it belongs to all of them with finitely many exceptions) then for x

$$\lim_{r\to 0} \frac{\log 1/\mu([x-r, x+r])}{(\log 1/r)^{1+\eta}} = 0 \tag{5.1}$$

holds. Therefore, to prove that the set of those points x for which (5.1) is satisfied for all $\eta > 0$ has capacity $\frac{1}{4}$ it is enough to show that

$$\liminf_{k\to\infty} E(M_k, \varepsilon_k) = \bigcup_{k=1}^{\infty} \bigcap_{m=k}^{\infty} E(M_m, \varepsilon_m)$$

has capacity $\frac{1}{4}$.

Applying Lemma 4.4.3 with $H = E(M_k, \varepsilon_k)$, $n = M_{k+1} = 2^{M_k}$, and $\alpha = \varepsilon_{k+1} = 2^{-5M_k\varepsilon_k}$, which satisfy

$$n > \alpha^{-3},$$
$$\alpha < 4^{-10}(2 \cdot 2^{-M_k\varepsilon_k})^4$$

provided n is sufficiently large (notice that $H = E(M_k, \varepsilon_k)$ consists of intervals of length $2 \cdot 2^{-M_k\varepsilon_k}$, which are of distance $\geq 1/2M_k$ apart), we get that

$$\begin{aligned} \operatorname{cap}(E(M_k; \varepsilon_k) \cap E(M_{k+1}, \varepsilon_{k+1})) &\geq \operatorname{cap}(E(M_k, \varepsilon_k)) \exp(-\varepsilon_{k+1}^{1/5}) \\ &\geq \tfrac{1}{4} \exp(-\varepsilon_k^{1/5} - \varepsilon_{k+1}^{1/5}) \end{aligned}$$

(see also Corollary 4.4.4).

We would like to repeat this process by setting

$$H = E(M_k, \varepsilon_k) \cap E(M_{k+1}, \varepsilon_{k+1}),$$

$n = M_{k+2}$, $\alpha = \varepsilon_{k+2}$. Everything works well if H consists of intervals of length $2^{-M_{k+1}\varepsilon_{k+1}}$ or larger because then we can apply Lemma 4.4.3. Unfortunately, H may contain intervals of length smaller than $2^{-M_{k+1}\varepsilon_{k+1}}$; however, there can be at most $2M_k$ such intervals, hence 1 divided by the logarithm of the reciprocal of their total capacity is at most (see (4.27))

$$\begin{aligned} 2M_k / \log(4 \cdot 2^{M_{k+1}\varepsilon_{k+1}}) &\leq 2M_k / (\log 2) M_{k+1}\varepsilon_{k+1} \\ &= 2M_k \exp(-M_k(\log 2 - 5\varepsilon_k)) / \log 2, \end{aligned}$$

which is much smaller than ε_{k+1}. Now if we drop these intervals from

$$H = E(M_k, \varepsilon_k) \cap E(M_{k+1}, \varepsilon_{k+1}),$$

we get a set H' consisting of intervals of length $2^{-M_{k+1}\varepsilon_{k+1}}$ or bigger for which

$$\operatorname{cap}(H') \geq \tfrac{1}{4} \exp(-\varepsilon_k^{1/5} - 2\varepsilon_{k+1}^{1/5})$$

(apply (4.27) again for $E = H$, $E_1 = H'$, $E_2 = H \setminus H'$). Thus, we can then continue with H' instead of H. However, to keep things simple, in what follows we shall assume $H' = H$.

Applying repeatedly Lemma 4.4.3 in this fashion to

$$H = \bigcap_{m=k}^{s} E(M_m, \varepsilon_m)$$

and $n = M_{s+1}$, $\alpha = \varepsilon_{s+1}$, we get in a similar fashion that for every $s > k$

$$\begin{aligned} \text{(5.2)} \quad \operatorname{cap}\left(\bigcap_{m=k}^{s} E(M_m, \varepsilon_m)\right) &> \tfrac{1}{4} \exp\left(-2 \sum_{m=k}^{s} \varepsilon_m^{1/5}\right) \\ &\geq \tfrac{1}{4} \exp\left(-2 \sum_{m=k}^{s} e^{-m}\right) \geq \tfrac{1}{4} \exp(-2/k), \end{aligned}$$

where we have used that for large

$$m\colon \varepsilon_m = \exp(-M_{m-1}(\log 2 - 5\varepsilon_{m-1})) < e^{-10m}.$$

Now the sets $E(M_m, \varepsilon_m)$ are closed, hence for $s \to \infty$ the left-hand side in (5.2) tends to

$$\text{cap}\left(\bigcap_{m=k}^{\infty} E(M_m, \varepsilon_m)\right).$$

This and (5.2) already prove our claim that

$$\text{cap}(\liminf_{k\to\infty} E(M_k, \varepsilon_k)) = \tfrac{1}{4}.$$

It is left to prove that μ is not regular. But this is easy: If

$$P_{M_k}(x) = \prod_{j=1}^{M_k-1}\left(x - \frac{j}{M_k}\right),$$

then P_{M_k} vanishes on the support of $\sum_{m=1}^{k} \mu_m$ and so

$$\begin{aligned}\left(\int |P_{M_k}|^2 d\mu\right)^{1/2M_k} &\le \left(\sum_{m=k+1}^{\infty} \mu_m\right)^{1/2M_k} \le \left(\sum_{m=k+1}^{\infty} M_m\delta_m\right)^{1/2M_k} \\ &= \left(\sum_{m=k+1}^{\infty} 2^{M_{m-1}} \cdot 2^{-mM_{m-1}}\right)^{1/2M_k} \\ &\le \left(2\cdot 2^{-kM_k}\right)^{1/2M_k} \to 0\end{aligned}$$

as $k \to \infty$, proving $\mu \notin \mathbf{Reg}$, because the preceding estimate gives

$$\lim_{k\to\infty} \gamma_{M_k}(\mu)^{1/M_k} = \infty. \qquad \square$$

Proof of Theorem 4.2.4. Let us suppose that $\gamma\colon (0,1) \to (0,1)$ decreases faster than any power x^L as $x \to 0+0$. Clearly, we can assume without loss of generality that γ is monotonically increasing. Then we can choose $\varepsilon_k > 0$, $\varepsilon_k \to 0$, and $l_k \nearrow \infty$ such that if $M_0 = 1$, $M_{k+1} = 2^{M_k}$ then $\gamma(2^{-M_k\varepsilon_k}) = 2^{-M_k l_k}$. Now let $\delta_k = 2^{-M_{k-1}l_{k-1}}$, and consider the measure μ constructed in the proof of Theorem 4.2.2 with this ε_k and δ_k. Exactly as there, we get for $x \in E(M_k, \varepsilon_k)$ and $2^{-M_k\varepsilon_k} \le r \le 2^{-M_{k-1}\varepsilon_{k-1}}$ that

$$\mu([x-r, x+r]) \ge \delta_k = \gamma(2^{-M_{k-1}\varepsilon_{k-1}}) \ge \gamma(r),$$

and so

$$\text{cap}(\{x \mid \mu([x-r; x+r]) \ge \gamma(r)\}) \ge \text{cap}(E(M_k; \varepsilon_k)) \to \tfrac{1}{4}$$

as $k \to \infty$, by Corollary 4.4.4 (clearly we may assume $\varepsilon_k > M_k^{-1/3}$ to be able to apply Corollary 4.4.4).

But μ is far from being regular because the same computation that was done at the end of the preceding proof shows (for the same P_{M_k} as there)

$$\left(\int |P_{M_k}|^2 d\mu\right)^{1/M_k} \leq (2 \cdot M_k 2^{-M_k l_k})^{1/M_k} \to 0$$

when $k \to \infty$, and then $\mu \notin \mathbf{Reg}$ follows. □

Proof of Theorem 4.2.6. Without loss of generality we may assume that $\gamma(r)/r$ is monotonically increasing. There is a sequence $\{l_n\}$ such that $l_n \nearrow \infty$ as $n \to \infty$ and $\gamma(2^{-n}) = l_n^{-2} 2^{-n}$. Let μ_n be the measure that assigns mass $\exp(-l_{n-1}2^{n-1})$ to each point $j/2^n$, $j = 1, \ldots, 2^n$, and set $\mu = \sum_{n=1}^{\infty} \mu_n$.

If $2^{-n} \leq r < 2^{-n+1}$ then for every $x \in [0,1]$

$$\begin{aligned}\mu([x-r, x+r]) &\geq \mu_n([x-r, x+r]) \geq \exp(-l_{n-1}2^{n-1}) \\ &= \exp(-(\gamma(2^{-n+1}))^{-1}/l_{n-1}) \geq \exp(-(\gamma(r))^{-1}/l_{n-1}),\end{aligned}$$

which proves

$$\lim_{r \to 0+0} \gamma(r) \log(\mu[x-r, x+r]) = 0.$$

On the other hand, μ is not regular because for the polynomial

$$P_{2^n}(x) = \prod_{j=1}^{2^n} \left(x - \frac{j}{2^n}\right)$$

of degree 2^n we have

$$\begin{aligned}\int |P_{2^n}|^2 d\mu &\leq \sum_{k=n+1}^{\infty} \|\mu_k\| = \sum_{k=n+1}^{\infty} 2^k \exp(-l_{k-1}2^{k-1}) \\ &\leq 2^{n+2} \exp(-l_n 2^n),\end{aligned}$$

showing that

$$\gamma_{2^n}(\mu)^{1/2^n} \geq e^{l_n/2}(2^{n+2})^{-1/2^{n+1}} \to \infty \qquad \text{as } n \to \infty.$$

□

Proof of Theorem 4.2.8. Let us suppose that for some $\eta > 0$ and $\alpha < \frac{1}{4}$ the sets

$$A_m = \left\{x \,\middle|\, \mu\left(\left[x - \frac{1}{m}, x + \frac{1}{m}\right]\right) \geq e^{-\eta m}\right\}$$

have capacity at most α for infinitely many ms, say, for $m \in \mathcal{N} \subseteq \mathbf{N}$. We have to show that then μ is not regular.

Let $m \in \mathcal{N}$ and

$$B_m = \left\{ x \,\middle|\, \mu\left(\left[x - \frac{1}{2m}, x + \frac{1}{2m}\right]\right) \geq e^{-\eta m} \right\}.$$

Clearly,

$$\bigcup_{x \in B_m} \left[x - \frac{1}{2m}, x + \frac{1}{2m}\right] \subseteq A_m. \tag{5.3}$$

Let H_m be a set of the form

$$H_m = \bigcup_{x \in B_m} (x - \varepsilon_m(x), x + \delta_m(x)), \tag{5.4}$$

where the functions ε_m, δ_m take only the values $1/2m$ or $1/4m$. We claim that we can choose $\varepsilon_m(x)$ and $\delta_m(x)$ in such a way that every interval contiguous to H_m is of length greater than or equal to $1/4m$. In fact, let us start with $\varepsilon_m \equiv \delta_m = 1/4m$, and let the corresponding set (5.4) be H_m^*. If $J_l = (\xi_1^{(l)}; \xi_2^{(l)})$, $l = 1, \ldots, l_0$, are those intervals contiguous to H_m^* that have length smaller than $1/4m$, then for each l there are $x_1^{(l)}, x_2^{(l)} \in B_m$ satisfying $x_1^{(l)} < \xi_1^{(l)} < \xi_2^{(l)} < x_2^{(l)}, x_2^{(l)} - x_1^{(l)} < 3/4m$. Thus, if we set $\varepsilon_m(x_2^{(l)}) = \delta_m(x_1^{(l)}) = 1/2m$, $l = 1, 2, \ldots, l_0$, and leave all the other $\varepsilon_m(x), \delta_m(x)$ unchanged $(= 1/4m)$, then for the resulting set (5.4) our claim is satisfied.

If $x \notin H_m$ then $x \notin B_m$; therefore

$$\mu\left(\left(x - \frac{1}{2m}, x + \frac{1}{2m}\right)\right) < e^{-\eta m}. \tag{5.5}$$

The compact set $[0, 1] \setminus H_m$ is covered by the open intervals $\{(x - 1/2m, x + 1/2m)\}_{x \in H_m}$ and by taking a minimal subcover we can conclude from (5.5) that

$$\mu([0, 1] \setminus H_m) \leq 4me^{-\eta m} \tag{5.6}$$

(note that in a minimal cover no three intervals can have common points, therefore the number of intervals $(x - 1/2m, x + 1/2m)$ in a minimal cover is at most $4m$).

We shall use (5.6) and

$$\operatorname{cap}(H_m) \leq \operatorname{cap}(A_m) \leq \alpha < \tfrac{1}{4} \tag{5.7}$$

(see (5.3)) to construct monic polynomials P_n with small $L^2(\mu)$ norm. The degree n of these polynomials will be $n = [\eta m/4]$, which ensures that the integral of $|P_n|^2$ over $[0,1]\setminus H_m$ with respect to μ will be much smaller than 4^{-2n} in view of (5.6). On the other hand, we shall construct P_n in such a way that on H_m the absolute value $|P_n|$ will not be much larger than α^n – this is where (5.7) enters the picture. Note, however, that the sets H_m change with m (or n), that is, we essentially have to solve the Chebyshev problem for the case when the set where the polynomials should be minimalized changes together with the degree of the polynomials.

Since H_m consists of certain intervals $I_j^{(m)} = (a_j^{(m)}, b_j^{(m)})$, $j = 1, \ldots, k(m)$, of length greater than equal to $1/2m$, and every interval contiguous to H_m is of length $\geq 1/4m$, we have $k(m) \leq 2m$. Lemma 4.4.2 tells us that if $d\omega_{H_m}(x) = v_m(x)dx$ is the equilibrium distribution of H_m then

$$v_m(t) \leq \frac{1}{\pi|(t-a_j^{(m)})(b_j^{(m)}-t)|^{1/2}}, \qquad t \in (a_j^{(m)}, b_j^{(m)}), \quad j = 1, \ldots, k(m),$$

from which one can obtain by easy calculation that

$$\|v_m\|_{L^{3/2}[0,1]} \leq 11m \tag{5.8}$$

(recall also that $k(m) \leq 2m$).

Fix now a $1 > \gamma > 0$ such that

$$e^{\gamma}\alpha < \tfrac{1}{4},$$

and let

$$n = [\eta m/4].$$

From (5.8) and Hölder's inequality we obtain for every interval $I \subseteq [0,1]$ and $x \in [0,1]$

(5.9)

$$\int_I |\log|x-t||v_m(t)dt = \int_{|x-t|\leq e^{-\gamma n}} + \int_{|x-t|>e^{-\gamma n}}$$

$$\leq \left(\int_{|x-t|\leq e^{\gamma n}} |\log|x-t||^3 dt\right)^{1/3} \|v_m\|_{L^{3/2}[0,1]} + \gamma n \int_I v_m(t)dt$$

$$\leq 11mne^{-\gamma n/3} + \gamma n \int_I v_m(t)dt \leq 50 \cdot \eta^{-1} n^2 e^{-\gamma n/3} + \gamma n \int_I v_m(t)dt.$$

Let us now consider the points

$$0 = y_{0,n}^{(n)} < y_{1,n}^{(n)} < \cdots < y_{n,n}^{(n)} < y_{n+1,n}^{(n)} = 1$$

with the property (set $v_m(t) = 0$ if $t \notin H_m$)

$$\int_{y_{j,n}^{(n)}}^{y_{j+1,n}^{(n)}} v_m(t)dt = \frac{1}{n+1}, \quad j = 0, 1, \ldots, n, \tag{5.10}$$

and form the monic polynomials

$$P_n(x) = \prod_{j=1}^{n} (x - y_{j,n}^{(n)})$$

of degree n. It was shown in [LuSa1, pp. 40–2] that this $P_n(x)$ satisfies

$$(5.11) \qquad (n+1)^{-1} \log |P_n(x)| \le -p(v_n; x) - \int_{y_{l,n}^{(n)}}^{y_{l+1,n}^{(n)}} \log |x - t| v_n(t) dt,$$

where $p(v_n; x)$ denotes the potential of the measure $v_n(x)dx$ and where l is chosen to satisfy $x \in [y_{l,n}^{(n)}, y_{l+1,n}^{(n)}]$ (see (6.13) and (6.14) in [LuSa1]). To be more precise [LuSa1] contains the previous estimate for $[-1, 1]$ instead of $[0, 1]$ but linear transformation does not introduce any new constant. Since for $x \in H_m$ we have $-p(v_n; x) = \log(\text{cap}(H_n))$, we get from (5.9), (5.11), and (5.7) that

$$(5.12) \qquad \begin{aligned} \|P_n\|_{H_m} &\le (\text{cap}(H_m))^{n+1} \exp((n+1)(50\eta^{-1} n^2 e^{-\gamma n/3} + \gamma)) \\ &\le 2(\alpha e^{\gamma})^n \end{aligned}$$

provided $n = [\eta m/4]$, $m \in \mathcal{N}$, is sufficiently large, say, $n \ge n_0$.

On $[0, 1] \setminus H_m$ we have $|P_n| \le 1$; therefore (5.6) yields

$$(5.13) \qquad \int_{[0,1] \setminus H_m} |P_n|^2 d\mu \le 4m e^{-\eta m} \le 20\eta^{-1} n e^{-4n} \le \left(\tfrac{1}{5}\right)^{2n}$$

for large n, say, $n \ge n_1$.

Now if $n \ge \max\{n_0, n_1\}$ and n is of the form $n = [\eta m/4]$, $m \in \mathcal{N}$, then (5.12) and (5.13) imply

$$\int |P_n|^2 d\mu \le 2(\alpha e^{\gamma})^{2n} \|\mu\| + \left(\tfrac{1}{5}\right)^{2n},$$

and so, in view of $\alpha e^{\gamma} < \frac{1}{4}$, we have

$$\liminf_{n \to \infty} \gamma_n(\mu)^{-1/n} < \tfrac{1}{4},$$

which proves that μ is not regular. □

4.6 Summary of Regularity Criteria and Their Relations

For the convenience of the reader we give a summary of the criteria given in Sections 4.1 and 4.2 for the regularity of measures and discuss their relative strength. Let us first recall their definitions.

Erdős–Turán. $\frac{d\mu}{d\omega_{S(\mu)}} > 0 \qquad \omega_{S(\mu)}$-*a.e.*

Ullman. $c_\mu = \mathrm{cap}(S(\mu))$.

Widom. *For every carrier C of μ there exist compact sets $V_n \subseteq S(\mu)$ with*

$$\mathrm{cap}(V_n) \to \mathrm{cap}(S(\mu)),\ \omega_{V_n}(C) \to 1 \quad \textit{as} \quad n \to \infty.$$

Λ. $\mathrm{cap}\left(\left\{x \,\middle|\, \limsup_{r\to 0+} \frac{\log 1/\mu(\Delta_r(z))}{\log 1/r} < \infty\right\}\right) = \mathrm{cap}(S(\mu))$.

Λ*. *For some constant $L > 0$*

$$\lim_{r\to 0} \mathrm{cap}(\{z \mid \mu(\Delta_r(z)) > r^L\}) = \mathrm{cap}(S(\mu)).$$

λ. $S(\mu) = [0,1]$ *and for almost every $x \in [0,1]$*

$$\liminf_{r\to 0} r \log \mu([x-r, x+r]) \geq 0.$$

λ*. $S(\mu) = [0,1]$ *and for every $\eta > 0$*

$$\lim_{n\to\infty} \mathrm{meas}\left(\left\{x \,\middle|\, \mu\left(\left[x - \frac{1}{n}, x + \frac{1}{n}\right]\right) \geq e^{-\eta n}\right\}\right) = 1.$$

All of these imply regularity (for Λ^* we have assumed that $S(\mu) = \mathrm{supp}(\mu)$ is regular with respect to the Dirichlet problems in Ω), and conversely, $S(\mu) = [0,1]$ and $\mu \in \mathbf{Reg}$ implies for every $\eta > 0$

$$\lim_{n\to\infty} \mathrm{cap}\left(\left\{x \,\middle|\, \mu\left(\left[x - \frac{1}{n}, x + \frac{1}{n}\right]\right) \geq e^{-\eta n}\right\}\right) = 1$$

(see Sections 4.1 and 4.2).

We have already seen in Section 4.1 that the Erdős–Turán criterion implies Ullman's criterion and the latter one is equivalent to Widom's criterion. Examples 1.5.1 and 4.1.5 show that in general Ullman's criterion does not imply the Erdős–Turán one.

Now we show that each of these imply critera Λ and Λ^*. Clearly, this follows from Theorem 4.6.1.

Theorem 4.6.1. *For any μ the set*

$$C = \left\{x \,\middle|\, \limsup_{r\to 0+} \frac{\log 1/\mu(\Delta_r(z))}{\log 1/r} \leq 2\right\}$$

is a carrier of μ.

Proof of Theorem 4.6.1. Suppose that C is not a carrier. Then there exists a compact set K in the complement of C which has positive μ

measure. We set $\nu = \mu|_K$, and it is enough to show that for some $z_0 \in K$ and $c > 0$

$$\nu(\Delta_r(z_0)) \geq cr^2 \tag{6.1}$$

for every $0 < r < 1$.

But this can be easily verified by the usual quartering process. In fact, if Q_0 is a closed square with one side equal to some number q and parallel to the x-axis containing K, and $Q_{00}, Q_{01}, Q_{02}, Q_{03}$ are the four closed quadrants of Q_0 (containing one vertex of Q_0), then for some $\varepsilon_1 \in \{0, 1, 2, 3\}$

$$\nu(Q_{0\varepsilon_1}) \geq \nu(Q_0)/4 = \nu(K)/4.$$

Dividing $Q_{0\varepsilon_1}$ into four equal squares $Q_{0\varepsilon_1 0}, Q_{0\varepsilon_1 1}, Q_{0\varepsilon_1 2}, Q_{0\varepsilon_1 3}$, we have an $\varepsilon_2 \in \{0, 1, 2, 3\}$ with

$$\nu(Q_{0\varepsilon_1\varepsilon_2}) \geq \nu(Q_{0\varepsilon_1})/4 \geq \nu(K)/4^2.$$

Proceeding this way we get nested closed squares $Q_0, Q_{0\varepsilon_1}, Q_{0\varepsilon_1\varepsilon_2}, \ldots$ with sides equal to $q, q/2, q/2^2, \ldots$, each containing the next one such that

$$\nu(Q_{0\varepsilon_1\varepsilon_2\cdots\varepsilon_k}) \geq \nu(K)/4^k.$$

If z_0 is the common point of these squares then (6.1) is satisfied and obviously we must have $z_0 \in K$ (recall that $\operatorname{supp}(\nu) \subseteq K$ and K is compact). $\square$

Next we consider criteria Λ and Λ^*. We mention first of all that Λ does not imply Λ^*. Based on criterion Λ^* it is not too difficult to construct a measure, say, on $[0, 1]$, which is regular and satisfies (2.1), but for which (2.4) does not hold (in fact, Theorem 4.2.3 implies that if

$$\lim_{r\to 0} \operatorname{cap}(\{x \in [0, 1 - \varepsilon_L] \mid \mu([x - r, x + r]) \geq r^L\}) = (1 - \varepsilon_L)/4$$

for some ε_L, where $\varepsilon_L \to 0$ as $L \to \infty$, then μ is regular and such a μ can be chosen in such a way that (2.1) is true but (2.4) fails to hold for a single L).

Conversely, Λ^* can be much weaker than Λ as is shown by the following theorem.

Theorem 4.6.2. *There exists a μ such that $S(\mu) = [0, 1]$, and for every $L > 0$ and $x \in [0, 1]$*

$$\lim_{r\to 0} \operatorname{cap}(\{x \mid \mu([x - r, x + r]) \geq r^L\}) = \tfrac{1}{4} \tag{6.2}$$

holds, but the set

$$\left\{x \,\middle|\, \limsup_{r\to 0+} \frac{\log 1/\mu([x - r, x + r])}{\log 1/r} < \infty\right\} \tag{6.3}$$

is of zero capacity.

Proof of Theorem 4.6.2. Let

$$E(n,\alpha)=\bigcup_{i=1}^{n-1}\left[\frac{i}{n}-e^{-\alpha n},\frac{i}{n}+e^{-\alpha n}\right]$$

be the sets introduced in Lemma 4.4.3, $S_n=\{\frac{i}{n}\}_{i=1}^{n-1}$, and μ_n the measure that assigns mass

$$\delta_n=\exp(-2^n/n^2)$$

to each point of S_{2^n}. We show that

$$\mu=\sum_{n=1}^{\infty}\mu_n$$

satisfies (6.2) and (6.3).

Let $L>0$ be fixed and

$$\exp(-2^n/n)<r\le\exp(-2^{n-1}/(n-1)).$$

Then for $x\in E(2^n,1/n)$ the interval $[x-r,x+r]$ contains at least one point of S_{2^n}; therefore for large n, say, $n\ge n_L$,

$$\mu([x-r,x+r])\ge\delta_n\ge(\exp(-2^{n-1}/(n-1)))^L\ge r^L,$$

and this proves (6.2) (making use of Corollary 4.4.4).

To prove that the set in (6.3) is of zero capacity, let $L>0$ be fixed and

$$\exp(-2^{k+1}/(k+1)^3)<r\le\exp(-2^k/k^3).$$

Since for large k, say, $k\ge k_L$,

$$\sum_{n=k+1}^{\infty}\|\nu_n\|\le\sum_{n=k+1}^{\infty}2^n\delta_n=\sum_{n=k+1}^{\infty}2^n\exp(-2^n/n^2)$$
$$\le(\exp(-2^{k+1}/(k+1)^3))^L<r^L,$$

the inequality

$$\mu([x-r,x+r])\ge r^L \tag{6.4}$$

can only hold when

$$[x-r,x+r]\bigcap\left(\bigcup_{n=1}^{k}S_{2^n}\right)\ne\emptyset.$$

Now taking into account that

$$\bigcup_{n=1}^{k}S_{2^n}=S_{2^k}\quad\text{and}\quad r<\exp(-2^k/k^3),$$

it follows that (6.4) implies

$$x \in E(2^k, k^{-3}).$$

Thus, the set in (6.3) is contained in

$$\liminf_{k\to\infty} E(2^k, k^{-3}) = \bigcup_{n=1}^{\infty} \bigcap_{k=n}^{\infty} E(2^k, k^{-3}).$$

It is easy to check that for $n < k < 2^{n/2}$

$$\begin{aligned} &E(2^n, n^{-3}) \bigcap E(2^k, k^{-3}) \\ &\qquad \subseteq \bigcup_{j=1}^{2^n-1} \left[\frac{j}{2^n} - \exp(-2^k/k^3), \frac{j}{2^n} + \exp(-2^k/k^3)\right]; \end{aligned}$$

therefore

$$\bigcap_{k=n}^{2^n} E(2^k, k^{-3}) \subseteq \bigcup_{j=1}^{2^n-1} \left[\frac{j}{2^n} - \varepsilon_n, \frac{j}{2^n} + \varepsilon_n\right] =: B_n,$$

where

$$\varepsilon_n = \exp(-2^{2^{n/4}}).$$

The inequality

$$\frac{1}{\log(1/\text{cap}(E))} \le \sum_j \frac{1}{\log(1/\text{cap}(E_j))},$$

valid for $E = \cup_{j=1}^{\infty} E_j$, $\text{diam}(E) \le 1$ (see (4.27) or [Ts, Chapter III, Theorem III.17]), when applied with $E = B_n$ and

$$E_j = \left[\frac{j}{2^n} - \varepsilon_n, \frac{j}{2^n} + \varepsilon_n\right], \quad j = 1, \dots, 2^n - 1,$$

gives

$$\text{cap}\left(\bigcap_{k=n}^{2^n} E(2^k, k^{-3})\right) \le \text{cap}(B_n) \le \exp(-2^{2^{n/4}}/2^n) \to 0$$

as $n \to \infty$, and together with this we have

$$\text{cap}\left(\bigcap_{k=n}^{\infty} E(2^k, k^{-3})\right) = 0,$$

proving

$$\text{cap}\left(\liminf_{k\to\infty} E(2^k, k^{-3})\right) = 0. \qquad \square$$

Finally, we consider criteria λ and λ^*. Thus, in what follows let $S(\mu) = [0,1]$. It is clear that $\lambda \Rightarrow \lambda^*$ but not vice versa. It is much less obvious that $\lambda^* \Rightarrow \Lambda^*$.

Theorem 4.6.3. *Condition λ^* implies Λ^*.*

We shall prove more, namely, that λ^* implies

$$\lim_{r\to 0+0} \operatorname{cap}(\{x \mid \mu([x-r,x+r]) \geq r^L\}) = \tfrac{1}{4}$$

for every $L > 1$.

Proof of Theorem 4.6.3. Let us fix an $L > 1$, $0 < \alpha < 4^{-10}(L-1)$, and set $\varepsilon = \alpha^2$. According to condition λ^* there is an n_0 such that for $n \geq n_0$ we have for the set

$$A_n = \left\{ x \,\middle|\, \mu\left(\left[x - \frac{1}{4n}, x + \frac{1}{4n}\right]\right) \geq e^{-\alpha^2 n} \right\}$$

the inequality

$$\operatorname{meas}(A_n) > 1 - \frac{\varepsilon}{2}.$$

If

$$J_n = \left\{ j \,\middle|\, 1 \leq j \leq n,\ A_n \cap \left[\frac{j}{n} - \frac{1}{4n}, \frac{j}{n} + \frac{1}{4n}\right] \neq \emptyset \right\}$$

then we must have $|J_n| > (1-\varepsilon)n$. But for each $j \in J_n$ the definition of A_n gives

$$\mu\left(\left[\frac{2j-1}{2n}, \frac{2j+1}{2n}\right]\right) \geq e^{-\alpha^2 n}.$$

This implies that there is an interval $I_{j,n}$ of length $4e^{-\alpha n}$ in $[(2j-1)/2n, (2j+1)/2n]$ for which

$$\mu(I_{j,n}) \geq \mu\left(\left[\frac{2j-1}{2n}, \frac{2j+1}{2n}\right]\right) e^{-\alpha n} \geq 4^{1+2\alpha} e^{-(\alpha+2\alpha^2)(n-1)}$$

provided n is sufficiently large, say, $n \geq n_1$ (just divide the interval $[(2j-1)/2n, (2j+1)/2n]$ into $[\exp(\alpha n)]$ equal parts; one will have μ–measure at least as large as

$$\mu([(2j-1)/2n, (2j+1)/2n])e^{-\alpha n};$$

this interval enlarged so as to have length $4e^{-\alpha n}$ will suffice as $I_{j,n}$). But then for

$$4e^{-\alpha(n-1)} > r \geq 4e^{-\alpha n}$$

and any $x \in I_{j,n}$, $j \in J_n$,

$$\mu([x-r, x+r]) \geq \mu(I_{j,n}) \geq 4^{1+2\alpha} e^{-(\alpha+2\alpha^2)(n-1)}$$
$$= (4e^{-\alpha(n-1)})^{1+2\alpha} > r^L \quad (n \geq n_2).$$

In summary, for large n ($n \geq \max(n_0, n_1, n_2)$) the set

$$E_\varepsilon^*(n,\alpha) = \bigcup_{j \in J_n} I_{j,n}$$

is contained in

$$B(r,L) = \{x \mid \mu([x-r,x+r]) \geq r^L\}.$$

To $E_\varepsilon^*(n,\alpha)$ we can apply Lemma 4.4.5 (note that each $I_{j,n}$ has capacity $\geq e^{-\alpha n}$) to conclude for large n

$$\operatorname{cap}(B(r,L)) \geq \operatorname{cap}(E_\varepsilon^*(n,L)) \geq \tfrac{1}{4}\exp(-\alpha^{1/5} - 2\varepsilon/\alpha)$$
$$= \tfrac{1}{4}\exp(-\alpha^{1/5} - 2\alpha).$$

Thus

$$\liminf_{r \to 0+} \operatorname{cap}(B(r,L) \geq \tfrac{1}{4}\exp(-\alpha^{1/5} - 2\alpha),$$

and since here $\alpha > 0$ is arbitrary, condition Λ^* holds. □

As a final summary we state the following.

Summary. *The order of the criteria discussed above with regard to decreasing amount of assumptions on* μ *(and hence increasing strength in applications) is roughly:*

Erdős–Turán criterion
Ullman's and Widom's criteria
λ
λ^*
Λ
Λ^*.

5

Localization

In this chapter we show that in many cases regularity of a measure is a local property. This is more remarkable if one considers that the definition itself of the regularity of a measure μ involves the whole support of μ. For example, if the support of μ is the unit circle, then it will turn out that μ is regular if and only if its restriction to every arc is regular, and, of course, the Green functions associated with subarcs (that appear in the definition of regularity) are hardly related to the Green function of the exterior of the unit disk. On the other hand, if we consider that regularity means that μ must be sufficiently dense around $\partial\Omega$, then we can see that localization is actually a natural procedure.

5.1 Global Versus Local Behavior

Although each orthonormal polynomial $p_n(\mu;\cdot)$ is determined by the whole measure μ, experience shows that their asymptotic behavior on certain subsets of $S(\mu)$ depends only on the restriction of μ to this subset. In this section we shall elaborate on this idea.

The upper estimates of Chapter 1 seemingly used the whole measure μ (cf. Definition 1.1 of the minimal-carrier Green function). But Theorem 1.1.4 easily implies that if we restrict μ to a subset K then for the measure $\mu_{|K}$ we also have

$$\limsup_{n\to\infty} |p_n(\mu;z)|^{1/n} \le e^{g_{\mu_{|K}}(z;\infty)}$$

locally uniformly on $\mathbb{C}$. In fact, this follows from the obvious inequality $g_\mu(z;\infty) \le g_{\mu_{|K}}(z;\infty)$. The situation is very different for the lower estimates in Theorem 1.1.4 (away from the support): Since the support of $\mu_{|K}$ is smaller than that of μ, the corresponding Green function $g_{\mathbb{C}\setminus\mathrm{Pc}(S(\mu_{|K}))}(z;\infty)$ is bigger than $g_{\mathbb{C}\setminus\mathrm{Pc}(S(\mu))}(z;\infty)$.

If we consider the definition of regular (nth-root) behavior from Section 3.1 and its equivalent formulations given in Section 3.2, we can see that the situation is similar to the lower estimates in Theorem 1.1.4: Everything, even the definitions, heavily relies on the whole measure μ, for example, the support of μ has played a prominent role. Despite these facts, in the next sections we will show that in many cases the regularity of μ is inherited by restriction.

In this section we only show the following sufficient local condition for regularity.

Theorem 5.1.1. *Let us suppose that $\{K_j\}_{j=1}^{\infty}$ are compact sets and quasi every point of $\partial\Omega$ is contained in the union of the supports $\{\mathrm{supp}(\mu|_{K_j})\}_{j=1}^{\infty}$. Then if every $\mu_{K_j} := \mu|_{K_j}$ is regular, then μ is also regular.*

Remark. It is not enough to assume that quasi every point of $\partial\Omega$ is contained in the sets $\{K_j\}_{j=1}^{\infty}$. Consider, for example, a nonregular measure μ that is the sum of a thin discrete measure with support in the unit disk plus the arc measure on a closed proper subarc I of the unit circumference. If each K_j is the unit circumference then each μ_{K_j} is regular but μ is not.

Proof of Theorem 5.1.1. If each μ_{K_j} is regular then by Theorem 3.2.1(v) for every j quasi every point z on $S(\mu_{K_j})$ satisfies

$$\limsup_{n\to\infty}\left(\frac{|P_n(z)|}{\|P_n\|_{L^2(\mu|_{K_j})}}\right)^{1/n}\le 1$$

whatever the polynomial sequence $\{P_n\}_{n=1}^{\infty}$ with $\deg(P_n)\le n$ is (see also Remark (5) after Theorem 3.2.1). But then

$$\limsup_{n\to\infty}\left(\frac{|P_n(z)|}{\|P_n\|_{L^2(\mu)}}\right)^{1/n}\le 1 \tag{1.1}$$

also holds for quasi every $z\in S(\mu_{K_j})$, and since these sets cover quasi every point of $\partial\Omega$, we can conclude (1.1) for quasi every $z\in\partial\Omega$. Now $\mu\in\mathbf{Reg}$ follows from the equivalence of (i) and (v) of Theorem 3.2.1. □

In Section 5.3 we shall consider a certain converse to Theorem 5.1.1; namely, we will conclude the regularity of restrictions μ_K of μ to compact sets K from the regularity of μ. That the set K for which we take the restriction cannot be arbitrary is shown by the following example.

Example 5.1.2. Let K be the classical Cantor set and μ_1 a measure with support $S(\mu_1)=K$ that is not regular. Since $\mathrm{cap}(K)>0$, the existence of such a μ_1 follows from Theorem 1.1.9 if we consider an arbitrary discrete measure with support K. With μ_0 the Lebesgue measure on $[0,1]$ we set $\mu = \mu_0 + \mu_1$. Then, of course, $\mu \geq \mu_0$ is regular but $\mu_{|K} = \mu_1$ is not. Note that in this case

$$\operatorname{cap}(\partial\Omega(\mu_K) \setminus \operatorname{Int}(K)) = \operatorname{cap}(K) > 0. \qquad \square$$

The preceding example showed that for localization we need the condition

$$\operatorname{cap}(\partial\Omega(\mu_K) \setminus \operatorname{Int}(K)) = 0.$$

Now we show that in general $\operatorname{cap}(\partial\Omega(\mu_K) \setminus \partial\Omega) = 0$ is also indispensable.

Example 5.1.3. The idea is that regularity of μ does not imply any denseness of the measure μ inside $\operatorname{Pc}(S(\mu))$. Based on this observation it is an easy task to construct a measure μ and a compact set K such that μ is regular, $\operatorname{cap}(\partial\Omega(\mu_K) \setminus \operatorname{Int}(K)) = 0$, and μ_K is not regular. Let, for example, μ_0 be the Lebesgue measure on the unit circle, μ_1 be a sufficiently sparse discrete measure with compact support equal to the right half of the unit disk, $\mu = \mu_0 + \mu_1$, and K be a square containing this half disk but not containing -1. Then μ is regular, but μ_K is not. In this example $\operatorname{cap}(\partial\Omega(\mu_K) \setminus \partial\Omega) \neq 0$. $\square$

5.2 Localization at a Single Point

We start with the following definition.

Definition 5.2.1. *A point z_0 is called a μ-*regular point *if*

$$\limsup_{\substack{n\to\infty \\ \deg P_n \leq n, P_n \not\equiv 0}} \left(\frac{|P_n(z_0)|}{(\int |P_n|^2 d\mu)^{1/2}} \right)^{1/n} \leq 1.$$

This is equivalent to

$$\lim_{n\to\infty} \left(\sup_{\substack{\deg P_n \leq n \\ P_n \not\equiv 0}} \frac{|P_n(z_0)|}{\|P_n\|_{L^2(\mu)}} \right)^{1/n} = 1,$$

and also to (see the proof of Theorem 3.2.1, (ii)⇒(iii))

$$\limsup_{n\to\infty} |p_n(\mu; z_0)|^{1/n} \leq 1.$$

The equivalence of (i) and (v) of Theorem 3.2.1 asserts that μ is regular if and only if quasi every point of $\partial\Omega$ is μ-regular. Here $\partial\Omega$ can be replaced by $S(\mu)$ or $\operatorname{Pc}(S(\mu))$ (see Theorem 3.2.2, Remark (5)). Our main result in this section says that under some mild condition μ-regularity of a point is a local property on $\partial\Omega$.

Theorem 5.2.2. *Let $z_0 \in \partial\Omega$ and U be a neighborhood of z_0. Suppose that $\mu|_U$ does not have a carrier of zero capacity. Then z_0 is μ-regular if and only if it is $\mu|_U$-regular.*

The condition is obviously equivalent to saying that every carrier of μ intersects U in a set of positive capacity.

Proof of Theorem 5.2.2. Obviously, $\mu|_U$-regularity of z_0 implies its μ-regularity, so we only have to prove the converse.

Without loss of generality we may assume that $S(\mu)$ has diameter at most 1. Since $\mu|_U$ does not have a carrier of zero capacity, it follows from Corollary 1.1.7 that $\{\gamma_n(\mu|_U)^{1/n}\}_{n=1}^{\infty}$ is a bounded sequence. Hence $\{|p_n(\mu|_U;\cdot)|^{1/n}\}_{n=1}^{\infty}$ is uniformly bounded on compact subsets of the plane. Expanding now an arbitrary polynomial P_n of degree at most n into its Fourier series in $\{p_k(\mu|_U;\cdot)\}$ we can see that there exists a constant C independent of n and P_n for which

(2.1)
$$\left\{\int |P_n|^2 d\mu\right\}^{1/2} \le C^n \left\{\int_U |P_n|^2 d\mu\right\}^{1/2}, \quad \deg P_n \le n,\ n = 1, 2, \ldots .$$

After this let us assume that z_0 is not a $\mu|_U$-regular point. Then there are an $\alpha > 1$, a subsequence $\mathcal{N} \subseteq \mathbf{N}$ of the natural numbers, and for each $n \in \mathcal{N}$ a polynomial P_n of degree at most n such that

$$\|P_n\|_{L^2(\mu|_U)} = 1 \tag{2.2}$$

but

$$|P_n(z_0)| > \alpha^n. \tag{2.3}$$

Choose an $1 \ge \varepsilon > 0$ such that $1 + 2\varepsilon < \alpha$ and let $\delta > 0$ be such that U contains the disk

$$D_\delta(z_0) := \{z \mid |z - z_0| < \delta\}.$$

Let now $\delta_0 = \delta/2C$. By Lemma 3.2.4 there is a polynomial Q of degree, say, k, such that $|Q(z_0)| = 1$, $|Q(z)| \le 1+\varepsilon$ for all $z \in \partial\Omega$, and $|Q(z)| < \delta/C$ for $z \in \partial\Omega$, $|z - z_0| \ge \delta_0$. We claim that

$$|Q(z)| \le 1/C \quad \text{for } z \in S(\mu) \setminus U. \tag{2.4}$$

In fact, consider the polynomial

$$Q^*(z) = (z - z_0)Q(z)$$

on the set $K := \mathrm{Pc}(S(\mu)) \setminus D_{\delta_0}(z_0)$. The boundary of K is contained in the union of $\partial D_{\delta_0}(z_0) \cap \mathrm{Pc}(S(\mu))$ and $\partial\Omega \setminus D_{\delta_0}(z_0)$. On the former set we have $|Q^*| \le 2\delta_0 = \delta/C$ because (by the maximum modulus theorem for $|Q|$) $|Q| \le 1 + \varepsilon \le 2$ on $\mathrm{Pc}(S(\mu))$, whereas on the latter set we have $|Q^*| \le |Q| \le \delta/C$ by the choice of Q. Applying the maximum modulus theorem for $|Q^*|$ we can conclude that $|Q^*| \le \delta/C$ on K and therefore

$$|Q(z)| \le \frac{1}{\delta}|Q^*(z)| \le \frac{1}{C}$$

when $z \in \mathrm{Pc}(S(\mu)) \setminus D_\delta(z_0)$. Finally, our claim follows from the fact that the latter set contains $S(\mu) \setminus U$.

Consider now the polynomials

$$R_{n(1+k)}(z) = P_n(z)Q^n(z)$$

of degree at most $n(1+k)$ for $n \in \mathcal{N}$. From (2.1), (2.2), and (2.4) we can derive

$$\left\{\int_{S(\mu)\setminus U} |R_{n(1+k)}|^2 d\mu\right\}^{1/2} \le \left\{\int_U |P_n|^2 d\mu\right\}^{1/2} = 1,$$

whereas $|Q| \le 1 + \varepsilon$ for $z \in S(\mu)$ yields

$$\left\{\int_U |R_{n(1+k)}|^2 d\mu\right\}^{1/2} \le (1+\varepsilon)^n.$$

Taking into account (2.3) and $|Q(z_0)| = 1$ we have

$$\limsup_{\substack{n\to\infty\\ n\in\mathcal{N}}} \left(\frac{|R_{n(1+k)}(z_0)|}{\|R_{n(1+k)}\|_{L^2(\mu)}}\right)^{1/n(1+k)} \ge \left(\frac{\alpha}{1+2\varepsilon}\right)^{1/(1+k)} > 1,$$

proving that z_0 is not a μ-regular point. □

As a supplement to Theorem 5.2.2 we show that Theorem 5.2.2 is not true without any further assumption on z_0 and U (such as the one in the formulation of the theorem).

Example 5.2.3. We construct a measure μ with support in $[-1,3]$ such that $z_0 = 2$ is μ-regular but not $\mu|_U$-regular, where U is the disk around z_0 of radius 1.

Let $n_k = k^k$, $a_k = 10^{-n_k}$, and μ_k be the point measure that has mass $1/n_k$ at $2 + a_k$. Finally, with Lebesgue measure μ_0 on $[-1,1]$ we set

$$\mu = \mu_0 + \sum_{k=1}^{\infty} \mu_k.$$

First we verify that $z_0 = 2$ is μ-regular. Indeed, if we assume that P_n, $n \in \mathcal{N}$, are polynomials with $\deg(P_n) \le n$,

(2.5) $$\|P_n\|_{L^2(\mu)} = 1, \quad n \in \mathcal{N},$$

and

(2.6) $$|P_n(2)| > \alpha^n, \quad n \in \mathcal{N},$$

for some $\alpha > 1$, then we get a contradiction as follows. Equation (2.5) implies

$$\|P_n\|_{L^2(\mu_0)} \le 1,$$

and so the Markov–Bernstein inequality (see also Nikolskii's inequality [Ti, 4.9.6.]

$$\|P_n'\|_{L^\infty[-1,1]} \le 3n\|P_n'\|_{L^2[-1,1]})$$

implies

(2.7) $$\|P_n'\|_{L^2(\mu_0)} = O(n^3).$$

μ_0 is certainly regular, so if we apply Theorem 3.2.1(iii) to P_n' we get from (2.7)

$$|P_n'(x)| \le 6^n$$

for $x \in [2,3]$ and large $n \in \mathcal{N}$ (this is also a trivial consequence of (2.7), Nikolskii's inequality [Ti, 4.9.6], and the Bernstein–Walsh lemma). This and (2.6) imply for $n_k \le n < n_{k+1}$

$$|P_n(2 + a_{k+1})| \ge \alpha^n - a_{k+1}6^n \ge \alpha^n - 10^{-n}6^n \ge \alpha^n - 1,$$

which in turn yields

$$\|P_n\|_{L^2(\mu)} \ge \|P_n\|_{L^2(\mu_{k+1})} \ge (\alpha^n - 1)\cdot n_{k+1}^{-1/2} \ge (\alpha^{n_k} - 1)n_{k+1}^{-1/2} > 1,$$

contradicting (2.5). This proves the μ-regularity of $z_0 = 2$.

With U as in the statement of the example we have

$$\mu_{|U} = \sum_{k=1}^{\infty} \mu_k.$$

Now let

$$P_k(x) = \prod_{s=1}^{k} (x - (2 + a_s)).$$

For this polynomial we have

$$|P_k(2)| = \prod_{s=1}^{k} a_s,$$

whereas for $x \in [2, 2 + a_{k+1}]$

$$|P_k(x)| \leq \prod_{s=1}^{k} a_s,$$

hence

$$\int |P_k|^2 d\mu_{|U} = \sum_{s=k+1}^{\infty} \int |P_k|^2 d\mu_s \leq \left(\prod_{s=1}^{k} a_s\right)^2 \left(\sum_{s=k+1}^{\infty} \frac{1}{n_s}\right)$$
$$\leq \left(\prod_{s=1}^{k} a_s\right)^2 2n_{k+1}^{-1},$$

and so

$$\left(\frac{|P_k(2)|}{(\int |P_k| d\mu_{|U})^{1/2}}\right)^{1/k} \geq \tfrac{1}{2} n_{k+1}^{1/2k} = \tfrac{1}{2} k^{1/2} \to \infty,$$

proving that 2 is not $\mu|_U$-regular. □

5.3 Localization Theorems

In this section we state and prove our main localization theorems as well as a localized version of the results of Section 3.2.

First we consider the problem of relating the regularity of the restriction $\mu_K = \mu|_K$ of μ to some compact set K to properties of $\{p_n(\mu; \cdot)\}$. That is, we ask if the regularity of μ_K is reflected in the original orthonormal polynomials associated with μ(!). Here K cannot be arbitrary (see Examples 5.1.2 and 5.1.3), what we need is that the part of μ carried by K should be essentially carried by the interior Int(K) as well. More precisely, we will assume

$$\text{(3.1)} \qquad \operatorname{cap}(\partial\Omega(\mu_K) \setminus (\operatorname{Int}(K) \cap \partial\Omega)) = 0.$$

Recall that $\Omega(\mu_K)$ is the unbounded component of $\mathbf{C} \setminus S(\mu_K) = C \setminus \operatorname{supp}(\mu_K)$. Typical examples of such Ks are closed disks when μ is supported on the real line or on the unit circumference (in these cases the set appearing in (3.1) has at most two points). Note also that for the case $S(\mu) = \partial\Omega$ condition (3.1) is equivalent to

$$\text{(3.2)} \qquad \operatorname{cap}(S(\mu_K) \setminus \operatorname{Int}(K)) = 0.$$

We have seen in Examples 5.1.2 and 5.1.3 that (3.1), in general, is indispensable for the regularity of μ_K.

Our first theorem is the following.

Theorem 5.3.1. *Let K be a compact set such that the support of $\mu_K := \mu|_K$ is an infinite set and (3.1) holds. Then the following statements are equivalent.*

(i) $\mu_K \in$ **Reg**, *that is, the sequence $\{p_n(\mu_K;\cdot)\}_{n=0}^{\infty}$ has regular (nth-root) asymptotic behavior.*

(ii) *We have*

$$\limsup_{n\to\infty} |p_n(\mu;z)|^{1/n} \le e^{g_{\Omega(\mu_K)}(z;\infty)} \tag{3.3}$$

locally uniformly for $z \in \mathbb{C}$.

(iii) *For any sequence $\{P_n\}$ of nonzero polynomials of degree* $\deg(P_n) \le n$ *we have*

$$\limsup_{n\to\infty} \left(\frac{|P_n(z)|}{\|P_n\|_{L^2(\mu)}} \right)^{1/n} \le e^{g_{\Omega(\mu_K)}(z;\infty)} \tag{3.4}$$

locally uniformly for $z \in \mathbb{C}$.

(iv) *The relation*

$$\limsup_{n\to\infty} |p_n(\mu;z)|^{1/n} \le 1 \tag{3.5}$$

holds quasi everywhere on $\partial\Omega(\mu_K)$.

(v) *For any sequence $\{P_n\}$ as in* (iii)

$$\limsup_{n\to\infty} \left(\frac{|P_n(z)|}{\|P_n\|_{L^2(\mu)}} \right)^{1/n} \le 1 \tag{3.6}$$

for quasi every $z \in \partial\Omega(\mu_K)$.

If in addition we have $S(\mu_K) \subseteq \partial\Omega(\mu_K)$, in particular, if

$$\mathrm{Int}(\mathrm{Pc}(S(\mu_K)) = \emptyset,$$

then each of (i)–(v) *is equivalent to the following statement.*

(vi) *For any infinite sequence $\{P_n\}$ of nonzero polynomials of degree* $\deg(P_n) \le n$ *and for any sequence of weight functions $g_n \ge 0$, $n \in \mathbb{N}$, defined on $S(\mu)$ such that $\lim g_n^{1/n} = g$ uniformly on $S(\mu_K)$, where $g > 0$ is continuous on $S(\mu_K)$, we have with $d\mu_n = g_n d\mu$*

$$\limsup_{n\to\infty} \left(\frac{g_n(z)^{1/2}|P_n(z)|}{\|P_n\|_{L^2(\mu_n)}} \right)^{1/n} \le 1$$

for quasi every $z \in \partial\Omega(\mu_K) = S(\mu_K)$.

If $S(\mu_K)$ is a regular set with respect to the solution of the Dirichlet problem in the domain $\Omega(\mu_K) = \mathbb{C} \setminus \mathrm{Pc}(S(\mu_K))$ then our estimates hold uniformly in the range described.

Remarks. (1) Note that the equivalence of (i)–(v) with μ replaced by $\mu|_K$ in (3.3)–(3.6) was stated in Theorem 3.2.1. Here the main point is that μ and not μ_K stands in (3.3)–(3.6). Of course, the equivalence of (i)–(vi) to, for example,

$$\limsup_{n\to\infty} \left(\frac{|P_n(z)|}{\|P_n\|_{L^2(\mu_K)}} \right)^{1/n} \leq 1 \quad \text{for qu.e. } z \in S(\mu_K)$$

is contained in Theorem 3.2.1.

(2) We have already mentioned that assumption (3.1) cannot be dropped. Without it one can only prove the pairwise equivalence of (ii)–(vi) and that each of them follows from (i).

(3) It is worth noting that the implication (i)⇒(vi) holds without the assumption (3.1), and the limit superior is uniform on $S(\mu_K)$ if the latter set is regular with respect to the Dirichlet problem in $\Omega(\mu_K)$. Moreover, the $L^2(\mu)$ norm can be replaced by any $L^p(\mu)$ norm for $1 \leq p < \infty$ (cf. Section 3.4).

Corollary 5.3.2. *Let K be as in* Theorem 5.3.1 *and assume that $S(\mu_K)$ is regular with respect to the Dirichlet problem in $\Omega(\mu_K)$. Then μ_K is regular if and only if*

$$\limsup_{n\to\infty} \|p_n(\mu;\cdot)\|_{\sup,S(\mu_K)}^{1/n} = 1.$$

Another equivalent condition is

$$\lim_{n\to\infty} \left(\sup_{\substack{\deg(P_n)\leq n \\ P_n\not\equiv 0}} \frac{\|P_n\|_{\sup,S(\mu_K)}}{\|P_n\|_{L^p(\mu)}} \right)^{1/n} = 1, \quad 1 \leq p < \infty.$$

Proof of Theorem 5.3.1. (i)⇒(ii) follows from (i)⇒(iii) of Theorem 3.2.1 applied to μ_K in place of μ and to $p_n(\mu;\cdot)$ in place of P_n (using also that the $L^2(\mu)$ norm is at least as large as the $L^2(\mu_K)$ norm).

In a similar manner (i)⇒(vi) follows from the corresponding implication in Theorem 3.2.1, furthermore (vi)⇒(v) is obvious. Now the proof of (ii)⇒(iii) and (iv)⇒(v) is the same as that of (ii)⇒(iii) in Theorem 3.2.1 (use Fourier expansion in $\{p_k(\mu;\cdot)\}$), and (iii)⇒(iv) is trivial because $g_{\Omega(\mu_K)}(z;\infty) = 0$ quasi everywhere on $\partial\Omega(\mu_K)$. Thus, all that remains to be proved is the implication (v)⇒(i) and this is the only place where we use condition (3.1).

We show (v)⇒(i) by making use of the localization Theorem 5.2.2. Suppose on the contrary that (v) holds but μ_K is not regular. Then there are a $0 < \lambda < \mathrm{cap}(S(\mu_K))$ and an infinite subsequence $\mathcal{N} \subseteq \mathbb{N}$ such that for each $n \in \mathcal{N}$ the monic orthogonal polynomial

$$q_n(z) = \frac{1}{\gamma_n(\mu_K)} p_n(\mu_K; z)$$

satisfies

$$\int |q_n|^2 d\mu_K < \lambda^{2n}.$$

Without loss of generality we may assume that $S(\mu_K)$ is contained in the disk $D_{1/2}(0) = \{z \mid |z| \le \frac{1}{2}\}$, and so Theorem 2.1.1 implies $|q_n(z)| < 1$ for $z \in S(\mu_K)$.

We choose an $\varepsilon > 0$ with the property

$$\lambda^{1/(1+\varepsilon)} < \mathrm{cap}(S(\mu_K)) \tag{3.7}$$

and set $\rho = \lambda^{2/\varepsilon}$. According to (3.1) the set $\mathrm{Int}(K) \cap S(\mu_K)$ is of positive capacity (otherwise $\mathrm{cap}(S(\mu_K)) = 0$ and so the regularity of μ_K would follow); therefore for some $z_0 \in \mathrm{Int}(K) \cap S(\mu_K)$ the set $\mathrm{Int}(K) \cap S(\mu_K) \cap D_\rho(z_0)$, where

$$D_\rho(z_0) = \{z \mid |z - z_0| < \rho\},$$

is of positive capacity, and hence it contains a compact set K_0 of positive capacity. Let

$$\nu = \mu_K + \omega_{K_0},$$

where ω_{K_0} is the equilibrium distribution of the set K_0. For the polynomials

$$Q_n(z) = q_n(z)(z - z_0)^{[\varepsilon n]}$$

of degree $n + [\varepsilon n] \le n(1+\varepsilon)$ we have for $n \in \mathcal{N}$

$$\begin{aligned} \int |Q_n|^2 d\nu &\le \int |q_n|^2 d\mu_K + \rho^{[\varepsilon n]} < \lambda^{2n} + \rho^{[\varepsilon n]} \\ &\le (\lambda^{1/(1+\varepsilon)})^{2n(1+\varepsilon)} + \frac{1}{\rho}(\rho^{\varepsilon/2(1+\varepsilon)})^{2n(1+\varepsilon)} \\ &= \left(1 + \frac{1}{\rho}\right)(\lambda^{1/(1+\varepsilon)})^{2n(1+\varepsilon)}, \end{aligned}$$

which, in view of (3.7) and $\mathrm{supp}(\nu) = S(\mu_K)$, shows that ν is not regular. But then, by Theorem 3.2.1(v) there is a set $E \subseteq \partial\Omega(\mu_K)$ of positive capacity such that no point of E is a ν-regular point. From (3.1) we can deduce that then

$$E_1 := E \cap \mathrm{Int}(K) \cap \partial\Omega(\mu)$$

also has positive capacity.

Set now

$$\tilde{\mu} = \mu|_{S(\mu)\setminus S(\mu_K)} + \nu.$$

Then $\tilde{\mu}|_{\mathrm{Int}(K)} = \nu|_{\mathrm{Int}(K)} \geq \omega_{K_0}$ does not have a carrier of zero capacity and no point of $E_1 \subseteq \mathrm{Int}(K)$ is $\tilde{\mu}|_{\mathrm{Int}(K)}$-regular (recall that since $\tilde{\mu}|_{\mathrm{Int}(K)} \leq \nu$, $\tilde{\mu}|_{\mathrm{Int}(K)}$-regularity of a point implies its ν-regularity, and the points in E_1 are not ν-regular). Thus we can apply Theorem 5.2.2 with $U = \mathrm{Int}(K)$ to every point of E_1 from which we get that no point of E_1 is $\tilde{\mu}$-regular, and since $\tilde{\mu} \geq \mu$ we finally conclude that no point of E_1 is μ-regular. But this contradicts (v) (note that $\mathrm{cap}(E_1) > 0$) and this contradiction proves the implication (v)$\Rightarrow$(i).

With this we have verified the equivalence of (i)–(vi). The statement concerning the uniformity of the estimates when $S(\mu_K)$ is regular can be easily checked based on our earlier proofs (cf. Theorem 3.2.3). □

The following theorem immediately follows from Theorems 5.1.1 and 5.3.1.

Theorem 5.3.3. *Let us suppose that $\{K_j\}_{j=1}^{\infty}$ are compact sets such that quasi every point of $\partial\Omega$ is contained in the union of the of the supports $\{S(\mu_{K_j})\}_{j=1}^{\infty}$ and that each K_j satisfies the condition*

$$\mathrm{cap}(\partial\Omega(\mu_{K_j}) \setminus (\mathrm{Int}(K_j) \cap \partial\Omega)) = 0. \tag{3.8}$$

Then μ is regular if and only if every $\mu_{K_j} = \mu|_{K_j}$ is regular.

When $\mathrm{Int}(\mathrm{Pc}(S(\mu))) = 0$, *in particular, if μ is supported on* $\mathbb{R}$, *or even when only $S(\mu) = \partial\Omega$, then condition* (3.8) *reads as*

$$\mathrm{cap}(S(\mu_{K_j}) \setminus \mathrm{Int}(K_j)) = 0. \tag{3.9}$$

Proof of Theorem 5.3.3. That the regularity of μ implies the regularity of each μ_{K_j} follows from Theorems 3.2.1 and 5.3.1. The converse statement is Theorem 5.1.1.

That (3.8) and (3.9) are the same conditions when $S(\mu) = \partial\Omega$ follows from the fact that then $\partial\Omega(\mu_{K_j}) \subseteq S(\mu_{K_j}) \subseteq S(\mu) = \partial\Omega$. □

6

Applications

In this chapter we give some applications of the results from Chapters 1–5. These amply illustrate that **Reg** is an important and natural class of measures that appears in different problems of mathematical analysis. In Section 6.1 we consider rational interpolants to Markov functions and see how regularity of the measure defining the Markov function in question is related to the rate of convergence of the interpolants. Section 6.2 characterizes the regularity of a measure generating a Markov function by an exact, maximal rate of convergence of best rational approximants on compact sets. In Section 6.3 we consider similar questions but for ray sequences of Padé approximants. An interesting feature of the proof is that the problem on the upper half of the Padé table is reduced to that on the lower half by the localization theorems of the preceding chapter. In Sections 6.4 and 6.5 we connect regularity to weighted polynomials and answer the question of where the L^p norm of these weighted polynomials lives. Section 6.6 is devoted to the relation of $\mu \in \mathbf{Reg}$ to Fourier coefficients and best L^2 polynomial approximation of analytic functions. Finally, in Section 6.7 we investigate sets E that have the property that orthonormal polynomials with respect to any weight that is positive on E have regular nth-root behavior.

6.1 Rational Interpolants to Markov Functions

Orthogonal polynomials are closely related to continued fractions, and one of the classical results in the analytic theory of continued fractions is Markov's theorem (see [Ma] or [Pe]). In this section we prove a generalization of this theorem to sequences of interpolating rational functions. With the help of results from Chapter 3 it is not only possible to prove convergence, but also to determine the exact rate of convergence for these approximants.

A function of the form

$$f(z) = \int \frac{d\mu(x)}{x - z} = \frac{c_0}{z} + \frac{c_1}{z^2} + \cdots \tag{1.1}$$

is called a Markov function if μ is a positive measure with compact support $S(\mu) \subseteq \mathbf{R}$; that is, Markov functions are Cauchy transforms of positive measures μ with compact support in $\mathbf{R}$. The coefficients c_j in the development of (1.1) can easily be recognized as the moments

$$c_j = -\int x^j \, d\mu(x), \quad j = 0, 1, \dots, \tag{1.2}$$

of the measure μ. For functions of type (1.1) A. Markov [Ma] proved in 1895 that the continued fraction development

$$\cfrac{b_1}{z - a_1 + \cfrac{b_2}{z - a_2 + \cdots}} \tag{1.3}$$

of f at infinity converges locally uniformly in $\overline{\mathbf{C}} \setminus \mathrm{Co}(S(\mu))$. Since $S(\mu) \subseteq \mathbf{R}$, the convex hull $\mathrm{Co}(S(\mu))$ is equal to the smallest interval $I(S(\mu))$ containing $S(\mu)$.

The numerators A_n and the denominators B_n of the convergents

$$\frac{A_n(z)}{B_n(z)} = \cfrac{b_1}{z - a_1 + \cfrac{b_2}{z - a_2 + \cdots + \cfrac{b_n}{z - a_n}}}, \quad n = 1, 2, \dots,$$

satisfy the recurrence relations

$$A_n(z) = (z - a_n)A_{n-1}(z) + b_n A_{n-2}(z), \quad A_0 \equiv 0, \ A_{-1} \equiv 1;$$

$$B_n(z) = (z - a_n)B_{n-1}(z) + b_n B_{n-2}(z), \quad B_0 \equiv 1, \ B_{-1} \equiv 0,$$

and it is well known that the nth convergent is the $[n - 1/n]$ Padé approximant to the function (1.1) (see [Pe]). Hence the convergents of (1.3) are rational interpolants with all interpolation points being identical to infinity.

We will consider here rational interpolants with more general systems of interpolation points. Basically, there will be only two restrictions: The interpolation points have to be chosen from outside of $I(S(\mu))$ and have to stay away from $S(\mu)$, and have to lie symmetric with respect to the real axis.

In order to avoid the compulsory zero which the function (1.1) has at infinity, we investigate a slightly more general class of Markov functions, namely, functions of the form

$$f(z) = c + \int \frac{d\mu(x)}{x - z}, \tag{1.4}$$

where c is a real constant and μ is again a positive measure with compact support $S(\mu) \subseteq \mathbb{R}$.

For every $n \in \mathbb{N}$ we select a set $A_n = \{x_{0,n}, \dots, x_{2n,n}\}$ of $2n+1$ interpolation points from $\overline{\mathbb{C}} \setminus I(S(\mu))$. The points need not to be distinct. Hence, A_n is in general a multiset. Put

$$\omega_n(z) := \prod_{\substack{j=0 \\ x_{jn} \neq \infty}}^{2n} (z - x_{jn}). \tag{1.5}$$

The degree of ω_n is equal to the number of finite points in A_n. In order that $\omega_n(x)$ be real and do not change its sign on $S(\mu)$, we assume that nonreal points in A_n appear only in conjugate pairs, that is, we assume that

$$A_n = \overline{A_n} := \{x \in \mathbb{C} \mid \overline{x} \in A\} \quad \text{and} \quad A_n \subseteq \overline{\mathbb{C}} \setminus I(S(\mu)). \tag{1.6}$$

Denote by $\mathcal{R}_{m,n} \equiv \mathcal{R}_{mn}$ the set of all rational functions with numerator degree $\leq m$, denominator degree $\leq n$, and complex coefficients.

Definition 6.1.1. *By $r_n = r_n(f, A_n, \cdot) \in \mathcal{R}_{nn}$, $n \in \mathbb{N}$, we denote the rational function that interpolates the function f of type* (1.4) *in the $2n+1$ points of the set $A_n = \{x_{0n}, \dots, x_{2n,n}\}$. If some of these points are confluent (identical), then the interpolation is understood in Hermite's sense, that is, if an interpolation point x_{jn} appears $k > 1$ times in A_n, then the value of f at x_{jn} together with the first $k-1$ derivatives has to be interpolated by r_n. It is easy to see that this is equivalent to the assertion that the left-hand side of*

$$\frac{f(z) - r_n(f, A_n; z)}{\omega_n(z)} = O(z^{-(2n+1)}) \quad \text{as} \quad |z| \to \infty \tag{1.7}$$

is bounded at every finite point of A_n and at infinity it has the indicated behavior. We note that interpolation at infinity has not been excluded.

Contrary to the polynomial case, the existence of a rational interpolant with the properties required in Definition 6.1.1 can in general not be guaranteed. However, because of the special structure of the Markov functions and because of the assumptions made in (1.6), the situation is different here. We have the following lemma, in which, besides the unique existence, some other important properties of the interpolant and a remainder formula are proved.

Lemma 6.1.2. (a) *For any set $A_n = \{x_{0n}, \dots, x_{2n,n}\}$, $n \in \mathbb{N}$, of $2n+1$ interpolation points from $\overline{\mathbb{C}} \setminus I(S(\mu))$ possessing the symmetry property* (1.6) *there exists uniquely a rational interpolant*

$$r_n(z) = r_n(f, A_n; z) = \frac{p_n(z)}{q_n(z)} \in \mathcal{R}_{nn} \tag{1.8}$$

to the function (1.4). *We assume that the denominator polynomial* q_n *is monic.*

(b) *The polynomial* q_n *is of degree* n, *all its zeros are simple and contained in* $I(S(\mu))$, *every component of* $I(S(\mu)) \setminus S(\mu)$ *contains at most one zero, and* q_n *satisfies the weighted orthogonality relation*

$$\int q_n(x) x^k \frac{d\mu(x)}{\omega_n(x)} = 0 \quad \text{for } k = 0, \ldots, n-1. \tag{1.9}$$

(c) *The two polynomials* p_n *and* q_n *in* (1.8) *have no common factors.*

(d) *The remainder term of the interpolant has the representation*

$$(f - r_n(f, A_n; \cdot))(z) = \frac{\omega_n(z)}{q_n^2(z)} \int \frac{q_n^2(x) d\mu(x)}{\omega_n(x)(x - z)} \tag{1.10}$$

for all $z \in \Omega$.

Remark. The assumptions made in (1.6) with respect to the location of the interpolation points are necessary for Lemma 6.1.2 to hold true. The least obvious is perhaps the assumption that no interpolation points are allowed to lie in the set $I(S(\mu)) \setminus S(\mu)$, not even if their number is even in every component of $I(S(\mu)) \setminus S(\mu)$, which would be compatible with the requirement that ω_n does not change its sign on $S(\mu)$. The next example shows that such a relaxation of (1.6) could lead to interpolation defects.

Example 6.1.3. Let μ be defined by $d\mu(x) = dx$ for $x \in S := [-2, -1] \cup [1, 2]$,

$$f(z) = \int \frac{d\mu(x)}{x - z} = \int_S \frac{dx}{x - z}, \tag{1.11}$$

$n = 1$, and the interpolation set $A_1 = \{0, 0, \infty\}$. We have $f(0) = f(\infty) = 0$ and $f'(0) = 1$. Since $\mathcal{R}_{1,1}$ contains only Möbius transforms or constants, that is, only functions that are univalent on $\overline{\mathbb{C}}$ or constants, the data cannot be interpolated. □

A proof of the assertions of the lemma can be found in [GoLo]. Since part (b) especially is basic for the understanding of the relationship between weighted orthogonality and rational interpolation and since a proof does not require too much space we shall give one in what follows after completing the statement of the main results of this section.

Since our interest here is the convergence problem, we have to make assumptions about the asymptotic distribution of the interpolation points x_{jn}, $j = 0, \dots, 2n$, as $n \to \infty$. A triangular matrix

$$(1.12) \qquad \mathbf{A} = \begin{pmatrix} A_0 \\ A_1 \\ \vdots \\ A_n \\ \vdots \end{pmatrix} = \begin{pmatrix} a_{00} & & & & \\ a_{01} & a_{11} & a_{21} & & \\ \vdots & & & \ddots & \\ a_{0n} & \cdots & \cdots & & a_{2n,n} \\ \vdots & & & & & \ddots \end{pmatrix}$$

of interpolation points is called an *interpolation scheme*. By $L(\mathbf{A})$ we denote the set of all limit points of $\{x_{jn}\}_{j=0}^{2n}$ as $n \to \infty$.

Definition 6.1.4. *We say that the scheme* $\mathbf{A}$ *has a probability measure* $\alpha = \alpha(\mathbf{A})$ *as its* asymptotic distribution *if*

$$(1.13) \qquad \frac{1}{2n+1}\sum_{j=0}^{2n} \delta_{x_{jn}} \to \alpha \quad \textit{as} \quad n \to \infty$$

in the weak topology of measures in* $\overline{\mathbb{C}}$. *Another formulation of this fact is the assertion that for any Borel set* $B \subseteq \overline{\mathbb{C}}$ *with* $\alpha(\partial B) = 0$ *we have*

$$(1.14) \qquad \lim_{n\to\infty} \frac{1}{2n+1}\operatorname{card}\{j \mid x_{jn} \in B\} = \alpha(B).$$

It is obvious that $\operatorname{supp}(\alpha) \subseteq L(\mathbf{A})$.

Definition 6.1.5. *We call an interpolation scheme* $\mathbf{A}$ admissible *if for every* $n \in \mathbb{N}$ *the set* $A_n \subseteq \overline{\mathbb{C}} \setminus I(S(\mu))$ *possesses the symmetry property* (1.6) *and if*

$$(1.15) \qquad L(\mathbf{A}) \subseteq \Omega = \overline{\mathbb{C}} \setminus S(\mu).$$

Assumption (1.15) means that the interpolation points have to stay away asymptotically from $S(\mu)$, although they may approach points in $I(S(\mu)) \setminus S(\mu)$.

The next theorem is the main result of this section. It is a generalization of Markov's classical theorem on the convergence of continued fractions.

Theorem 6.1.6. *Let* f *be a Markov function of type* (1.4) *and* $\mathbf{A}$ *an admissible interpolation scheme with asymptotic distribution* $\alpha = \alpha(\mathbf{A})$.

(a) *We have*

$$(1.16) \qquad \limsup_{n\to\infty} |f(z) - r_n(f, A_n; z)|^{1/2n} \le e^{-g(\alpha,\Omega;z)}$$

locally uniformly in $\overline{\mathbb{C}} \setminus I(S(\mu))$, *where*

$$g(\alpha, \Omega; z) = \int g_\Omega(z, x) d\alpha(x) \tag{1.17}$$

is the Green potential of α *in* $\Omega = \mathbb{C} \setminus S(\mu)$.

(b) *The following three assertions are equivalent.*

(i) $\mu \in \mathbf{Reg}$.

(ii) *We have*

$$\lim_{n\to\infty} |f(z) - r_n(f, A_n; z)|^{1/2n} = e^{-g(\alpha,\Omega;z)} \tag{1.18}$$

locally uniformly in $\overline{\mathbb{C}} \setminus (I(S(\mu)) \cup L(\mathbf{A}))$.

(iii) *For any infinite subsequence* $N \subseteq \mathbb{N}$ *we have*

$$\limsup_{n\to\infty, n\in N} |f(z) - r_n(f, A_n; z)|^{1/2n} = e^{-g(\alpha,\Omega;z)} \tag{1.19}$$

for quasi every $z \in \overline{\mathbb{C}} \setminus I(S(\mu))$.

Remark. If $\operatorname{cap}(S(\mu)) = 0$, then $g(\alpha, \Omega; z) \equiv \infty$.

Corollary 6.1.7. *If the interpolation scheme* $\mathbf{A}$ *is admissible, then the sequence* $\{r_n(f, A_n; \cdot)\}_{n=1}^{\infty}$ *converges locally uniformly in* $\mathbb{C} \setminus I(S(\mu))$ *to* f.

Remark. If all interpolation points a_{jn} in (1.12) are chosen to equal infinity, then the scheme is admissible, and Corollary 6.1.7 specializes to Markov's classical theorem.

Proof of Corollary 6.1.7. From Helly's selection theorem it follows that any infinite subsequence $N \subseteq \mathbb{N}$ contains an infinite subsequence, which we continue to denote by N, such that the weak* limit (1.13) exists in the weak* topology on $\overline{\mathbb{C}}$. For the natural numbers n not belonging to N we can choose interpolation sets $A_n \subseteq \overline{\mathbb{C}} \setminus I(S(\mu))$ of $2n + 1$ points in such a way that (1.6) is satisfied and that the newly chosen sets have also α as their asymptotic distribution. For the completed interpolation scheme Theorem 6.1.6 is applicable. Since $\exp(-g(\alpha, \Omega; z)) < 1$ for all $z \in \Omega$, the corollary follows from (1.16). □

Proof of Lemma 6.1.2. (a) First we prove the uniqueness of $r = r_n(f, A_n; \cdot)$, $n \in \mathbb{N}$. Let $r_1, r_2 \in \mathcal{R}_{nn}, r_j = p_j/q_j$, $j = 1, 2$, be two interpolants of f in the $2n + 1$ points of A_n. Multiplying $r_1 - r_2$ by $q_1 q_2 \omega_n$ gives with (1.7) that

$$(q_2 p_1 - q_1 p_2)(z) = O(z^{\deg(\omega_n)-1}) \quad \text{as} \quad |z| \to \infty.$$

Hence, here the left-hand side is a polynomial of one degree less than ω_n. But since this polynomial interpolates the zero function at all finite points of A_n, it has to be identically zero, and therefore $q_2p_1 \equiv q_1p_2$, and furthermore $r_1 \equiv r_2$.

Next we prove the existence of the interpolants. Let $q \in \Pi_n^*$ be the monic orthogonal polynomial with respect to the orthogonality relation (1.9). From this orthogonality and the identities

(1.20)

$$\int \frac{q(z)x^l - z^l q(x)}{x-z} \frac{d\mu(x)}{\omega_n(x)} = \int \left(\frac{q(z)}{q(x)}x^l - z^l\right)(x-z)^{-1}q(x)\frac{d\mu(x)}{\omega_n(x)}$$

$$= \int \left[z^{l-1} + z^{l-2}x + \cdots + z^{n+1}x^{l-n-2} + \frac{\frac{q(z)}{q(x)}x^l - z^{n+1}x^{l-n-1}}{x-z}\right]$$

$$\times\, q(x)\frac{d\mu(x)}{\omega_n(x)}$$

it follows that for all $2n+1 \geq l > n$ the expressions in (1.20) are polynomials of degree at most n. From this we immediately derive that the polynomial

$$p(z) = \int \frac{q(z)\omega_n(x) - \omega_n(z)q(x)}{x-z}\frac{d\mu(x)}{\omega_n(x)} + cq(z)$$

$$= \int \left[q(z) - \frac{\omega_n(z)}{\omega_n(x)}q(x)\right]\frac{d\mu(x)}{x-z} + cq(z)$$

is also of degree at most n, and from here and the orthogonality relation (1.9) we derive

(1.21)

$$q(z)f(z) - p(z)$$

$$= \omega_n(z)\int \frac{q(x)d\mu(x)}{\omega_n(x)(x-z)}$$

$$= \omega_n(z)\int \left[-\frac{1}{z} - \frac{x}{z^2} - \cdots - \frac{x^{n-1}}{z^n} + \frac{x^n}{z^n(x-z)}\right]q(x)\frac{d\mu(x)}{\omega_n(x)}$$

$$= \omega_n(z)O(z^{-(n+1)})$$

as $|z| \to \infty$. A comparison with (1.7) then shows that p/q is a rational function interpolating f in the points of the interpolation set A_n that was used in (1.4) to define ω_n.

(b) First we prove that the denominator polynomial q_n satisfies the orthogonality relation (1.9). Multiplying (1.7) by q_n and performing some elementary transformations we get

$$\text{(1.22)}\quad \begin{aligned} \frac{q_n f - p_n}{\omega_n}(z) &= \int \left[\frac{q_n(z)}{\omega_n(z)} - \frac{q_n(x)}{\omega_n(x)}\right] \frac{d\mu(x)}{x-z} + \frac{cq_n(z) - p_n(z)}{\omega_n(z)} \\ &+ \int \frac{q_n(x)d\mu(x)}{\omega_n(x)(x-z)} = O(z^{-(2n+1)+\deg(q_n)}) \quad \text{as} \quad |z| \to \infty. \end{aligned}$$

The first two terms on the right-hand side of (1.22) are rational functions with poles only at zeros of ω_n or at infinity. Since both the last term on the right-hand side and the term on the left-hand side of (1.22) are bounded at the zeros of ω_n and are zero at infinity, it follows that the first two terms on the right-hand side of (1.22) cancel out, and we have

$$\text{(1.23)}\quad \frac{q_n f - p_n}{\omega_n}(z) = \int \frac{q_n(x)d\mu(x)}{\omega_n(x)(x-z)} = O(z^{-(n+1)}) \quad \text{as} \quad |z| \to \infty.$$

Expanding the integrand in the second term in powers of z as in (1.21), it is easy to see that (1.23) implies that q_n satisfies the orthogonality relation (1.9).

Since ω_n does not change its sign on $S(\mu)$, it follows from Theorem 2.1.1 that $\deg(q_n) = n$ and all zeros of q_n are contained in $I(S(\mu))$. In Remark (1) to Lemma 1.1.3 it has been noted that in the case of $S(\mu) \subseteq \mathbb{R}$ we further know that all zeros are simple and every component of $I(S(\mu)) \setminus S(\mu)$ contains at most one of them.

(c) If p_n and q_n in (1.8) had common factors, there would exist a reduced representation of $r_n(f, A_n; z)$ with a denominator polynomial $\tilde{q}_n$ of degree less than n. But since $\tilde{q}_n$ satisfies (1.9) this is not possible.

(d) From the orthogonality (1.9) it follows that for every $l = 1, \ldots, n$ we have the identity

(1.24)

$$\begin{aligned} \int x^l \frac{q_n(x)d\mu(x)}{\omega_n(x)(x-z)} &= z^l \int \left[-\frac{1}{z} - \cdots - \frac{x^{l-1}}{z^l} + \frac{x^l}{z^l(x-z)}\right] \frac{q_n(x)}{\omega_n(x)} d\mu(x) \\ &= z^l \int \frac{q_n(x)d\mu(x)}{\omega_n(x)(x-z)}, \end{aligned}$$

which implies that

$$\text{(1.25)}\quad \int \frac{q_n(x)d\mu(x)}{\omega_n(x)(x-z)} = \frac{1}{q_n(z)} \int \frac{q_n(x)^2 d\mu(x)}{\omega_n(x)(x-z)}.$$

Together with (1.23) this proves the remainder formula (1.10). □

Next we come to the proof of Theorem 6.1.6. It is based on the following theorem, which is, from the point of view of the subject of this book, even more interesting than Theorem 6.1.6 itself. The theorem shows that there exists a one-to-one correspondence between the asymptotic behavior of certain weighted orthonormal polynomials and the convergence behavior of rational interpolants.

Theorem 6.1.8. *Let f be a Markov function of type (1.4), let $\mathbf{A}$ be an interpolation scheme with interpolation sets possessing the symmetry property (1.6), and let μ_n be the weight measure defined by*

$$d\mu_n := \frac{d\mu}{\omega_n}, \quad n \in \mathbb{N}. \tag{1.26}$$

Then

$$\lim_{n\to\infty} \left| (f(z) - r_n(f, A_n; z)) \frac{p_n(\mu_n; z)^2}{\omega_n(z)} \right|^{1/n} = 1 \tag{1.27}$$

locally uniformly in $\overline{\mathbb{C}} \setminus I(S(\mu))$, where $p_n(\mu_n; z)$, $n \in \mathbb{N}$, denote the orthonormal polynomials with respect to the weight measure μ_n.

Remarks. (1) It is not necessary to assume that condition (1.15) holds true for the interpolation scheme $\mathbf{A}$.
(2) Because of the symmetry property (1.6) the polynomials ω_n are real and do not change their sign on $S(\mu)$. We can therefore assume without loss of generality that μ_n is positive for all $n \in \mathbb{N}$.

Proof of Theorem 6.1.8. Set, as in Lemma 6.1.2, $r_n(f, A_n; \cdot) = p_n/q_n$ with monic q_n. From the orthogonality (1.9) we know that

$$q_n(z) = \|q_n\|_{L^2(\mu_n)} p_n(\mu_n; z), \tag{1.28}$$

and therefore the remainder formula (1.10) yields that

$$\left(\frac{f - r_n(f, A_n; \cdot)}{\omega_n} \right)(z) p_n(\mu_n; z)^2 = \frac{1}{\|q_n\|^2_{L^2(\mu_n)}} \int \frac{q_n(x)^2 d\mu(x)}{(x - z)\omega_n(x)}. \tag{1.29}$$

For $x \in S(\mu)$ and $z \notin \mathbb{R}$ we have

$$\left| \operatorname{Im}\left(\frac{1}{x - z} \right) \right| \geq \frac{|\operatorname{Im}(z)|}{\operatorname{diam}(\{z\} \cup S(\mu))^2}, \tag{1.30}$$

for $z \in \{x \in \mathbb{R} \mid x < \inf S(\mu)\}$

$$\frac{1}{x - z} \geq \frac{1}{\operatorname{diam}(\{z\} \cup S(\mu))}, \tag{1.31}$$

for $z \in \{x \in \mathbb{R} \mid x > \sup S(\mu)\}$

$$\frac{-1}{x-z} \geq \frac{1}{\operatorname{diam}(\{z\} \cup S(\mu))}, \tag{1.32}$$

for all $z \notin I(S(\mu))$

$$\left|\frac{1}{x-z}\right| \leq \frac{1}{\operatorname{dist}(z, S(\mu))}. \tag{1.33}$$

From (1.30)–(1.33) it follows that there exist two positive continuous functions $d_1(z) > 0$ and $d_2(z) < \infty$ on $\mathbb{C} \setminus I(S(\mu))$ independent of n such that

$$d_1(z)\|q_n\|^2_{L^2(\mu)} \leq \left|\int q_n(x)^2 \frac{d\mu_n(x)}{x-z}\right| \leq d_2(z)\|q_n\|^2_{L^2(\mu_n)} \tag{1.34}$$

for all $z \in \overline{\mathbb{C}} \setminus I(S(\mu))$. From these inequalities together with (1.29) the limit (1.27) follows immediately. □

Finally, we turn to Theorem 6.1.6.

Proof of Theorem 6.1.6. With Theorem 6.1.8 the ground work of the proof is done. We start with a consideration of the interpolation points equal to infinity.

Without loss of generality we can assume in Theorem 6.1.6 that $\infty \notin L(\mathbf{A})$. For if $\infty \in L(\mathbf{A})$, then by a Möbius transformation the original problem can be transformed into an equivalent one with $\infty \notin L(\mathbf{A})$. More precisely: There exists $a \in \mathbb{R} \setminus (L(\mathbf{A}) \cup I(S(\mu)))$; we consider the new variable w defined by

$$w = \varphi(z) = \frac{1}{z-a}. \tag{1.35}$$

In this variable the function f has the form

$$\tilde{f}(w) := (f \circ \varphi^{-1})(w) = f(a) - \int \frac{d\tilde{\mu}(y)}{y-w} \tag{1.36}$$

with $d\tilde{\mu}(y) := y^2 d\mu(\varphi^{-1}(y))$, $y \in \varphi(S(\mu))$, and the interpolants $\tilde{r}_n$ and the polynomials $\tilde{\omega}_n$, $\tilde{q}_n$, and $\tilde{p}_n$ are given by

$$\tilde{r}_n(\tilde{f}, \tilde{A}_n; w) = r_n(f, A_n; \varphi^{-1}(w)) \in \mathcal{R}_{nn} \tag{1.37}$$

with $\tilde{A}_n = \varphi(A_n)$, and

$$\begin{aligned} \tilde{\omega}_n(w) &= w^{2n+1}(\omega_n \circ \varphi^{-1})(w) \in \Pi_{2n+1}, \\ \tilde{q}_n(w) &= w^n(q_n \circ \varphi^{-1})(w) \in \Pi_n, \\ \tilde{p}_n(w) &= w^n(p_n \circ \varphi^{-1})(w) \in \Pi_n. \end{aligned} \tag{1.38}$$

The function $\tilde{f}$ is, except for its sign, again of the type (1.4). To verify (1.36) we consider the identity

(1.39)
$$\tilde{f}\left(\frac{1}{z-a}\right) = c + \int \frac{d\mu(x)}{x-a} - \int \left(\frac{1}{x-a} - \frac{1}{z-a}\right)^{-1} \left(\frac{1}{x-a}\right)^2 d\mu(x)$$
$$= c + \int \left[\frac{1}{x-a} - \frac{z-a}{(x-a)(z-x)}\right] d\mu(x) = f(z).$$

After this preparation we come to the actual proof, which consists of a combination of the asymptotic results from Theorems 3.3.2 and 3.3.3 with Theorem 6.1.8. Since we have assumed that $\infty \notin L(\mathbf{A})$, we know that $\operatorname{supp}(\alpha) \subseteq L(\mathbf{A})$ has compact support in $\Omega \setminus \{\infty\}$, and Theorems 3.3.2 and 3.3.3 are applicable with the weight functions

(1.40)
$$g_n(x) := \frac{1}{\omega_n(x)}, \quad n \in \mathbb{N}.$$

It is easy to see that (1.13) and (1.15) imply $g_n^{1/n} \to g := \exp(2p(\alpha;\cdot))$ uniformly on $S(\mu)$ as $n \to \infty$. The measure σ in Theorems 3.3.2 and 3.3.3 is now α and the constant $\gamma = 2$.

(a) From (1.13) and the lower envelope theorem it follows that

(1.41)
$$\limsup_{n\to\infty} |\omega_n(z)|^{1/n} \le e^{-2p(\alpha;z)}$$

locally uniformly for $z \in \mathbb{C}$ and equality holds true in (1.41) quasi everywhere.

From (3.3.20) of Theorem 3.3.2 with $\gamma = 2$ and $\sigma = \alpha$ together with (1.41) it follows that

(1.42)
$$\limsup_{n\to\infty} \left| \frac{\omega_n(z)}{p_n(\mu_n; z)^2} \right|^{1/2n} \le e^{-g(\alpha,\Omega;z)}$$

locally uniformly for $z \in \mathbb{C} \setminus I(S(\mu))$. By Theorem 6.1.8 this implies (1.16).

(b) We prove the chain of implications: assertion (iii) of Theorem 3.3.3 $\Rightarrow$ Theorem 6.1.6(iii) $\Rightarrow$ Theorem 6.1.6(ii) $\Rightarrow$ assertion (iii) of Theorem 3.3.3. Since in Theorem 3.3.3 it has been shown that Theorem 3.3.3(iii) is equivalent to $\mu \in \mathbf{Reg}$, part (b) of Theorem 6.1.6 is then proved.

If $\operatorname{cap}(S(\mu)) = 0$ then (i) holds as well as (ii) and (iii) by (1.16). Hence we may assume in what follows that $\operatorname{cap}(S(\mu)) > 0$.

Assertion (iii) of Theorem 3.3.3 $\Rightarrow$ Theorem 6.1.6(iii): From (iii) in Theorem 3.3.3 we know that

(1.43)
$$\lim_{n\to\infty} \left| \frac{1}{p_n(\mu_n; z)} \right|^{1/n} = e^{p(\alpha;z) - g(\alpha,\Omega;z)}$$

locally uniformly for $z \in \mathbb{C} \setminus I(S(\mu))$. From the limit (1.13) it follows by the lower envelope theorem that (1.41) holds true also for any infinite subsequence $N \subseteq \mathbb{N}$, and we again have equality quasi everywhere. Therefore we have

$$\limsup_{n\to\infty, n\in N} \left| \frac{\omega_n(z)}{p_n(\mu_n; z)^2} \right|^{1/2n} = e^{-g(\alpha,\Omega;z)} \tag{1.44}$$

for quasi every $z \in \overline{\mathbb{C}} \setminus I(S(\mu))$. With Theorem 6.1.8 this implies (1.19).

(iii)⇒(ii): Since the rational interpolant $r_n(f, A_n; \cdot)$ has all its poles in $I(S(\mu))$, it follows from the remainder formula (1.10) that the functions

$$\frac{1}{2n} \log |f(z) - r_n(f, A_n; z)| \tag{1.45}$$

are harmonic on every compact set in $\overline{\mathbb{C}} \setminus (I(S(\mu)) \cup L(\mathbf{A}))$ for n sufficiently large, and they are bounded because of (1.16). Hence, we can deduce from the pointwise convergence quasi everywhere in (1.19), which is true for any infinite subsequence $N \subseteq \mathbb{N}$, that we have the proper limit (1.18) locally uniformly in $\overline{\mathbb{C}} \setminus (I(S(\mu)) \cup L(\mathbf{A}))$. The conclusion can be drawn by local considerations of the sequence of harmonic functions.

Theorem 6.1.6(ii) ⇒ assertion (iii) of Theorem 3.3.3: By Theorem 6.1.8 the limit (1.18) is equivalent to

$$\lim_{n\to\infty} \left| \frac{\omega_n(z)}{p_n(\mu_n; z)^2} \right|^{1/2n} = e^{-g(\alpha,\Omega;z)} \tag{1.46}$$

locally uniformly for $z \in \overline{\mathbb{C}} \setminus (I(S(\mu)) \cup L(\mathbf{A}))$. From (1.13) it then follows that

$$\lim_{n\to\infty} e^{p(\alpha;z)-g(\alpha,\Omega;z)} |p_n(\mu_n; z)|^{1/n} = 1 \tag{1.47}$$

also holds locally uniformly for $z \in \overline{\mathbb{C}} \setminus (I(S(\mu)) \cup L(\mathbf{A}))$. Since the logarithm of the function on the left-hand side of (1.47) is harmonic in $\mathbb{C} \setminus I(S(\mu))$ (cf. the harmonicity of (3.3.37) in the proof of Theorem 3.3.2), the convergence extends locally uniformly to $\mathbb{C} \setminus I(S(\mu)) = \mathbb{C} \setminus \mathrm{Co}(S(\mu))$. This proves assertion (iii) of Theorem 3.3.3 with $\alpha = \sigma$ and $\gamma = 2$. □

6.2 Best Rational Approximants to Markov Functions

We continue the study of rational approximants to Markov functions by investigating the convergence of best rational approximants. The type of result is similar to that in the last section. First we prove a general estimate for the rate of convergence, and then we show that the rate is sharp in the sense of geometric convergence if and only if the measure μ in the definition of the function (2.1) is regular.

The function f to be approximated is, as in (1.4), of the form

$$f(z) = \int \frac{d\mu(x)}{x - z} + c, \tag{2.1}$$

where μ is a positive measure with compact support $S(\mu) \subseteq \mathbb{R}$ and $c \in \mathbb{R}$ is a constant. We approximate f on a compact set $V \subseteq \overline{\mathbb{C}} \setminus S(\mu)$ that is assumed to be symmetric with respect to $\mathbb{R}$, that is,

$$V = \overline{V} := \{z \in \mathbb{C} \mid \bar{z} \in V\}. \tag{2.2}$$

Definition 6.2.1. *By* $r_n^* = r_n^*(f, V; \cdot) \in \mathcal{R}_{nn}$, $n \in \mathbb{N}$, *we denote any rational function that satisfies*

$$\|f - r_n^*\|_{\sup,V} = \inf_{r \in \mathcal{R}_{nn}} \|f - r\|_{\sup,V}. \tag{2.3}$$

It is called the best rational approximant *of degree n to f on V.*

Under our rather general assumptions the best rational approximant $r_n^*(f, V; \cdot)$ may not exist. We know that $r_n^*(f, V; \cdot)$ exists for every $n \in \mathbb{N}$ if V is a real interval or the union of finitely many real intervals (see, e.g., [Br, Section III]), but it may not exist if V contains isolated points. In any case there exists a near best rational approximant $r_n^*(f, V; \cdot)$ in the sense that

$$\|f - r_n^*\|_{\sup,V} \leq 2 \inf_{r \in \mathcal{R}_{nn}} \|f - r\|_{\sup,V}.$$

For simplicity we use the convenient terminology "best approximant" for $r_n^*(f, V; \cdot)$s in the case of near best approximants. The results to follow are not touched by this change in the definition. The question of uniqueness of best rational approximants is even more difficult to answer than the existence problem. The uniqueness of $r_n^*(f, V; \cdot)$ is known only in the case when V is a real interval and the best approximant is selected from $\mathcal{R}_n^{\mathbb{R}}$, the set of all rational functions of degree at most n with real coefficients. In this special case uniqueness can be proved by considering alternation points (see [Br]). On the other hand, no example of a Markov function (2.1) and a symmetric compact set $V \subseteq \overline{\mathbb{C}} \setminus S(\mu)$ is known for which the best rational approximant $r_n^*(f, V; \cdot)$ is not unique. In general, that is, for arbitrary analytic functions f, uniqueness of best rational functions in $\mathcal{R}_{nn}$ is not true. (For more references concerning the nonuniqueness problem see Section 6.2 in the historical notes.)

In the statement of our main result we need the *condenser capacity* $\text{cap}(V, S)$ of two disjoint compact sets $V, S \subseteq \overline{\mathbb{C}}$, which is defined in Section VIII of the Appendix.

Theorem 6.2.2. *Let f be a Markov function of type* (2.1) *and $V \subseteq \overline{\mathbb{C}} \setminus S(\mu)$ a compact set with the symmetry property* (2.2).

(a) *We have*

$$\limsup_{n\to\infty} \|f - r_n^*(f, V; \cdot)\|_{\sup,V}^{1/2n} \le e^{-1/\operatorname{cap}(V,S(\mu))} \tag{2.4}$$

(b) *If* $\operatorname{cap}(V) > 0$, *then*

$$\lim_{n\to\infty} \|f - r_n^*(f, V; \cdot)\|_{\sup,V}^{1/2n} = e^{-1/\operatorname{cap}(V,S(\mu))} \tag{2.5}$$

if and only if $\mu \in$ **Reg**.

Remark. If $\operatorname{cap}(V) = 0$, then $\operatorname{cap}(V, S(\mu)) = 0$, and by (2.4) we have

$$\lim_{n\to\infty} \|f - r_n^*(f, V; \cdot)\|_{\sup,V}^{1/2n} = 0$$

independently of whether $\mu \in$ **Reg** or not. Hence, in part (b) the assumption $\operatorname{cap}(V) > 0$ is necessary.

Since the proof of Theorem 6.2.2 is very long and technically quite involved, we present the following outline of it.

The outline of the proof of Theorem 6.2.2. Parts (a) and (b) are separately proved.

Two lemmas, Lemmas 6.2.3 and 6.2.4, prepare the proof of part (a). Lemma 6.2.3 is a simple technical one on approximating measures by discrete ones. Part (a) of Lemma 6.2.4 easily follows from the results in the preceding section: It gives an estimate on rational interpolants via Green potentials of limit distributions of the interpolation points (cf. (2.10), (2.11)). The most important step is part (b) of Lemma 6.2.4 which gives a technique as to how to get good rational approximants on V from rational interpolants. The idea is to cancel in the partial-fraction representation of the interpolants all those terms that contain poles closer to V than δ_0/n, where n is the degree in question (cf. (2.17)). This yields rational approximants with rate of approximation expressed by suprema of potentials that are majorized by the above Green potentials (cf. (2.12)). Then the proof of part (a) of Theorem 6.2.2 is obtained by taking interpolation schemes with asymptotic distribution equal to the equilibrium distribution on V of the condenser $(V, S(\mu))$. Much of the technical details are caused by the irregular points of V.

In the necessity part of Theorem 6.2.2(b) we use the same rational interpolants. The idea is that they yield an approximation that is in between the optimum rate ($\|f - r_n^*\|_V$) and the rate expressed by the condenser capacity (this is what is proved in Theorem 6.2.2(a)); hence if these two rates collapse (i.e., we have equality in (2.4)) then we know sufficient information on the rate of approximation given by the rational approximants to be able to apply the results of the preceding section.

The most severe proof is that of the sufficiency part in Theorem 6.2.2(b), that is, when we want to conclude equality in (2.4) from the regularity of μ. The method is rather ad hoc, and most of the difficulties arise because V may not be regular with respect to the Dirichlet problems in $\mathbb{C}\setminus V$. $\mu \in$ **Reg** is applied in the form that is guaranteed by the equivalence of (i) and (vi) of Theorem 3.2.1, where the weight functions g_n are associated with the denominator polynomials of the best approximants (cf. (2.56), (2.59), (2.60)). Two lemmas, Lemmas 6.2.5 and 6.2.6, contain essential steps of the proof. The first one simplifies the structure of V. The second, and technically most difficult, one is concerned with moving the zeros of polynomials Q_n from $\mathbb{C}\setminus V$ onto V in such a way that the so-obtained polynomials behave on V in a fashion similar to that of the original polynomials Q_n (cf. (2.66) and (2.67)). This is done by selecting symmetric, weighted Fekete points with respect to weights $1/Q_n$. □

We prepare the proof of Theorem 6.2.2(a) by stating two lemmas. In both of them we use the equilibrium distribution ω_{V,S_0} on V of the condenser (V, S_0) (see Appendix VIII).

Lemma 6.2.3. *Let V be as in* Theorem 6.2.2, *and assume in addition that $\partial V \cap I(S(\mu))$ has only finitely many points,* $\text{cap}(V) > 0$, *and* $\infty \notin V$. *Let further $S_0 \supseteq S(\mu)$ be a compact set with $S_0 \subseteq \mathbb{R}$, $V \cap S_0 = \emptyset$, and* $\text{cap}(S_0) > 0$. *Then there exists an admissible interpolation scheme $\mathbf{A} = (A_1, \ldots, A_n, \ldots)^t$ with all its interpolation points in $\partial V \setminus I(S(\mu))$ and ω_{V,S_0} as its asymptotic distribution $\alpha(\mathbf{A})$.*

Recall that $I(S(\mu))$ is the smallest interval on $\mathbb{R}$ containing $S(\mu)$, and the admissibility of $\mathbf{A}$ means that $\mathbf{A}$ is symmetric onto $\mathbb{R}$, each interpolation set A_n is disjoint from $I(S(\mu))$, and no point of $L(\mathbf{A})$, the set of limit points of the sets A_n, belongs to $S(\mu) = \text{supp}(\mu)$.

Remark. In the sequel a compact set S_0 different from $S(\mu)$ will be used if $\text{cap}(S(\mu)) = 0$, otherwise we can set $S_0 = S(\mu)$.

Proof of Lemma 6.2.3. Since V is symmetric with respect to $\mathbb{R}$, the same is true for ω_{V,S_0}. It is easy to verify that we can select for each $n \in \mathbb{N}$ an interpolation set A_n with $2n+1$ points $x_{0,n}, \ldots, x_{2n,n} \in \partial V$ possessing the symmetry property (1.6) and having $\omega_{V,S(\mu)}$ as their asymptotic distribution, that is, the limit (1.13) holds true with $\alpha = \omega_{V,S(\mu)}$. Since we have assumed that $\partial V \cap I(S(\mu))$ consists of only finitely many points, the $\omega_{V,S(\mu)}$-measure of these points is zero, and we can therefore select all points $x_{jn}, 0 \le j \le n$, $n \in \mathbb{N}$, from $\partial V \setminus I(S(\mu))$. Hence, the interpolation scheme $\mathbf{A} := (A_1, A_2, \dots)^t$ is admissible in the sense of Definition 6.1.5.

□

If $\mathbf{A} = (A_1, A_2, \dots)^t$ is an admissible interpolation scheme with finite interpolation points then define, as in (1.5) and (1.26),

$$\begin{aligned} \omega_n(z) &:= \prod_{x \in A_n} (z - x), \\ d\mu_n(x) &:= \frac{d\mu(x)}{\omega_n(x)}, \quad x \in S(\mu),\ n \in \mathbb{N}. \end{aligned} \tag{2.6}$$

Since $\mathbf{A}$ is admissible, we can assume without loss of generality that μ_n is a positive measure. (If ω_n is not positive on $S(\mu)$, then it has to be negative everywhere on $S(\mu)$, and we can multiply it by -1.)

Lemma 6.2.4. *Let V be as in* Theorem 6.2.2, *and let $\mathbf{A}$ be an admissible interpolation scheme with finite interpolation points and asymptotic distribution $\alpha = \alpha(\mathbf{A})$.*

(a) *Let q_n be the monic denominator polynomial of the rational interpolant $r_n = r_n(f, A_n; \cdot)$, $n \in \mathbb{N}$, and let $N \subseteq \mathbb{N}$ be an infinite subsequence such that the two limits*

$$\frac{1}{n}\nu_{q_n} \xrightarrow{*} \nu_0, \quad \frac{1}{n}\log \|q_n\|_{L^2(\mu_n)} \to c_0 \in \mathbb{R}\cup\{-\infty\} \quad \text{as} \quad n \to \infty, \quad n \in N, \tag{2.7}$$

exist. Then

$$\operatorname{supp}(\nu_0) \subseteq S(\mu), \tag{2.8}$$

$$\limsup_{n\to\infty, n\in N} \frac{1}{2n} \log |f(z) - r_n(z)| \le p(\nu_0 - \alpha; z) + c_0 \tag{2.9}$$

locally uniformly for $z \in \overline{\mathbb{C}} \setminus I(S(\mu))$, with equality quasi everywhere in $\overline{\mathbb{C}} \setminus I(S(\mu))$,

$$\lim_{n\to\infty, n\in N} \frac{1}{2n} \log |f(z) - r_n(z)| = p(\nu_0 - \alpha; z) + c_0 \tag{2.10}$$

locally uniformly for $z \in \overline{\mathbb{C}} \setminus (I(S(\mu)) \cup L(\mathbf{A}))$, and

$$p(\nu_0 - \alpha; z) + c_0 \le -g(\alpha, \Omega; z) \tag{2.11}$$

for all $z \in \mathbb{C}$ (the Green potential $g(\alpha, \Omega; z)$ was defined in (1.17)).

(b) *There exist rational functions $\tilde{r}_n \in \mathcal{R}_{nn}$, $n \in \mathbb{N}$, such that for every infinite subsequence $N \subseteq \mathbb{N}$ for which the two limits (2.7) exist we have*

$$\limsup_{n\to\infty, n\in N} \|f - \tilde{r}_n\|_{\sup,V}^{1/2n} \le \exp\left(\sup_{x\in V} p(\nu_0 - \alpha; x) + c_0\right). \tag{2.12}$$

Remark. The functions $\tilde{r}_n$ are obtained from the rational interpolants $r_n = r_n(f, A_n; \cdot)$ by removing all poles that the functions r_n have on V and in certain neighborhoods of V.

Proof of Lemma 6.2.4. (a) It has been shown in Lemma 6.1.2 that the denominator polynomial q_n is the nth orthogonal polynomial with respect to the measure μ_n, and therefore it has at most one zero in every component of $I(S(\mu)) \setminus S(\mu)$. This implies (2.8).

Since q_n is assumed to be monic, we have

$$q_n(z) = \|q_n\|_{L^2(\mu_n)} p_n(\mu_n; z),$$

where $p_n(\mu_n; \cdot)$ is the orthonormal polynomial with respect to μ_n. From the assumption that α is the asymptotic distribution of the scheme $\mathbf{A}$, that is,

$$\frac{1}{2n+1}\nu_{A_n} \overset{*}{\longrightarrow} \alpha \quad \text{as} \quad n \to \infty, \tag{2.13}$$

and from the assumed limits in (2.7) we immediately deduce by the principle of descent and the lower envelope theorem that

$$\liminf_{n\to\infty, n\in N} \frac{1}{2n} \log \left| \frac{p_n(\mu_n; z)^2}{\omega_n(z)} \right| \geq p(\alpha - \nu_0; z) - c_0 \tag{2.14}$$

locally uniformly for $z \in \mathbb{C} \setminus I(S(\mu))$, and equality holds quasi everywhere in $\mathbb{C} \setminus I(S(\mu))$. From Theorem 6.1.8 we then know that (2.14) is equivalent to (2.9).

In the same way we deduce (2.10) from (2.7) and (2.13) by Theorem 6.1.8.

From (1.16) in Theorem 6.1.6 together with (2.9) we know that (2.11) holds true for quasi every $z \in \mathbb{C} \setminus I(S(\mu))$. Since both sides of (2.11) are continuous in the fine topology (see Appendix II), the inequality in (2.11) extends to all $z \in \mathbb{C}$.

(b) It has to be shown that there exist rational functions $\tilde{r}_n \in \mathcal{R}_{nn}$, $n \in \mathbb{N}$, satisfying (2.12). This part of the proof causes most of the technical difficulties.

We start from the rational interpolants $r_n = r_n(f, A_n; \cdot)$ and remove the poles of these functions from V and from certain neighborhoods of V. Since r_n has only simple poles, it can be represented as

$$r_n(z) = \sum_{j=1}^{n} \frac{a_{jn}}{z - y_{jn}} + c_n$$

with $a_{jn}, c_n \in \mathbb{C}, y_{jn} \in I(S(\mu))$. Set

$$\begin{aligned} \delta_n &:= \frac{1}{n}\delta_0, \\ V_n &:= \{z \in \mathbb{C} \mid \operatorname{dist}(z, V) \leq \delta_n\}, \\ V_{2,n} &:= \{z \in \mathbb{C} \mid \operatorname{dist}(z, V) \leq 2\delta_n\}, \quad n \in \mathbb{N}, \end{aligned} \tag{2.15}$$

where we assume $\delta_0 > 0$ small enough to ensure that $V_{2,1} \cap S(\mu) = \emptyset$, and that both $V_{2,1}$ and V intersect exactly the same components (intervals) of $I(S(\mu)) \setminus S(\mu)$. Let

$$Y_n := \{y_{1n}, \dots, y_{m(n)n}\} \tag{2.16}$$

be the set of all poles of r_n in V_n. The numbers $m(n)$ are bounded by the number of components of $I(S(\mu)) \setminus S(\mu)$ that intersect V. We define

$$\begin{aligned} \tilde{r}_n(z) &:= r_n(f, A_n; z) - r_{1n}(z) \in \mathcal{R}_{nn}, \\ r_{1n}(z) &:= \sum_{j=1}^{m(n)} \frac{a_{jn}}{z - y_{jn}} \in \mathcal{R}_{m(n)-1,m(n)}, \\ h_n(z) &:= \prod_{j=1}^{m(n)} (z - y_{jn}). \end{aligned} \tag{2.17}$$

First, we derive asymptotic estimates for h_n and $h_n(f - r_n)$ as $n \to \infty$. Since the degree $m(n)$ of h_n is bounded, and since all zeros of h_n are contained in $V_n \subseteq V_1$, we have

$$\limsup_{n\to\infty} |h_n(z)|^{1/n} \le 1 \tag{2.18}$$

locally uniformly in $\mathbb{C}$ and

$$\lim_{n\to\infty} |h_n(z)|^{1/n} = 1 \tag{2.19}$$

locally uniformly in $\mathbb{C} \setminus (I(S(\mu)) \cap V)$. Then (2.18) together with (2.9) yields

$$\limsup_{n\to\infty} \frac{1}{2n} \log |h_n(f - r_n)(z)| \le p(\nu_0 - \alpha; z) + c_0 \tag{2.20}$$

locally uniformly in $\mathbb{C} \setminus I(S(\mu))$.

We need a version of the estimate (2.20) that holds true not only in $\mathbb{C} \setminus I(S(\mu))$, but also on $I(S(\mu)) \cap V$, and asymptotically on the points of the sets Y_n as $n \to \infty$. To get such an estimate, we introduce a sequence of compact sets $U_n, n \in \mathbb{N}$, with

$$V \subseteq U_n \subseteq V_{2,n}.$$

Let $Z(q_n)$ denote the set of zeros of the polynomial q_n. Then for every $n \in \mathbb{N}$ we can choose U_n such that

$$\begin{aligned} &\operatorname{dist}(\partial U_n, Z(q_n)) \ge \frac{1}{2}\delta_n = \frac{1}{2n}\delta_0, \\ &U_n \cap Z(q_n) = Y_n, \quad n \in \mathbb{N}. \end{aligned} \tag{2.21}$$

Indeed, in every component of $I(S(\mu)) \setminus S(\mu)$ there is at most one zero of q_n. If this zero is in V_n, then it belongs to Y_n, and we will choose ∂U_n close to $\partial V_{2,n}$ in the critical region. This is always possible since $\partial V_{2,n}$ has a distance $\frac{1}{n}\delta_0$ from V_n. If, however, the zero in question does not belong to V_n, then it does not belong to V, and therefore we choose ∂U_n close to V, which is again always possible.

Next, we show that it follows from (2.21) that for any sequence of points $z_n \in \partial U_n$, $n = 1, 2, \ldots$, we have

$$\lim_{n\to\infty} |h_n(z_n)|^{1/n} = 1,$$

(2.22)

$$\limsup_{n\to\infty, n\in N} \frac{1}{2n} \log|h_n(f - r_n)(z_n)| \leq \max_{x\in V} p(\nu_0 - \alpha; x) + c_0.$$

In the proof of (2.22) we can restrict ourselves to convergent sequences $z_n \in \partial U_n$, $n \in N_0 \subseteq \mathbb{N}$, with $z_n \to z_0$ as $n \to \infty, n \in N_0$. Clearly $z_0 \in V$.

For sufficiently large $n \in N_0$ every polynomial h_n can have only one zero close to z_n since all the other zeros belong to components of $I(S(\mu)) \setminus S(\mu)$ that do not contain z_0, and their distance from z_n is therefore bounded away from zero. But also the zero of h_n that asymptotically comes close to z_n has a distance from z_n not smaller than $\frac{1}{2n}\delta_0$ because of (2.21). These considerations show that there exists a constant $a > 0$ independent of n such that

$$|h_n(z_n)| \geq \frac{1}{2n}\delta_0 a^{m(n)-1}, \tag{2.23}$$

which proves the first limit of (2.22).

The proof of the second relation in (2.22) starts from the modified remainder formula

$$[h_n(f - r_n)](z) = \frac{\omega_n(z)}{(\frac{q_n}{h_n})(z)^2} \int \frac{[q_n \frac{q_n}{h_n}](x) d\mu(x)}{\omega_n(x)(x - z)}, \tag{2.24}$$

which follows from (1.23) and (1.24). By the same arguments as were applied in the derivation of (2.23) we can show that

$$\lim_{n\to\infty, n\in N_0} \frac{1}{n} \log \left| \frac{q_n}{h_n}(z_n) \right| = -p(\nu_0; z_0) \tag{2.25}$$

for any sequence $z_n \in \partial U_n, n \in N_0 \subseteq N$, with $z_n \to z_0$ as $n \to \infty, n \in N_0$ (also making use of the first limit in (2.7)). From (2.13) and the principle of descent it follows that

$$\limsup_{n\to\infty} \frac{1}{2n} \log |\omega_n(z)| \leq -p(\alpha; z) \tag{2.26}$$

locally uniformly on $\mathbb{C}$. Since the limit (2.19) holds true uniformly on $S(\mu)$, we can repeat the estimate (1.34) in the nth-root sense for the integral in (2.24). With the second limit in (2.7) this yields

$$(2.27) \qquad \lim_{n\to\infty, n\in N} \frac{1}{2n} \log \left| \int \frac{[q_n \frac{q_n}{h_n}](x) d\mu(x)}{\omega_n(x)(x-z)} \right| = c_0$$

uniformly for $z \in V_1 \supseteq V$. Formula (2.24) together with (2.25), (2.26), and (2.27) proves that

$$(2.28) \qquad \limsup_{n\to\infty, n\in N_0} \frac{1}{2n} \log |h_n(f - r_n)(z_n)| \le p(\nu_0 - \alpha, z_0) + c_0$$

for all sequences $z_n \in \partial U_n, n \in N_0 \subseteq N$, with $z_n \to z_0$ as $n \to \infty, n \in N_0$, where N is the same subsequence of $\mathbb{N}$ as was used in (2.7).

From (2.28) the second relation of (2.22) immediately follows.

With the two limits in (2.22) we are prepared to prove (2.12). First, we derive an estimate for the sequence $\{r_{1n}\}$ (cf. (2.17)) on ∂V_n. From the definition of h_n and $r_{1,n}$ in (2.17) it follows that

$$P_{1n}(z) := h_n(z) r_{1n}(z) \in \Pi_{m(n)-1}, \quad n \in \mathbb{N},$$

and furthermore that

$$(2.29) \qquad P_{1n}(y_{jn}) = h_n'(y_{jn}) a_{jn} = -[h_n(f - r_n)(z)]_{z=y_{jn}} \quad \text{for} \quad j = 1, \ldots, m(n),$$

where $\{y_{jn}\}$ are the $m(n)$ zeros of h_n. Since the last expression in (2.29) is analytic on U_n, it follows from the maximum principle and the second relation in (2.22) that

$$\limsup_{n\to\infty, n\in N} \max_{j=1,\ldots,m(n)} \frac{1}{2n} \log |P_{1n}(y_{jn})| \le \max_{x\in V} p(\nu_0 - \alpha; x) + c_0.$$

Using Lagrange's interpolation formula

$$P_{1n}(z) = \sum_{j=1}^{m(n)} \frac{P_{1n}(y_{jn}) h_n(z)}{h_n'(y_{jn})(z - y_{jn})},$$

we derive

$$|P_{1n}(z)| \le \sum_{j=1}^{m(n)} \left| \frac{P_{1n}(y_{jn}) h_n(z)}{h_n'(y_{jn})(z - y_{jn})} \right|, \quad z \in \mathbb{C}.$$

Since h_n has at most one zero in every component of $I(S(\mu)) \setminus S(\mu)$, we have

$$\limsup_{n\to\infty, n\in N} \frac{1}{2n} \log \sum_{j=1}^{m(n)} \frac{|h_n(z)|}{|h_n'(y_{jn})(z - y_{jn})|} \le 0$$

uniformly on every compact set in $\mathbb{C}$. Hence, we have shown that

$$\limsup_{n\to\infty, n\in N} \frac{1}{2n} \log |P_{1n}(z)| \leq \max_{x\in V} p(\nu_0 - \alpha; x) + c_0 \tag{2.30}$$

locally uniformly for $z \in \mathbb{C}$.

From the definitions in (2.17) we immediately derive the identity

$$f - \tilde{r}_n = \frac{1}{h_n}[h_n(f - r_n + r_{1n})] = \frac{1}{h_n}[h_n(f - r_n) + P_{1n}],$$

and with both limits in (2.22), it follows from (2.30) that

$$\limsup_{n\to\infty} \frac{1}{2n} \log |f(z_n) - \tilde{r}_n(z_n)| \leq \max_{x\in V} P(\nu_0 - \alpha; x) + c_0 \tag{2.31}$$

for any sequence of points $z_n \in \partial U_n, n \in \mathbb{N}$. Since $f - \tilde{r}_n$ is analytic on U_n, (2.31) proves (2.12) and the proof of Lemma 6.2.4 is complete. □

Proof of Theorem 6.2.2(a). Using, if necessary, the same transformation as in (1.35)–(1.39), we can assume without loss of generality that $\infty \notin V$, that is, V is a compact subset of $\mathbb{C}$.

Since we do not know whether V is regular with respect to Dirichlet problems in $\overline{\mathbb{C}} \setminus V$, nor that $\operatorname{cap}(S(\mu)) > 0$, we introduce auxiliary sets V_δ, which are defined for $\delta > 0$ by

$$\begin{aligned} \tilde{V}_\delta &:= \{z \in \mathbb{C} \mid \operatorname{dist}(z, V) \leq \delta\}, \\ V_\delta &:= \tilde{V}_\delta \cup \{z \in \mathbb{C} \mid \operatorname{dist}(z, \tilde{V}_\delta \cap I(S(\mu))) \leq \delta\}. \end{aligned} \tag{2.32}$$

The set V_δ is regular with respect to Dirichlet problems in $\overline{\mathbb{C}} \setminus V_\delta$, $\operatorname{cap}(V_\delta) > 0$, and $\partial V_\delta \cap I(S(\mu))$ consists of only finitely many points. From the outer continuity of the condenser capacity (see Appendix VIII) it follows that for every $\varepsilon > 0$ there exists a $\delta > 0$ and a compact set S_0 with $\mathbb{R} \supseteq S_0 \supseteq S(\mu)$, $\operatorname{cap}(S_0) > 0$, and $V_\delta \cap S_0 = \emptyset$ such that

$$e^{-1/\operatorname{cap}(V_\delta, S_0)} \leq e^{-1/\operatorname{cap}(V, S(\mu))} + \varepsilon. \tag{2.33}$$

Let ω_{V_δ, S_0} be the equilibrium distribution of the condenser (V_δ, S_0) on V_δ (see Appendix VIII). Since V_δ is regular, we have

$$g(\omega_{V_\delta, S_0}, \overline{\mathbb{C}} \setminus S_0; z) = \frac{1}{\operatorname{cap}(V_\delta, S_0)} \tag{2.34}$$

for all $z \in V_\delta$.

Since the set V_δ satisfies the assumptions of Lemma 6.2.3, we therefore know that there exists an admissible interpolation scheme $\mathbf{A}$ with asymptotic distribution $\alpha(\mathbf{A}) = \omega_{V_\delta, S_0}$. We apply Lemma 6.2.4 to the set V_δ and set $\alpha = \omega_{V_\delta, S_0}$. For any infinite subsequence $N \subseteq \mathbb{N}$ for which the two limits ν_0 and c_0 in (2.7) exist, we deduce from (2.12) and (2.11) in Lemma 6.2.4 together with (2.33) and (2.34) that

(2.35)
$$\begin{aligned}\limsup_{n\to\infty, n\in N} \|f - r_n^*\|_{\sup,V_\delta}^{1/2n} &\leq \sup_{x\in V_\delta} \exp(p(\nu_0 - \omega_{V_\delta,S_0}; x) + c_0) \\ &\leq \sup_{x\in V_\delta} \exp(-g(\omega_{V_\delta,S_0}, \Omega; x)) \\ &\leq \sup_{x\in V_\delta} \exp(-g(\omega_{V_\delta,S_0}, \mathbb{C}\setminus S_0; x)) \\ &= \exp\left(-\frac{1}{\operatorname{cap}(V_\delta, S_0)}\right) \leq e^{-1/\operatorname{cap}(V,S(\mu))} + \varepsilon.\end{aligned}$$

Since the right-hand side of (2.35) is independent of the subsequence N, and since $\varepsilon > 0$ is arbitrary, (2.35) proves (2.4). □

In the proof of Theorem 6.2.2(b) we shall use the following lemma that simplifies the structure of the set V.

Lemma 6.2.5. *Let* $V \subseteq \mathbb{C}\setminus S(\mu)$, $\operatorname{cap}(V) > 0$, *be a compact set as in* Theorem 6.2.2(b). *Then for every* $\varepsilon > 0$ *there exists a compact set* $V^* \subseteq V$ *such that* V^* *is symmetric with respect to* $\mathbb{R}$,

$$\operatorname{dist}(V^*\setminus\mathbb{R}, \mathbb{R}) > 0,$$

$\mathbb{C}\setminus V^*$ *is connected,* $\operatorname{Int}(V^*) = \emptyset$, V^* *is regular with respect to the Dirichlet problem in* $\overline{\mathbb{C}}\setminus V^*$, *and*

$$\operatorname{cap}(V^*, S(\mu)) \geq \operatorname{cap}(V, S(\mu)) - \varepsilon.$$

Remark. Applying twice the lemma we get the following: If $V_1, V_2 \subseteq \overline{\mathbb{C}}$ are disjoint compact sets then for every $\varepsilon > 0$ there are compact subsets $V_1^* \subseteq V_1$ and $V_2^* \subseteq V_2$ such that $\operatorname{Int}(V_1^* \cup V_2^*) = \emptyset$, $\mathbb{C}\setminus(V_1^*\cup V_2^*)$ is connected and regular with respect to the Dirichlet problem, and

$$\operatorname{cap}(V_1^*, V_2^*) \geq \operatorname{cap}(V_1, V_2) - \varepsilon.$$

Proof of Lemma 6.2.5. Since $\operatorname{cap}(V, S(\mu)) = \operatorname{cap}(\partial V, S(\mu))$, we can assume from the beginning that $\operatorname{Int}(V) = \emptyset$.

First we remove two small strips from V around $\mathbb{R}$: For $\eta > 0$ set

$$V^\eta = \{z \in V \mid \operatorname{dist}(z, \mathbb{R}) \geq \eta\} \cup (V \cap \mathbb{R}).$$

We claim that

$$\lim_{\eta\to 0} \operatorname{cap}(V^\eta, S(\mu)) = \operatorname{cap}(V, S(\mu)).$$

In fact, in the opposite case there is an $\alpha > 1/\text{cap}(V, S(\mu))$ such that with the equilibrium distributions $\omega^\eta := \omega_{V^\eta, S(\mu)}$ and $\omega_\eta := \omega_{S(\mu), V^\eta}$ of the condenser $(V, S(\mu))$ (see Appendix VIII) we have for some sequence $\{\eta_j\}_{j=1}^\infty$ tending to 0

$$p(\omega^\eta - \omega_\eta; z) - p(\omega^\eta - \omega_\eta; w) \geq \alpha, \quad \eta = \eta_1, \eta_2, \dots,$$

for quasi every $z \in V^\eta$ and quasi every $w \in S(\mu)$. Without loss of generality we can assume that $\{\omega_{\eta_j}\}$ and $\{\omega^{\eta_j}\}$ converge to some measures ω_* and ω^* in the weak* topology. Since each ω^η is supported on V and each ω_η is supported on $S(\mu)$, and $\cup_{j=1}^\infty V^{\eta_j} = V$, we get from the lower envelope theorem that

$$p(\omega^* - \omega_*; z) - p(\omega^* - \omega_*; w) \geq \alpha$$

for quasi every $z \in V$ and $w \in S(\mu)$, which implies $\alpha \leq 1/\text{cap}(V, S(\mu))$ (see (A.25) in Appendix VIII), and this contradicts the choice of α.

Thus, if we set $V_1 = V^\eta$ then for some small $\eta > 0$ we will have

$$\text{cap}(V_1, S(\mu)) \geq \text{cap}(V, S(\mu)) - \varepsilon/3.$$

Next we erase some vertical strips from V_1 to obtain a compact set with connected complement. Let $G_1, \dots, G_k, \dots$ be an enumeration of the components of $\mathbb{C} \setminus V_1$, and let $z_k \in G_k$, $k = 1, 2, \dots$. We are going to construct nested compact sets $V_1 = V_1^0 \supseteq V_1^1 \supseteq V_1^2 \supseteq \cdots$ in such a way that for each k

$$\text{cap}(V_1^k, S(\mu)) > \text{cap}(V_1, S(\mu)) - \varepsilon/3 \tag{2.36}$$

and V_1^k does not have any point in a strip

$$\{z \mid |\text{Re}(z) - \text{Re}(z_k)| < a_k\} \tag{2.37}$$

for some $a_k > 0$. In fact, if $V_1^{k-1}, k = 1, 2, \dots,$ has already been constructed with the property (2.36), then by the argument that we used in the selection of V_1 we can omit all points of a strip (2.37) from V_1^{k-1} so that the remaining points form a compact set V_1^k with the property (2.36).

Now let

$$V_2 = \bigcap_{k=1}^\infty V_1^k.$$

By the upper continuity of the Green capacity (see [He, Lemma 7.18]) it easily follows then that

$$\text{cap}(V_2, S(\mu)) = \lim_{k\to\infty} \text{cap}(V_1^k, S(\mu)) \geq \text{cap}(V_1, S(\mu)) - \varepsilon/3.$$

Clearly, $\mathbb{C} \setminus V_2$ is already connected.

Finally, we turn to the construction of V^* itself. By [An1,2] for each $m \in \mathbb{N}$ there is a regular compact set $V_2^m \subseteq V_2$ such that

$$\operatorname{cap}(V_2^m) > \operatorname{cap}(V_2) - \frac{1}{m}.$$

Replacing V_2^m by

$$\left(\bigcup_{k=1}^{m} V_2^k\right) \cup \left(\bigcup_{k=1}^{m} \overline{V_2^k}\right)$$

if necessary, we may assume that each V_2^m is symmetric with respect to $\mathbb{R}$ (note that V_2 has this property) and the sequence $\{V_2^m\}$ is increasing. We claim that for large m the set $V^* = V_2^m$ satisfies the requirements in the lemma. By the previous construction it has only remained to prove that

$$\operatorname{cap}(V_2^m) > \operatorname{cap}(V_2) - \varepsilon/3 \tag{2.38}$$

for all large m.

Let $g_m(z;w) = g_{\mathbb{C}\setminus V_2^m}(z,w)$ be the Green function of the domain $\mathbb{C}\setminus V_2^m$ with pole at $w \in \mathbb{C} \setminus V_2^m$. It is verified in the Appendix during the proofs of Theorems A.1 and A.2 that

$$g_m(z;\infty) \downarrow g_{\mathbb{C}\setminus V_2}(z;\infty), \quad z \in \mathbb{C} \setminus V_2,$$

locally uniformly in $z \in \mathbb{C} \setminus V_2$ as $m \to \infty$. Since $g_m(z,w) \geq g_{\mathbb{C}\setminus V_2}(z,w)$, we get from Harnack's inequality (Appendix III) and the preceding relation that

$$\lim_{m\to\infty} g_m(z,w) = g_{\mathbb{C}\setminus V_2}(z,w) \tag{2.39}$$

locally uniformly and monotonically for $z, w \in \mathbb{C}\setminus V_2$. If $\overline{\nu}$ is the probability measure on $S(\mu)$ that minimizes the Green energy

$$\int\int_{S(\mu)} g_{\mathbb{C}\setminus V_2}(z,w)d\overline{\nu}(z)d\overline{\nu}(w) = \frac{1}{\operatorname{cap}(V_2, S(\mu))},$$

then by (2.39) and the monotone convergence theorem we get

$$\frac{1}{\operatorname{cap}(V_2, S(\mu))} = \lim_{m\to\infty} \int\int_{S(\mu)} g_m(z,w)d\overline{\nu}(z)d\overline{\nu}(w)$$

$$\geq \limsup_{m\to\infty} \frac{1}{\operatorname{cap}(V_2^m, S(\mu))},$$

from which (2.38) follows for all large m. □

Proof of Theorem 6.2.2(b). Without loss of generality we can assume in the proof of part (b) that $\operatorname{cap}(S(\mu)) > 0$ since otherwise both the limit in (2.5) and $\mu \in \mathbf{Reg}$ hold for μ.

Proof of the necessity of the condition: $\mu \in$ Reg in (b). We assume the limit in (2.5) to hold true and prove that this implies $\mu \in \mathbf{Reg}$. Most of the technical difficulties here are due to the fact that V may contain points from $I(S(\mu)) \setminus S(\mu)$ or that V may not be regular with respect to Dirichlet problems in $\overline{\mathbb{C}} \setminus V$.

Let $z_0 \in \mathbb{C} \setminus (I(S(\mu)) \cup V)$ be fixed, and let $\varepsilon_m > 0$ be a sequence of numbers with $\varepsilon_m \to 0$ as $m \to \infty$. From the outer continuity of the condenser capacity it follows, as after (2.32), that there exists a sequence of numbers $\delta_m > 0$ with $\delta_m \to 0$ as $m \to \infty$ such that the sets $V_m := V_{\delta_m}$, $m \in \mathbb{N}$, defined in (2.32) satisfy

$$0 \le \frac{1}{\operatorname{cap}(V, S(\mu))} - \frac{1}{\operatorname{cap}(V_m, S(\mu))} \le \varepsilon_m \tag{2.40}$$

and

$$|g(\omega_{V,S(\mu)} - \omega_m, \Omega; z_0)| \le \varepsilon_m, \tag{2.41}$$

where $\omega_m = \omega_{V_m,S(\mu)}$, $m \in \mathbb{N}$, denotes the equilibrium distribution on V_m of the condenser $(V_m, S(\mu))$. As in the proof of part (a), it follows from Lemma 6.2.3 that for every $m \in \mathbb{N}$ there exists an admissible interpolation scheme $\mathbf{A}^m = (A_1^m, A_2^m, \dots)^t$ on ∂V_m with asymptotic distribution

$$\alpha(\mathbf{A}^m) = \omega_m, \quad m \in \mathbb{N}. \tag{2.42}$$

For every $m \in \mathbb{N}$, let $N_m \subseteq \mathbb{N}$ be an infinite subsequence such that the two limits

$$\frac{1}{n}\nu_{q_n^m} \xrightarrow{*} \nu_m, \quad \frac{1}{n}\log \|q_n^m\|_{L^2(\mu_n^m)} \to c_m \in \mathbb{R} \cup \{-\infty\} \tag{2.43}$$

exist as $n \to \infty$, $n \in N_m$, where q_n^m is the denominator polynomial of the interpolant $r_n(f, A_n^m; \cdot)$ and μ_n^m is the measure defined as in (2.6) by the interpolation set A_n^m. We know from (2.10) of Lemma 6.2.4 that for sufficiently large m

$$\lim_{n\to\infty, n\in N_m} \frac{1}{2n}\log|f(z_0) - r_n(f, A_n^m; z_0)| = p(\nu_m - \omega_m; z_0) + c_m. \tag{2.44}$$

From (2.11) of Lemma 6.2.4 it follows that with $\Omega = \mathbb{C} \setminus S(\mu)$

$$p(\nu_m - \omega_m; z) + c_m \le -g(\omega_m, \Omega; z) \tag{2.45}$$

for all $z \in \mathbb{C}$. Since we have assumed that the best rational approximants $r_n^* = r_n^*(f, V; \cdot)$ have a rate of convergence given by (2.5), it follows from (2.12) of Lemma 6.2.4 that

$$\sup_{x\in V} p(\nu_m - \omega_m; x) + c_m \ge \frac{-1}{\operatorname{cap}(V, S(\mu))}. \tag{2.46}$$

Indeed, if (2.46) were false, then by (2.12) applied to V_m instead of V there would exist rational approximants that were better than assumed in (2.5).

With the two equilibrium distributions $\omega_m = \omega_{V_m,S(\mu)}$ and $\omega^m = \omega_{S(\mu),V_m}$ of the condenser $(S(\mu), V_m)$ we have the representation

$$g(\omega_m, \Omega; z) = p(\omega_m - \omega^m; z) + \tau_m \tag{2.47}$$

for all $z \in \mathbb{C}$ with a constant τ_m (see (A.22)–(A.23) in the Appendix). Subtraction of the Green potential (2.47) from (2.45) and (2.46) gives the inequalities

$$0 \leq p(\omega^m - \nu_m; z) - c_m - \tau_m \quad \text{for } z \in \Omega, \tag{2.48}$$

and

(2.49)
$$\begin{aligned} p(\omega^m - \nu_m; z_m) - c_m - \tau_m &\leq \frac{1}{\operatorname{cap}(V, S(\mu))} - g(\omega_m, \Omega; z_m) \\ &= \frac{1}{\operatorname{cap}(V, S(\mu))} - \frac{1}{\operatorname{cap}(V_m, S(\mu))} \leq \varepsilon_m \end{aligned}$$

for some $z_m \in V$. The second to last inequality follows from the regularity of V_m, whereas the last one follows from (2.40).

Since both measures ν_m and ω^m have their support on $S(\mu)$, the potentials in (2.48) and (2.49) are harmonic in Ω. From Harnack's inequality (see Appendix III) we know that there exists a constant $C = C(z_0, V, \Omega)$ such that any function h positive and harmonic in Ω satisfies

$$h(z_0) \leq C|h(z_m)|$$

independently of $z_m \in V$. With this inequality we deduce from (2.48) and (2.49) that

$$0 \leq p(\omega^m - \nu_m; z_0) - c_m - \tau_m \leq C\varepsilon_m,$$

and with (2.43) and (2.47) further that

$$|p(\nu_m - \omega_m; z_0) + c_m + g(\omega_m, \Omega; z_0)| \leq C\varepsilon_m \tag{2.50}$$

for all large $m \in \mathbb{N}$ (for which $z_0 \notin V_m$). From (2.44) it then follows together with (2.41) and (2.50) that for every $m \in \mathbb{N}$ there exists an n_m such that

$$\left| \frac{1}{2n} \log |f(z_0) - r_n(f, A_n^m; z_0)| + g(\omega_{V,S(\mu)}, \Omega; z_0) \right| \leq (C+2)\varepsilon \tag{2.51}$$

for $n \geq n_m$.

Now let the interpolation scheme $\mathbf{A} = (A_1, A_2, \dots)^t$ be defined by the sets

$$A_n := A_n^m \quad \text{for } n_m < n \leq n_{m+1}.$$

Then it is easy to verify that, because of (2.42), $\mathbf{A}$ has the asymptotic distribution

(2.52) $$\alpha(\mathbf{A}) = \omega_{V,S(\mu)}$$

if $\{n_m\}$ tends fast enough to infinity. From this and (2.51) it follows that for the sequence of rational interpolants $r_n(f, A_n; \cdot)$, $n \in \mathbb{N}$, we have the convergence

(2.53) $$\lim_{n\to\infty} |f(z_0) - r_n(f, A_n; z_0)|^{1/2n} = \exp(-g(\omega_{V,S(\mu)}, \Omega; z_0)).$$

It is easy to see that all these can be done uniformly in z_0 provided $|z_0|$ is large enough, that is, one and the same sequence $\{A_n\}$ can be chosen for all $|z_0| > R$ provided the disk $\{z \mid |z| < R\}$ contains $S(\mu) \cup V$, in such a way that the limit in (2.53) is uniform in $|z_0| > R$. Since $\mathbf{A}$ is an admissible interpolation scheme, it follows from the equivalence of assertions (i) and (ii) in part (b) of Theorem 6.1.6 that (2.52) and (2.53) imply $\mu \in \mathbf{Reg}$. To be more precise, to make this conclusion we have to remark that the proof of (ii)$\Rightarrow$(i) in Theorem 6.1.6(b) works word for word if we only know (1.18) uniformly for $|z| > R$ and not locally uniformly in $\overline{\mathbb{C}} \setminus (I(S(\mu)) \cup L(\mathbf{A}))$ as was assumed in Theorem 6.1.6(ii).

Thus, we have shown that (2.5) implies $\mu \in \mathbf{Reg}$, which is the first half of the claimed equivalence.

Proof of the sufficiency of the condition: $\mu \in$ Reg in (b). It will be shown indirectly that $\mu \in \mathbf{Reg}$ implies (2.5). We assume $\mu \in \mathbf{Reg}$ and

(2.54) $$\liminf_{n\to\infty} \|f - r_n^*\|_{\sup,V}^{1/2n} =: R_0 < e^{-1/\mathrm{cap}(V,S(\mu))},$$

and show that this leads to a contradiction.

Replacing V by $V^* \subseteq V$ from Lemma 6.2.5 with some small $\varepsilon > 0$, without loss of generality we can assume besides (2.54) that $\mathrm{Int}(V) = \emptyset$, $\mathbb{C} \setminus V$ is connected and regular with respect to the Dirichlet problem, and

(2.55) $$\mathrm{dist}(V \setminus \mathbb{R}, V \cap \mathbb{R}) > 0.$$

Let $R_n^* = r_n^*(f, V; \cdot) = p_n/q_n$ with p_n and q_n relatively prime. For every $n \in \mathbb{N}$ we define

(2.56)
$$Q_n(z) := q_n(z)\overline{q_n(\bar{z})} =: q_n(z)\bar{q}_n(z), \quad k = k_n = \deg(Q_n) + 2 \le 2(n+1).$$

The normalization of q_n and Q_n, $n \in \mathbb{N}$, is still open. We have $Q_n(z) = |q_n(z)|^2$, $z \in \mathbb{R}$. Let the positive measures ν_n and the constants $c_n \in \mathbb{R}$ be defined by

(2.57) $$p(\nu_n; z) + c_n = \log \frac{1}{|Q_n(z)|}, \quad n \in \mathbb{N}.$$

We have $\|\nu_n\| = \deg(Q_n)$ and $\mathrm{supp}(\nu_n) \subseteq \mathbb{C} \setminus V$.

Let $\widehat{\nu_n}$ be the balayage measure of ν_n defined by sweeping out ν_n from $\overline{\mathbb{C}} \setminus V$ onto V. The polynomials Q_n and q_n will now be normalized by choosing the constants c_n so that

$$p(\nu_n - \widehat{\nu_n}; z) + c_n = 0 \tag{2.58}$$

for every $z \in V$. We have $\|\widehat{\nu_n}\| = \|\nu_n\|$, $\operatorname{supp}(\widehat{\nu_n}) \subseteq \partial V$, and

$$\int g_{\mathbb{C}\setminus V}(x; \infty) d\nu_n(x) = c_n \geq 0.$$

For every $n \in \mathbb{N}$ we are going to select k_n distinct points $x_{jn} \in V$, $j = 1, \ldots, k_n$, that are symmetric with respect to $\mathbb{R}$. As in (2.6), we define

$$\begin{aligned} \omega_n(z) &= \prod_{j=1}^{k_n} (z - x_{jn}), \\ d\mu_n(x) &= \frac{d\mu(x)}{\omega_n(x)}, \quad x \in S(\mu). \end{aligned} \tag{2.59}$$

For the time being we assume that μ_n is positive for every $n \in \mathbb{N}$. At the end of the proof we shall indicate the necessary changes if this is not the case.

Next we verify that

$$\limsup_{n\to\infty} \left(\left| \frac{Q_n(x)}{\omega_n(x)} \right| \frac{1}{\|q_n\|^2_{L^2(\mu_n)}} \right)^{1/2n} \leq 1 \tag{2.60}$$

for quasi every $x \in S(\mu)$. Since all $x_{jn} \in V$, the set $\{|\omega_n|^{-1/2n}|_{S(\mu)}\}_{n=1}^{\infty}$ is a compact subset of $C_+(S(\mu))$. Hence, for every $\varepsilon > 0$ there is an $l = l_\varepsilon$, such that for each $\omega_n^{-1/2n}$ we can find a polynomial $T_{l_n} \geq 0$ of degree $l_n \leq l$ for which

$$1 - \varepsilon \leq |\omega_n^{-1/2n}(x)|/T_{l_n}(x) \leq 1 + \varepsilon, \quad x \in S(\mu).$$

This implies for the polynomial $R_n(z) = q_n(z) T_{l_n}(z)^n$ of degree at most $n(1+l)$ that

$$\left| \frac{Q_n(x)}{\omega_n(z)} \right| \frac{1}{\|q_n\|^2_{L^2(\mu_n)}} \leq \frac{(1+\varepsilon)^{2n} |R_n(x)|^2}{(1-\varepsilon)^{2n} \|R_n\|^2_{L^2(\mu)}}, \quad x \in S(\mu),$$

where we also have used that $Q_n(x) = |q_n(x)|^2$ for $x \in S(\mu)$. By Theorem 3.2.1, (i)$\Rightarrow$(v), we can conclude from $\mu \in \mathbf{Reg}$ that

$$\limsup_{n\to\infty} \left(\frac{|R_n(x)|}{\|R_n\|_{L^2(\mu)}} \right)^{1/n(1+l)} \leq 1$$

for quasi every $x \in S(\mu)$. Since here l is independent of n, this and the preceding relation imply (2.60) with 1 replaced by $(1+\varepsilon)/(1-\varepsilon)$ on the right, and for $\varepsilon \to 0$ we finally get (2.60) for quasi every $x \in S(\mu)$.

Set $e_n = f - r_n^*$, $n \in \mathbb{N}$, and let γ be a connected integration path that separates $S(\mu)$ from V. Since $\mathbb{C} \setminus (S(\mu) \cup V)$ is connected, there is such a γ. Then we have with appropriate orientation of γ

(2.61)

$$\begin{aligned}\sum_{j=1}^{k_n} \frac{Q_n(x_{jn})}{\omega_n'(x_{jn})} e_n(x_{jn}) &= \frac{1}{2\pi i} \oint_\gamma \left[\frac{q_n\overline{q}_n}{\omega_n}(f - r_n^*)\right](z)dz \\ &= \frac{-1}{2\pi i} \oint_\gamma \frac{p_n\overline{q}_n}{\omega_n}(z)dz + \frac{1}{2\pi i} \oint_\gamma \frac{q_n\overline{q}_n}{\omega_n}(z) \int \frac{d\mu(x)}{x-z} dz \\ &= \int \frac{-1}{2\pi i} \oint_\gamma \frac{q_n\overline{q}_n}{\omega_n}(z) \frac{dz}{z-x} d\mu(x) = \int \frac{|q_n(x)|^2}{\omega_n(x)} d\mu(x) \\ &= \|q_n\|^2_{L^2(\mu_n)}.\end{aligned}$$

The first equality in (2.61) follows from the residuum theorem since all the x_{jn} are chosen to be distinct. The other equalities in (2.61) can easily be verified. Note that the integrand is analytic "inside" γ, and also that $1/\omega_n$ has singularities only on V and $\deg(\omega_n) \geq \deg(q_n\overline{q}_n) + 2$.

The identity (2.61) implies

$$(2.62) \qquad \|q_n\|^{1/n}_{L^2(\mu_n)} \leq k_n^{1/2n} \left(\max_{j=1,\dots,k_n} \left|\frac{Q_n(x_{jn})}{\omega_n'(x_{jn})}\right|\right)^{1/2n} \|f - r_n^*\|^{1/2n}_{\sup,V}.$$

Our next aim is to get rid of the large zeros in Q_n. To this end let R be a large positive number and

$$q_n^R(z) = q_n(z) \prod_{\substack{|w|>R \\ q_n(w)=0}} \frac{z-R}{z-w} \exp(g_{\mathbb{C}\setminus V}(w;\infty) - g_{\mathbb{C}\setminus V}(R;\infty)),$$

$$\begin{aligned}(2.63) \qquad Q_n^R(z) &= q_n^R(z)\overline{q_n^R(\overline{z})} \\ &= Q_n(z) \prod_{\substack{|w|>R \\ Q_n(z)}} \frac{z-R}{z-w} \exp(g_{\mathbb{C}\setminus V}(w;\infty) - g_{\mathbb{C}\setminus V}(R;\infty)),\end{aligned}$$

that is, replace each zero of Q_n and q_n of modulus greater than R by R. The normalizing constants $\exp(g_{\mathbb{C}\setminus V}(w;\infty) - g_{\mathbb{C}\setminus V}(R;\infty))$ have been chosen so that (2.58) is still valid for Q_n^R; that is, if we set

$$p(\nu_n^R; z) + c_n^R = \log \frac{1}{|Q_n^R(z)|},$$

then

$$p(\nu_n^R - \widehat{\nu_n^R}; z) + c_n^R = 0 \tag{2.64}$$

holds for every $z \in V$. In fact, by Appendix VII the constant c_n in (2.58) is given by $\int g_{\mathbf{C}\setminus V}(x;\infty)d\nu_n(x)$ and so c_n^R equals $\int g_{\mathbf{C}\setminus V}(x;\infty)d\nu_n^R(x)$, which, again by Appendix VII, yields (2.64).

Seeing that the ratio

$$\frac{z-R}{z-w}\exp(g_{\mathbf{C}\setminus V}(w;\infty) - g_{\mathbf{C}\setminus V}(R;\infty)), \quad |w| \geq R,$$

tends to 1 uniformly on compact subsets of $\mathbf{C}$ as $R \to \infty$, by (2.62) and (2.63) we can choose for every $\eta > 0$ an $R = R_\eta$ such that

(2.65)

$$\|q_n^R\|_{L^2(\mu_n)}^{1/n} \leq (1+\eta)k_n^{1/2n}\left(\max_{j=1,\dots,k_n}\left|\frac{Q_n^R(x_{jn})}{\omega_n'(x_{jn})}\right|\right)^{1/2n} \|f - r_n^*\|_{\sup,V}^{1/2n}.$$

In Lemma 6.2.6 it will be shown that the k_n points $x_{jn} \in V$, $j = 1,\dots,k_n$, can be chosen for every $n \in \mathbf{N}$ in such a way that

$$\limsup_{n\to\infty}\left(\sup_{j=1,\dots,k_n}\left|\frac{Q_n^R(x_{jn})}{\omega_n'(x_{jn})}\right|\right)^{1/2n} \leq 1 \tag{2.66}$$

and

$$\liminf_{n\to\infty}\left|\frac{Q_n^R(z)}{\omega_n(z)}\right|^{1/2n} \geq 1 \tag{2.67}$$

holds uniformly for $z \in V$.

Let $N \subseteq \mathbf{N}$ be an infinite subsequence such that the lower limit in (2.54) is assumed through N, that is,

$$\lim_{n\to\infty, n\in N}\frac{1}{2n}\log\|f - r_n^*\|_{\sup,V} = \log R_0 < \frac{-1}{\operatorname{cap}(V, S(\mu))}, \tag{2.68}$$

and furthermore that the limits

$$\frac{1}{2n}\nu_n^R \xrightarrow{*} \nu, \quad \frac{1}{2n}\widehat{\nu_n^R} \xrightarrow{*} \hat{\nu}, \quad \frac{1}{2n}\nu_{\omega_n} \xrightarrow{*} \nu^*, \quad \frac{1}{2n}c_n^R \to c \in \mathbf{R} \tag{2.69}$$

exist as $n \to \infty$, $n \in N$. Recall that here ν_{ω_n} denotes the measure that has mass 1 at every zero of ω_n. In (2.69) the second limit follows from the first one because the regularity of V with respect to the Dirichlet problem in $\mathbf{C}\setminus V$ implies that the balayage operator $\nu \to \hat{\nu}$ is continuous in the weak* topology (see [La, Chapter IV, Section 3/17]).

From (2.65) together with (2.66) and (2.68) it follows that for some subsequence $N_0 \subseteq N$

$$\begin{aligned}
&\limsup_{n\to\infty, n\in N_0} \|q_n^R\|_{L^2(\mu_n)}^{1/n} \\
&\leq \limsup_{n\to\infty, n\in N_0} (1+\eta)(k_n)^{1/2n} \left(\max_{j=1,\dots,k_n} \left| \frac{Q_n^R(x_{jn})}{\omega_n'(x_{jn})} \right| \right)^{1/2n} \|f - r_n^*\|_{\sup,V}^{1/2n} \\
&\leq (1+\eta)R_0.
\end{aligned}$$

The proof of (2.60) shows that

$$\limsup_{n\to\infty} \left(\left| \frac{Q_n^R(x)}{\omega_n(x)} \right| \frac{1}{\|q_n^R\|_{L^2(\mu_n)}^2} \right)^{1/2n} \leq 1$$

is also true for quasi every $x \in S(\mu)$ (just replace Q_n by Q_n^R everywhere in that proof), and hence from the preceding relation it follows that

$$\limsup_{n\to\infty, n\in N} \frac{1}{2n} \log \left| \frac{Q_n^R(z)}{\omega_n(z)} \right| \leq \log R_0 + \log(1+\eta)$$

for quasi every $z \in S(\mu)$. With the lower envelope theorem and three of the four limits in (2.69) this proves that

$$p(\nu^* - \nu; z) - c \leq \log R_0 + \log(1+\eta) \tag{2.70}$$

for quasi every $z \in S(\mu)$. Note that ω_n has all its zeros in V, and furthermore the measures $\{\nu_n^R\}$ have support in $\{z \mid |z| \leq R\}$, so the lower envelope theorem can be applied.

From the limits in (2.69) and the lower envelope theorem we also deduce

$$\begin{aligned}
&\liminf_{n\to\infty, n\in N_0} \frac{1}{2n} \log \left| \frac{Q_n^R(z)}{\omega_n(z)} \right| \\
&\leq \limsup_{n\to\infty, n\in N_0} \frac{1}{2n} \log |Q_n^R(z)| + \liminf_{n\to\infty, n\in N} \frac{1}{2n} \log \frac{1}{|\omega_n(z)|} \\
&= p(\nu^* - \nu; z) - c
\end{aligned}$$

for quasi every $z \in \mathbb{C}$. With (2.67) this proves that

$$p(\nu^* - \nu; z) - c \geq 0 \tag{2.71}$$

for quasi every $z \in V$.

In the function $p(\nu^* - \nu; \cdot)$ the two measures ν and ν^* are positive and $\|\nu^*\| = \|\nu\| \leq 1$. Hence, it follows from (2.70) and (2.71), and the definition of the condenser capacity (see Appendix VIII) that

$$-\log R_0 - \log(1+\eta) \leq \frac{1}{\operatorname{cap}(V, S(\mu))}. \tag{2.72}$$

Since here $\eta > 0$ is arbitrary, this inequality contradicts the assumption made in (2.54), and we have therefore proved that $\mu \in \mathbf{Reg}$ implies (2.5).

In the last proof we assumed that $d\mu_n$ was a positive measure for all n, that is, $\omega_n(z)$ does not change sign on $S(\mu)$. Now we indicate the necessary changes if this is not the case.

Let $I_1, \dots, I_l$ be those intervals of $\mathbb{R}\setminus S(\mu)$ that have common points with V and let us fix in each of them one point $Y_1, \dots, Y_l$. We set $k_n = \deg(Q_n) + l + 2$ and choose k_n points $x_{jn} \in V$, $j = 1, \dots, k_n$, as before (2.59). The proof of Lemma 6.2.6 easily yields a choice of x_{jn} with properties (2.66) and (2.67).

Now if $\omega_n(z)$, which now has degree $k_n = \deg(Q_n) + l + 2$, does not change sign on $S(\mu)$ then the last proof is unchanged. If, however, for some ns ω_n does change its sign on the intervals $I_{l_1^{(n)}}, \dots, I_{l_{t_n}^{(n)}}$, $t_n \le l$, then we set

$$d\mu_n(x) = \frac{d\mu(x)}{|\omega_n(x)|},$$

and consider the polynomials

$$s_n(x) = \prod_{j=1}^{t_n} (x - Y_{l_j^{(n)}})$$

of degree at most l. Clearly, $s_n(x)/\omega_n(x)$ does not change its sign on $S(\mu)$ and $\deg(\omega_n) \ge \deg(s_n q_n \overline{q}_n) + 2$; hence with

$$d_1 = \min_{1 \le j \le l} \operatorname{dist}(Y_j, S(\mu))$$

we get exactly as in (2.61)

$$\begin{aligned} d_1^l \|q_n\|^2_{L^2(\mu_n)} &\le \left| \int \frac{|q_n(x)|^2 s_n(x)}{\omega_n(x)} d\mu(x) \right| = \cdots \\ &= \left| \sum_{j=1}^{k_n} \frac{Q_n(x_{jn})}{\omega_n'(x_{jn})} s_n(x_{jn}) e_n(x_{jn}) \right|, \end{aligned}$$

which leads to the following analogue of (2.62):

$$\|q_n\|^{1/n}_{L^2(\mu_n)} \le M^{1/2n} \left(\max_{j=1,\dots,k_n} \left| \frac{Q_n(x_{jn})}{\omega_n'(x_{jn})} \right| \right)^{1/2n} \|f - r_n^*\|^{1/2n}_{\sup,V}$$

with

$$M = k_n^{1/2n} \left(\max_{1 \le j \le l, x \in V} |Y_j - x| + 1 \right)^l \cdot d_1^{-l}.$$

The rest of the proof is the same as before.

For simplicity, in Lemma 6.2.6 we retain the case $k_n = \deg(Q_n) + 2$.

□

The last part of the proof of Theorem 6.2.2 has made essential use of the following lemma, in which we write Q_n instead of Q_n^R.

Lemma 6.2.6. *Let $V \subseteq \mathbb{C} \setminus S(\mu)$ be a compact set as in* Theorem 6.2.2 *with the simplifications carried out in the sufficiency proof for* Theorem 6.2.2(b), *that is, we assume that* $\operatorname{cap}(V) > 0$, *$V$ is symmetric with respect to $\mathbb{R}$,* $\operatorname{Int}(V) = \emptyset$, *$\overline{\mathbb{C}} \setminus V$ is connected and regular with respect to the Dirichlet problem, and*

$$d := \operatorname{dist}(V \setminus \mathbb{R}, \mathbb{R}) > 0. \tag{2.73}$$

Let further Q_n, $n \in \mathbb{N}$, be polynomials of degree $\leq 2n$ with the properties: Q_n is symmetric on $\mathbb{R}$, Q_n has zeros only outside V, there is a common compact subset of $\mathbb{C}$ containing all the zeros of every Q_n, and Q_n is normalized according to (2.58) *and* (2.64), *that is, if we set*

$$p(\nu_n; z) + c_n = \log \frac{1}{|Q_n(z)|}, \tag{2.74}$$

then

$$p(\nu_n - \widehat{\nu_n}; z) + c_n = 0 \tag{2.75}$$

for all $z \in V$.

Let for every $n \in \mathbb{N}$ the set $X_n = \{x_1, \ldots, x_k\} = \{x_{1n}, \ldots, x_{k_n n}\} \subseteq V$ of $k = k_n =$ $\deg(Q_n) + 2$ *points be selected so that X_n is symmetric on $\mathbb{R}$ ($\overline{X}_n = X_n$) and it maximizes the expression*

$$V_{mn}(x_1, \ldots, x_k) = \frac{\prod_{1 \leq i < j \leq k} |x_i - x_j|}{\prod_{j=1}^{k} Q_n(x_j)} \tag{2.76}$$

among all symmetric subsets of V.

With the polynomials

$$\omega_n(z) = \prod_{j=1}^{k_n} (z - x_{jn})$$

we then have

$$\limsup_{n \to \infty} \left(\max_{j=1,\ldots,k_n} \left| \frac{Q_n(x_{jn})}{\omega_n'(x_{jn})} \right| \right)^{1/2n} \leq 1 \tag{2.77}$$

and

$$\liminf_{n \to \infty} \left| \frac{Q_n(z)}{\omega_n(z)} \right|^{1/2n} \geq 1 \tag{2.78}$$

uniformly on $z \in V$.

Remarks. (1) We call the reader's attention to the fact that the present Q_ns are not the same as in the preceding proof; they correspond to the Q_n^R that we obtained by modifying the large zeros of Q_n there.

(2) The points $x_1, \ldots, x_k$ can be called weighted, symmetric Fekete points, since the concept of their definition is a generalization of that of Fekete points. It is easy to verify that a selection is always possible and that all points $x_1, \ldots, x_k$ are necessarily distinct.

(3) The result (at least in a weak form) holds true for more general sets V than stated in Lemma 6.2.6. Since the proof is quite involved even in the simplified case given in Lemma 6.2.6, we will not go into details in connection with possible generalizations.

(4) The proof would be considerably simpler without the symmetry assumption on X_n.

Proof of Lemma 6.2.6. We shall first assume that both sets $V \setminus \mathbf{R}$ and $V \cap \mathbf{R}$ are of positive capacity. The other two cases are simpler and are discussed at the end of the proof.

Let the enumeration of the points $\{x_1, \ldots, x_n\}$ be such that the first m pairs of the points (x_{2l-1}, x_{2l}), $l = 1, 2, \ldots, m$, are contained in $V \setminus \mathbf{R}$ and they are pairwise conjugate, that is, $x_{2l-1} = \overline{x_{2l}}$, and the remaining $k - 2m$ points x_j, $j = 2m+1, \ldots, k$, are contained in $V \cap \mathbf{R}$. Then $0 \le m \le k/2$ and temporarily we assume $0 < m < k/2$ for all $n \in \mathbf{N}$, that is, there always exist points in $V \setminus \mathbf{R}$ and $V \cap \mathbf{R}$. The opposite case is simpler and are discussed at the end of the proof.

We consider the rational functions

$$\begin{aligned} R_j(z) &= \frac{\omega_n(z)}{Q_n(z)(z - x_j)}, \quad j = 1, \ldots, k; \\ R_{ij}(z, w) &= R_i(z) R_j(w) \frac{(z - w)}{(z - x_j)(w - x_i)}, \quad i, j = 1, \ldots, k. \end{aligned} \tag{2.79}$$

For special values of z and w we have

$$|R_{ij}(z, \overline{z})| = \left| \frac{\omega_n(z)}{Q_n(z)} \right|^2 \frac{|z - \overline{z}|}{(|z - x_j||z - x_i|)^2}$$

for $i = 2l-1, j = 2l, l = 1, \ldots, m,$ and $2m < i \le j \le k$;

$$\begin{aligned} |R_j(x_j)| &= \left| \frac{\omega_n'(x_j)}{Q_n(x_j)} \right| \quad \text{for } j = 1, \ldots, k; \\ |R_{2l-1,2l}(x_{2l-1}, x_{2l})| &= \left| \frac{\omega_n'(x_{2l-1})}{Q_n(x_{2l-1})} \right|^2 \frac{1}{|x_{2l-1} - x_{2l}|} \\ &= \left| \frac{\omega_n'(x_{2l})}{Q_n(x_{2l})} \right|^2 \frac{1}{|x_{2l-1} - x_{2l}|} \end{aligned} \tag{2.80}$$

for $l = 1, \ldots, m$.

In the first identity we have used that $Q_n = \overline{Q}_n$ and $\omega_n = \overline{\omega}_n$. From the definition of V_{mn} in (2.76) it follows that

(2.81)

$$V_{mn}(x_1, \dots, x_{2(l-1)}, z, \overline{z}, x_{2l+1}, \dots, x_k) = \text{const.} \times |R_{2l-1,2l}(z, \overline{z})| \quad \text{for } l = 1, \dots, m;$$

$$V_{mn}(x_1, \dots, x_{j-1}, z, x_{j+1}, \dots, x_k) = \text{const.} \times |R_j(z)| \quad \text{for } j = 2m+1, \dots, k;$$

$$V_{mn}(x_1, \dots, x_{i-1}, z, x_{i+1}, \dots, x_{j-1}, w, x_{j+1}, \dots, x_k) = \text{const.} \times |R_{ij}(z, w)| \quad \text{for } 2m < i \le j \le k.$$

From the extremality of V_{mn} it then follows that for all $j = 1, \dots, 2m$ and for all $z \in V \setminus \mathbf{R}$ we have

$$\left|\frac{\omega_n'(x_j)}{Q_n(x_j)}\right| \ge \left|\frac{\omega_n(z)}{Q_n(z)}\right| \frac{(|z-\overline{z}||x_j-\overline{x}_j|)^{1/2}}{|z-x_j||z-\overline{x}_j|} \ge \left|\frac{\omega_n(z)}{Q_n(z)}\right| \frac{d}{\operatorname{diam}(V)}, \tag{2.82}$$

where d is the constant from (2.73), and for all $j = 2m+1, \dots, k$ and all $z \in V \cap \mathbf{R}$ we have

$$\left|\frac{\omega_n'(x_j)}{Q_n(x_j)}\right| \ge \left|\frac{\omega_n(z)}{Q_n(z)}\right| \frac{1}{|z-x_j|} \ge \left|\frac{\omega_n(z)}{Q_n(z)}\right| \frac{1}{\operatorname{diam}(V)}. \tag{2.83}$$

The inequality (2.82) follows from the first and last identities in (2.80) and the first identity in (2.81) by varying the pair of points $(x_j, \overline{x}_j)$ in $V \setminus \mathbf{R}$. In the same way (2.83) follows from the second identity in (2.80) and (2.81) by varying the point x_j in $V \cap \mathbf{R}$.

If a pair of points $(x_j, \overline{x}_j)$ is moved from $V \setminus \mathbf{R}$ to $V \cap \mathbf{R}$, and thereby m is altered to $m-1$, or if this operation is done in the opposite direction, then the next two inequalities follow similarly to the way (2.82) and (2.83) have been derived from the extremality of V_{mn}. For all $j = 1, \dots, 2m$ and all $z, w \in V \cap \mathbf{R}$ we have

$$\begin{aligned} \left|\frac{\omega_n'(x_j)}{Q_n(x_j)}\right|^2 &\ge \left|\frac{\omega_n(z)\omega_n(w)}{Q_n(z)Q_n(w)}\right| \frac{|z-w||x_j-\overline{x}_j|}{|z-x_j|^2|w-x_j|^2} \\ &\ge \left|\frac{\omega_n(z)\omega_n(w)}{Q_n(z)Q_n(w)}\right| \frac{d|z-w|}{\operatorname{diam}(V)^4}, \end{aligned} \tag{2.84}$$

and for all $2m < i \le j \le k$ and all $z \in V \setminus \mathbf{R}$ we have

(2.85)

$$\left|\frac{\omega_n'(x_i)}{Q_n(x_i)}\right|\left|\frac{\omega_n'(x_j)}{Q_n(x_j)}\right| \ge \left|\frac{\omega_n(z)}{Q_n(z)}\right|^2 \frac{|z-\overline{z}||x_i-x_j|}{|z-x_i|^2|z-x_j|^2} \ge \left|\frac{\omega_n(z)}{Q_n(z)}\right|^2 \frac{d|x_i-x_j|}{\operatorname{diam}(V)^4}.$$

Since $\deg(Q_n) < \deg(\omega_n)$ and since all zeros of ω_n are simple, Lagrange's interpolation formula implies that

$$Q_n(z) = \sum_{j=1}^{k} \frac{Q_n(x_j)\omega_n(z)}{\omega_n'(x_j)(z - x_j)},$$

and we have the upper estimate

$$\left|\frac{Q_n(z)}{\omega_n(z)}\right| \le \sum_{j=1}^{k} \left|\frac{Q_n(x_j)}{\omega_n'(x_j)}\right| \frac{1}{|z - x_j|} \quad \text{for all } z \in \mathbb{C}. \tag{2.86}$$

Now set

$$\begin{aligned} M_{1n} &:= \max_{j=1,\dots,2m} \left|\frac{Q_n(x_j)}{\omega_n'(x_j)}\right|, \\ M_{2n} &:= \max_{j=2m+1,\dots,k} \left|\frac{Q_n(x_j)}{\omega_n'(x_j)}\right| \\ M_{0n} &:= \max(M_{1n}, M_{2n}), \quad n \in \mathbb{N}. \end{aligned} \tag{2.87}$$

Then it follows from (2.82), (2.83), and (2.86) that

$$\begin{aligned} \left|\frac{Q_n(z)}{\omega_n(z)}\right| &\le M_{0n} \sum_{j=1}^{k} \frac{1}{|z - x_j|} \quad \text{for } z \in \mathbb{C}, \\ M_{1n} \frac{d}{\operatorname{diam}(V)} &\le \left|\frac{Q_n(z)}{\omega_n(z)}\right| \quad \text{for } z \in V \setminus \mathbb{R}, \\ M_{2n} \frac{1}{\operatorname{diam}(V)} &\le \left|\frac{Q_n(z)}{\omega_n(z)}\right| \quad \text{for } z \in V \cap \mathbb{R}. \end{aligned} \tag{2.88}$$

Most of the remaining investigation consists of systematic exploitation of the information contained in these three inequalities.

We have

$$\lim_{n\to\infty} \left(\sum_{j=1}^{k} \frac{1}{|z - x_j|}\right)^{1/2n} = 1 \tag{2.89}$$

locally uniformly for $z \in \mathbb{C} \setminus V$.

Let $N \subseteq \mathbb{N}$ be an infinite subsequence. Then we can select a subsequence of N, which we continue to denote by N, so that the limits

$$\begin{aligned} &\frac{1}{2n}\nu_n \xrightarrow{\;*\;} \nu, \quad \frac{1}{2n}\widehat{\nu_n} \xrightarrow{\;*\;} \hat{\nu}, \quad \frac{1}{2n}\nu_{\omega_n} \xrightarrow{\;*\;} \nu^*, \\ &\frac{1}{2n}c_n \to c \in \mathbb{R}, \quad \frac{1}{2n}\log M_{jn} \to m_j \in \mathbb{R} \cup \{-\infty, \infty\}, \; j = 0, 1, 2, \end{aligned} \tag{2.90}$$

exist as $n \to \infty$, $n \in N$, where the measures ν_n and the constants c_n have been defined in (2.74) and (2.75). Here $\hat{\nu}$ is the balayage measure that one gets by sweeping out the measure ν from $\overline{\mathbb{C}} \setminus V$ onto V. In fact, the only thing in (2.90) that needs explanation is the compatibility of the first two limits, but this follows from the fact that the balayage operator $\nu \to \hat{\nu}$ is continuous in the weak* topology if the set V, onto which we take the balayage, is regular with respect to the Dirichlet problem in $\mathbb{C} \setminus V$ (see [La, Chapter IV, Section 3/17]).

Let us also mention that, since all Q_n have zeros in a fixed compact set, with the measures ν_n that are the counting measures on the zeros of Q_n, we have that the sequence $\{c_n\}$ with terms

$$c_n = \int g_{\mathbb{C}\setminus V}(x;\infty)d\nu_n(x)$$

(cf. (2.75) and Appendix VII) is bounded, hence c in (2.90) is a finite number. Furthermore, the regularity of V, which is equivalent to the continuity of $g_{\mathbb{C}\setminus V}$, implies that

$$c = \int g_{\mathbb{C}\setminus V}(x;\infty)d\nu(x). \tag{2.91}$$

At this moment the only thing we know about the limits m_j in (2.90) is that $m_0 = \max(m_1, m_2)$, which follows immediately from the definition of the numbers M_j in (2.87). We shall see that actually $m_0 = m_1 = m_2 = 0$ is true.

The first estimate in (2.88) and (2.89) and the last limits in (2.90) yield

$$\limsup_{n\to\infty, n\in N} \frac{1}{2n} \log \left| \frac{Q_n(z)}{\omega_n(z)} \right| \le m_0 \quad \text{for } z \in \mathbb{C} \setminus V,$$

which, together with the first and fourth limits in (2.90), implies via the lower envelope theorem

$$p(\nu^* - \nu; z) - c \le m_0 \tag{2.92}$$

for quasi every $z \in \mathbb{C} \setminus V$. By continuity in the fine topology of the left-hand side we get (2.92) for all $z \in \mathbb{C} \setminus V$ and since $\mathbb{C} \setminus V$ is a domain, also for all $z \in \partial V = V$. Making use of (2.91) and formula (A.13) from the Appendix we can conclude from (2.92)

$$p(\nu^* - \hat{\nu}; z) \le m_0 \quad \text{for all } z \in V. \tag{2.93}$$

Next, we draw conclusions from the second and third inequalities in

(2.88). We deduce by the lower envelope theorem from the normalization (2.74)–(2.75) of Q_n and the limits in (2.90) that

$$\begin{aligned}\liminf_{n\to\infty,n\in N}\frac{1}{2n}\log\frac{|Q_n(z)|}{|\omega_n(z)|} &\le \limsup_{n\to\infty,n\in N}\frac{1}{2n}\log|Q_n(z)| \\ &\quad + \liminf_{n\to\infty,n\in N}\frac{1}{2n}\log\frac{1}{|\omega_n(z)|} \\ &= \limsup_{n\to\infty,n\in N}\frac{1}{2n}p(-\widehat{\nu_n};z) \\ &\quad + p(\nu^*;z) = p(\nu^*-\hat{\nu};z)\end{aligned} \tag{2.94}$$

for quasi every $z \in \mathbb{C}$, and analoguously

$$\limsup_{n\to\infty,n\in N}\frac{1}{2n}\log\frac{|Q_n(z)|}{|\omega_n(z)|} \ge p(\nu^*-\hat{\nu};z) \tag{2.95}$$

for quasi every $z \in \mathbb{C}$. From the second and third inequalities in (2.88) and the last limit in (2.90) it follows that

$$m_1 \le \liminf_{n\to\infty,n\in N}\frac{1}{2n}\log\left|\frac{Q_n(z)}{\omega_n(z)}\right| \tag{2.96}$$

uniformly for $z \in V \setminus \mathbb{R}$ and

$$m_2 \le \liminf_{n\to\infty,n\in N}\frac{1}{2n}\log\left|\frac{Q_n(z)}{\omega_n(z)}\right| \tag{2.97}$$

uniformly for $z \in V \cap \mathbb{R}$. With (2.94) this implies that

$$m_1 \le p(\nu^*-\hat{\nu};z) \quad \text{for quasi every } z \in V \setminus \mathbb{R}, \tag{2.98}$$

$$m_2 \le p(\nu^*-\hat{\nu};z) \quad \text{for quasi every } z \in V \cap \mathbb{R}. \tag{2.99}$$

Making use of the inequalities (2.84) and (2.85) we will improve (2.97) and (2.98). For this aim we consider the cases $m_1 \ge m_2$ and $m_1 < m_2$ separately.

We assume first that $m_1 \ge m_2$. From (2.84) and the definition of M_{1n} in (2.87), it follows that

$$\left|\frac{Q_n(z)Q_n(w)}{\omega_n(z)\omega_n(w)}\right| \ge \frac{d|z-w|}{\operatorname{diam}(V)^4}M_{1n}^2 \tag{2.100}$$

for all $z, w \in V \cap \mathbb{R}$. Because of (2.93), (2.94), and the assumption that $\operatorname{cap}(V \cap \mathbb{R}) > 0$, there exist $w_0 \in V \cap \mathbb{R}$ and an infinite subsequence $N_0 \subseteq N$ such that

$$\lim_{n\to\infty,n\in N_0}\frac{1}{2n}\log\left|\frac{Q_n(w_0)}{\omega_n(w_0)}\right| = p(\nu^*-\hat{\nu};w_0) \le m_0 = m_1.$$

From this and the inequality in (2.100) with $w = w_0$, it follows that

$$m_1 \le \liminf_{n\to\infty, n\in N_0} \frac{1}{2n} \log \left| \frac{Q_n(z)}{\omega_n(z)} \right| \tag{2.101}$$

uniformly on compact subsets of $z \in (V \cap \mathbf{R}) \setminus \{w_0\}$. However, since it is possible to select not only one point w_0 but two different points $w_0, w_1 \in V \cap \mathbf{R}$ with the aforegoing properties, it follows that (2.101) holds true also uniformly for $z \in V \cap \mathbf{R}$. From (2.101) and (2.94) with N replaced by N_0 we deduce that

$$m_0 = m_1 \le p(\nu^* - \hat{\nu}; z) \quad \text{for quasi every } z \in V \cap \mathbf{R} \tag{2.102}$$

in the case $m_1 \ge m_2$.

We next assume that $m_1 < m_2$. As a first step we derive a lower asymptotic estimate for

$$\max_{2m < i \le j \le k} \left(\left| \frac{Q_n(x_i)}{\omega_n'(x_j)} \right| \left| \frac{Q_n(x_j)}{\omega_n'(x_j)} \right| |x_i - x_j| \right)$$

as $n \to \infty$, $n \in N$. Let $j_n \in \{2m+1, \ldots, k\}$ be the index such that

$$\left| \frac{Q_n(x_{j_n})}{\omega_n'(x_{j_n})} \right| = M_{2n},$$

and set

$$M_{3n} := \max_{i \ne j_n} \left(\left| \frac{Q_n(x_i)}{\omega_n'(x_i)} \right| |x_i - x_{j_n}| \right). \tag{2.103}$$

Since $\deg(Q_n) + 1 < \deg(\omega_n)$, it follows from Lagrange's interpolation formula that

$$Q_n(z)(z - x_{j_n}) = \sum_{i=1}^{k} \frac{Q_n(x_i)(x_i - x_{j_n})\omega_n(z)}{\omega_n'(x_i)(z - x_i)},$$

and as in (2.86) and (2.88) we deduce

$$\left| \frac{Q_n(z)}{\omega_n(z)} \right| |z - x_{j_n}| \le M_{3n} \sum_{j=1}^{k} \frac{1}{|z - x_j|} \quad \text{for all } z \in \mathbf{C}.$$

In the same way as (2.93) was proved we can get that this implies

$$p(\nu^* - \hat{\nu}; z) \le m_3 \quad \text{for all } z \in \partial V = V, \tag{2.104}$$

where

$$m_3 = \limsup_{n\to\infty, n\in N} \frac{1}{2n} \log M_{3n}. \tag{2.105}$$

Combining (2.99) with (2.104) and the assumption $\operatorname{cap}(V \cap \mathbb{R}) > 0$, it follows that $m_2 \le m_3$. From the definition of M_{3n}, M_{2n}, m_3, and m_2 it follows on the other hand that $m_3 \le m_2$, which proves that $m_3 = m_2$. In particular, since we have assumed $m_1 < m_2$, the i in (2.103) for which the maximum is attained must be one of $2m+1, 2m+2, \ldots, k$ for all large n.

Now let $N_0 \subseteq N$ be an infinite subsequence such that we have a proper limit in (2.105). With the definition of M_{3n} and the last limit in (2.90) it then follows that

$$\lim_{n\to\infty, n\in N_0} \max_{2m<i\le j\le k} \frac{1}{2n} \log \left| \frac{Q_n(x_i)Q_n(x_j)}{\omega_n'(x_i)\omega_n'(x_j)}(x_i - x_j) \right| = 2m_2.$$

From the inequality (2.85) we deduce

$$m_2 \le \liminf_{n\to\infty, n\in N_0} \frac{1}{2n} \log \left| \frac{Q_n(z)}{\omega_n(z)} \right| \tag{2.106}$$

uniformly for $z \in V \setminus \mathbb{R}$. With (2.94) (applied for N_0 instead of N) this proves that

$$m_0 = m_2 \le p(\nu^* - \hat{\nu}; z) \quad \text{for quasi every } z \in V \setminus \mathbb{R} \tag{2.107}$$

in the case $m_1 < m_2$.

Summarizing the results in (2.96), (2.97), (2.101), and (2.106), we can see that

$$m_0 \le \liminf_{n\to\infty, n\in N_0} \frac{1}{2n} \log \left| \frac{Q_n(z)}{\omega_n(z)} \right| \tag{2.108}$$

uniformly on V. Furthermore from (2.98), (2.99), (2.102), and (2.107) we can conclude

$$m_0 \le p(\nu^* - \hat{\nu}; z) \tag{2.109}$$

for quasi every $z \in V$.

Now we use again the regularity of V. It is equivalent to the fact that V is dense in itself in the fine topology (see Appendix VI), which implies that the points $z \in V$ with property (2.109) are also dense in V in this topology (cf. the equivalence of the statements (a) and (d) in Appendix II), and hence (2.109) extends to all $z \in V$. Combining this with (2.93) we get

$$p(\nu^*; z) = p(\hat{\nu}; z) + m_0 \tag{2.110}$$

on V.

From the formula

$$\log \frac{1}{|x - z|} = p(\widehat{\delta_z}; x) - g_{\mathbb{C}\setminus V}(z; \infty)$$

(see (A.13) in Appendix VII) valid for $x \in V$ and $z \notin V$ we obtain by integration with respect to a given σ, where σ is a measure of compact support on V with total mass $\|\sigma\| = 1$, the so-called Riesz formula

$$\begin{aligned} p(\sigma; z) &= \int p(\widehat{\delta}_z; x)\,d\sigma(x) - g_{\mathbf{C}\setminus V}(z;\infty) \\ &= \int p(\sigma; x)\,d\widehat{\delta}_z(x) - g_{\mathbf{C}\setminus V}(z;\infty). \end{aligned}$$

On applying this with $\sigma = \nu^*$ and $\sigma = \hat{\nu}$, respectively, we can derive from (2.110) that

$$p(\nu^*; z) = p(\hat{\nu}; z) + m_0$$

for all $z \in \mathbf{C} \setminus V$. For $z \to \infty$ this yields $m_0 = 0$, and then $\nu^* = \hat{\nu}$ follows because, as we have just seen, (2.110) is true for every $z \in \mathbf{C}$.

Recalling the definition of m_0 in (2.90) this verifies (2.77), since our starting subsequence $N \subseteq \mathbf{N}$ before (2.90) was arbitrary.

In a similar fashion, since $m_0 = 0$ in (2.108), we get (2.78) from (2.108) again because N was arbitrary before (2.90).

At the beginning of the proof it has been assumed that for all $n \in \mathbf{N}$ the parameter m satisfies $m \neq 0$ and $m \neq k/2$. It is rather easy to check that a violation of this assumption only simplifies the proof since then all points $x_1, \ldots, x_k$ are either contained in $V \cap \mathbf{R}$ or in $V \setminus \mathbf{R}$.

We have also assumed in the beginning that $\operatorname{cap}(V \setminus \mathbf{R}) > 0$ and $\operatorname{cap}(V \cap \mathbf{R}) > 0$. But if either of these capacities is zero then the corresponding set ($V \setminus \mathbf{R}$ or $V \cap \mathbf{R}$) is actually empty because of the regularity of V and the assumption (2.73). Hence the preceding remark applies and the proof with considerable simplifications runs through. □

6.3 Nondiagonal Padé Approximants to Markov Functions

The convergence and divergence of nondiagonal sequences of Padé approximants to a Markov function f will be investigated.

We assume that the Padé approximants

$$[m/n](z) = \frac{p_{mn}(\frac{1}{z})}{q_{mn}(\frac{1}{z})}, \quad m, n \in \mathbf{N}, \tag{3.1}$$

to a Markov function f are developed at infinity, that is, $p_{mn} \in \Pi_m$, $q_{mn} \in \Pi_n$, $q_{mn} \not\equiv 0$, and the pair (p_{mn}, q_{mn}) satisfies

$$f(z)q_{mn}\left(\frac{1}{z}\right) - p_{mn}\left(\frac{1}{z}\right) = O(z^{-(m+n+1)}) \quad \text{as} \quad |z| \to \infty. \tag{3.2}$$

It is easy to see that polynomials p_{mn}, q_{mn} exist for every pair of indexes $(m, n) \in \mathbb{N}^2$. These polynomials are not unique, but the Padé approximant (3.1) is uniquely determined by (3.2). (For a general discussion of Padé approximants see [Bak] and [BaGM].)

We will primarily consider the convergence and divergence of ray sequences

$$\{[m/n] \mid (m,n) \in N(\lambda)\}, \quad 0 < \lambda < \infty, \tag{3.3}$$

which are defined by the property that the sequence of indexes $N = N(\lambda) \subseteq \mathbb{N}^2$ satisfies

$$\lim_{m+n\to\infty,(m,n)\in N} \frac{n}{m} = \lim_{N} \frac{n}{m} = \lambda. \tag{3.4}$$

For $\lambda \neq 1$ the degrees of the numerators and denominators grow with a different rate, and the corresponding $[m/n(z)]$ approximants are called nondiagonal Padé approximants. The value of λ describes the angle in the Padé table along which the sequence tends to infinity. If $\lambda > 1$, the approximants are asymptotically contained in the upper triangle, and if $\lambda < 1$, they are eventually in the lower triangle of the Padé table.

Since the Padé approximants are developed at infinity, there is no advantage in considering the extended class of Markov functions defined in (1.4) or (2.1), and we assume in this section the more classical definition

$$f(z) = \int \frac{d\mu}{x - z} \tag{3.5}$$

of a Markov function, where, as usual, μ is a positive measure with compact support $S(\mu) \subseteq \mathbb{R}$. In order to facilitate the analysis and to have a more explicit determination of the convergence and divergence domains, we make the assumption that

$$S(\mu) \subseteq [0, 1]. \tag{3.6}$$

The main result of this section is the following theorem.

Theorem 6.3.1. *Let $0 < \lambda < \infty$ be fixed and set*

$$\begin{aligned}
&\theta := \frac{1-\lambda}{1+\lambda}, \\
&J = J(\theta) := [\theta^2, 1], \\
&h(z) = h(\theta; z) := g_{\overline{\mathbb{C}}\setminus J}(z; \infty) - |\theta| g_{\overline{\mathbb{C}}\setminus J}(z; 0), \\
&D_\lambda := \{z \in \mathbb{C} \mid h\left(\frac{1-\lambda}{1+\lambda}; z\right) < 0\}, \\
&C_\lambda := \overline{\mathbb{C}} \setminus (\overline{D}_\lambda \cup [0, 1]).
\end{aligned} \tag{3.7}$$

Then for h we have the representation

$$(3.8)\qquad \begin{aligned} h(\theta; z) = \frac{-1}{\pi}\int_{\theta^2}^{1} \log\frac{1}{|z-x|}\sqrt{\frac{x-\theta^2}{x^2(1-x)}}\,dx - |\theta|\log\frac{1}{|z|} \\ + \log\frac{4}{1-\theta^2} - |\theta|\log\frac{1+|\theta|}{1-|\theta|}, \quad z \in \mathbb{C}. \end{aligned}$$

Let further f be a Markov function of type (3.5) *satisfying* (3.6).

(a) *Ray sequences* (3.3) *converge locally uniformly in the domain* C_λ *to* f. *More precisely: We have*

$$(3.9)\qquad \limsup_{m+n\to\infty, n\in N(\lambda)} |f(z) - [m/n](z)|^{1/(m+n)} \leq \exp\left(-h\left(\frac{1-\lambda}{1+\lambda}; z\right)\right)$$

locally uniformly for $z \in C_\lambda$.

(b) *Let* μ_λ *be the restriction* $\mu|_J$. *If* $\operatorname{supp}(\mu_\lambda) = J$ *and* $\mu_\lambda \in \mathbf{Reg}$, *then*

(i) *we have equality and a proper limit in* (3.9) *locally uniformly for* $z \in C_\lambda$, *and*

(ii) *for* $\lambda \neq 1$, *we have*

(3.10)

$$\lim_{m+n\to\infty, (m,n)\in N(\lambda)} |[m/n](z)|^{1/(m+n)} = \exp\left(\operatorname{sign}(\lambda-1)h\left(\frac{|1-\lambda|}{1+\lambda}; z\right)\right)$$

locally uniformly for $z \in D_\lambda$.

Remarks. (1) It is easy to verify that C_λ and D_λ are domains. They are called convergence and divergence domains. We have $C_\lambda = C_{1/\lambda}$ and $D_\lambda = D_{1/\lambda}$ for $0 \leq \lambda \leq \infty$, and $D_\lambda \neq \emptyset$ if and only if $\lambda \neq 1$.

(2) The right-hand side of (3.10) is smaller than 1 for all $z \in D_\lambda$ and $\lambda > 1$, and it is larger than 1 for all $z \in D_\lambda$ and $\lambda < 1$. Hence, in the first case the approximants $[m/n]$ converge to 0 and in the second case to ∞ for all $z \in D_\lambda$. Since $f(z) \notin \{0, \infty\}$ for all $z \in D_\lambda \setminus [0,1]$, we get from (3.10) that the approximants do not converge to f in $D_\lambda \setminus [0,1]$. This shows that for the case $\mu_\lambda \in \mathbf{Reg}$, the domain C_λ is the largest possible convergence domain for ray sequences (3.3).

(3) The proof of Theorem 6.3.1 is rather different for sequences in the lower and the upper triangles of the Padé table. For the transfer of results from one to the other half of the table, we use the localization results from Chapter 5 in a decisive way.

Corollary 6.3.2. *Let f and C_λ be as in* Theorem 6.3.1 *and let* $1 \le \lambda^* < \infty$ *be fixed. Then any sequence*

$$\{[m_j/n_j] \mid m_j, n_j \in \mathbb{N}\}$$

of Padé approximants to f with

$$m_j + n_j \to \infty \quad as \quad j \to \infty,$$

$$\frac{1}{\lambda^*} \le \liminf_{j\to\infty} \frac{n_j}{m_j} \le \limsup_{j\to\infty} \frac{n_j}{m_j} \le \lambda^* \tag{3.11}$$

converges to f locally uniformly in C_{λ^*}.

The Padé approximants $[m/n]$ are rational interpolants with all interpolation points identical to infinity. Because of the special properties of the function f there are no interpolation defects. The next lemma is in one direction a generalization of Lemma 6.1.2 since now we no longer have $m = n$, but on the other hand it is more special since now each interpolation point equals infinity.

Lemma 6.3.3. *Let $m, n \in \mathbb{N}$ with $m \ge n$.*

(a) *If $(p_{mn}, q_{mn}) \in \Pi_m \times \Pi_n$ are a pair of polynomials satisfying* (3.2), *we assume q_{mn} to be normalized so that the inverse polynomial*

$$Q_{mn}(z) = z^n q_{mn}\left(\frac{1}{z}\right) \in \Pi_n \tag{3.12}$$

is monic. Then Q_{mn} is unique and it satisfies

$$\int x^j Q_{mn}(x) x^{m-n} d\mu(x) = 0 \quad for\ j = 0, \ldots, n-1. \tag{3.13}$$

We have $\deg(Q_n) = n$, *all zeros of Q_{mn} are simple, they are contained in $I(S(\mu))$, and every component of $I(S(\mu)) \setminus S(\mu)$ contains at most one zero.*

(b) *We have the remainder formula*

$$(f - [m/n])(z) = \frac{1}{Q_{mn}^2(z) z^{m-n}} \int \frac{Q_{mn}^2(x) x^{m-n}}{x - z} d\mu(x) \tag{3.14}$$

for all $z \in \overline{\mathbb{C}} \setminus [0, 1]$.

Proof. (a) Let $h \in \Pi_n$ be an arbitrary polynomial. It follows from multiplying (3.2) by $z^m h(z)$ that

$$(hfQ_{mn})(z) z^{m-n} - h(z) z^m p_{mn}\left(\frac{1}{z}\right) = O(z^{-1-n+\deg(h)})$$

as $|z| \to \infty$. This gives the identity

$$
\begin{aligned}
&(hfQ_{mn})(z)z^{m-n} \\
&\quad= \int \frac{(hQ_{mn})(z)z^{m-n} - (hQ_{mn})(x)x^{m-n}}{x-z} d\mu(x) \\
&\qquad+ \int \frac{(hQ_{mn})(x)x^{m-n}}{x-z} d\mu(x) \qquad\qquad (3.15) \\
&\quad= h(z)p_{mn}\left(\frac{1}{z}\right) z^m + O(z^{-1-n+\deg(h)}) \quad \text{as} \quad |z| \to \infty.
\end{aligned}
$$

The first integral in (3.15) is a polynomial of degree at most $m-1+\deg(h)$, and the second integral is analytic and equal to zero at infinity. This implies

(3.16)
$$
h(z)p_{mn}\left(\frac{1}{z}\right) z^m = \int \frac{(hQ_{mn})(z)z^{m-n} - (hQ_{mn})(x)x^{m-n}}{x-z} d\mu(x).
$$

With $h \equiv 1$ it follows from (3.15) and (3.16) that

$$
\begin{aligned}
&\int \frac{Q_{mn}(x)x^{m-n}}{x-z} d\mu(x) \\
&\quad= \int \left[-\frac{1}{z} - \cdots - \frac{x^{n-1}}{z^n} + \frac{x^n}{z^n(x-z)} \right] Q_{mn}(x)x^{m-n} d\mu(x) \\
&\quad= O(z^{-n-1}) \quad \text{as} \quad |z| \to \infty.
\end{aligned}
$$

A comparison of the coefficients of $1/z, \ldots, 1/z^n$ then proves the orthogonality relation (3.13).

Since Q_{mn} is the monic orthogonal polynomial with respect to the positive measure $x^{m-n}d\mu(x)$, we know from Theorem 2.1.1 that $\deg(Q_{mn}) = n$ and all zeros of Q_{mn} belong to $I(S(\mu))$. In Remark 1 to Lemma 1.1.3 it has been noted that for the case $S(\mu) \subseteq \mathbb{R}$ we know further that all zeros are simple and every component of $I(S(\mu)) \setminus S(\mu)$ contains at most one of these zeros.

(b) If in (3.15) and (3.16) we set $h = Q_{mn}$, then it follows from these two identities that

$$
\begin{aligned}
Q_{mn}^2(z)z^{m-n}(f - [m/n])(z) &= (Q_{mn}^2 f)(z) - Q_{mn}(z)p_{mn}\left(\frac{1}{z}\right) z^m \\
&= \int \frac{Q_{mn}^2(x)x^{m-n}}{x-z} d\mu(x),
\end{aligned}
$$

which proves the remainder formula (3.14). □

Analogously to Theorem 6.1.8 we can prove the next lemma. We have only to substitute the remainder formula (1.10) by (3.14).

Lemma 6.3.4. *For $m \geq n$ we define the positive measure μ_{mn} by*

$$d\mu_{mn}(x) = x^{m-n} d\mu(x). \tag{3.17}$$

For any sequence (3.3) with $\lambda \leq 1$ we have

$$\lim_{m+n\to\infty,(m,n)\in N(\lambda)} |(f(z) - [m/n](z))p_n(\mu_{mn}; z)^2 z^{m-n}|^{1/(m+n)} = 1 \tag{3.18}$$

locally uniformly for $z \in \overline{\mathbb{C}} \setminus (I(S(\mu)) \cup \{0\})$, where $p_n(\mu_{mn}; \cdot)$ is the orthonormal polynomial with respect to the weight measure μ_{mn}.

The proof of Theorem 6.3.1 and Corollary 6.3.2 will be broken down in four pieces.

Proof of representation (3.8). With $J = J(\theta) = [\theta^2, 1]$, $0 \leq \theta < 1$, we have

$$g_{\overline{\mathbb{C}}\setminus J}(z; \infty) = -\int_{\theta^2}^{1} \log \frac{1}{|z-x|} \frac{dx}{\pi\sqrt{(x-\theta^2)(1-x)}} + \log \frac{4}{1-\theta^2} \tag{3.19}$$

for all $z \in \mathbb{C}$, which follows from a linear transformation of the domain $\overline{\mathbb{C}} \setminus [-2, 2]$ onto $\overline{\mathbb{C}} \setminus J$. Let J^* denote the interval $[1, \theta^{-2}]$, which is the inversion of J on the unit circle. Then we have

$$g_{\overline{\mathbb{C}}\setminus J}(z; 0) = g_{\overline{\mathbb{C}}\setminus J^*}\left(\frac{1}{z}; \infty\right).$$

For $g_{\overline{\mathbb{C}}\setminus J^*}(\frac{1}{z}; \infty)$ a representation like (3.19) gives

$$g_{\overline{\mathbb{C}}\setminus J^*}\left(\frac{1}{z}; \infty\right) = -\int_{1}^{\theta^{-2}} \log \frac{1}{|\frac{1}{z} - x|} \frac{dx}{\pi\sqrt{(x-1)(\theta^{-2}-x)}} + \log \frac{4}{\theta^{-2}-1}.$$

If we transform the integral by the substitution $x \mapsto 1/x$, we arrive at

$$\begin{aligned} g_{\overline{\mathbb{C}}\setminus J}(z; 0) &= g_{\overline{\mathbb{C}}\setminus J^*}\left(\frac{1}{z}; \infty\right) \\ &= \log \frac{1}{|z|} - \int_{\theta^2}^{1} \log \frac{1}{|z-x|} \frac{(\frac{\theta}{x})dx}{\pi\sqrt{(x-\theta^2)(1-x)}} + g_{\overline{\mathbb{C}}\setminus J^*}(0; \infty). \end{aligned} \tag{3.20}$$

Let φ denote the conformal mapping of $\overline{\mathbb{C}} \setminus J^*$ onto $\{z \mid |z| > 1\}$ with $\varphi(\infty) = \infty$. Then it is easy to verify that

$$\varphi(z) = \frac{2\theta^2}{1-\theta^2}\left(z - \frac{1+\theta^2}{2\theta^2} + \sqrt{(z-1)(z-\theta^{-2})}\right).$$

Hence,

$$g_{\overline{\mathbb{C}}\setminus J^*}(0;\infty) = \log|\varphi(0)| = \log\frac{(1+\theta)^2}{1-\theta^2} = \log\frac{1+\theta}{1-\theta}.$$

With (3.19) and (3.20) this proves (3.8).

We shall also need that

$$h(\theta; z) \le 0 \quad \text{for } z \in [0,\theta^2]. \tag{3.21}$$

This follows from $h(\theta;\theta^2) = 0$ (in fact, $h(\theta;z) = 0$ for all $z \in [\theta^2, 1]$) and from $h'(\theta;z) \ge 0$ for $z \in [0,\theta^2]$. To prove the latter inequality we observe that for $z \in (0,\theta^2]$

$$\begin{aligned} h'(\theta;z) &= |\theta|\frac{1}{z} - \frac{1}{\pi}\int_{\theta^2}^{1}\frac{1}{x-z}\sqrt{\frac{x-\theta^2}{x^2(1-x)}}dx \\ &\ge \frac{1}{|\theta|} - \frac{1}{|\theta|}\int_{\theta^2}^{1}\frac{\theta}{x}\frac{1}{\pi\sqrt{(x-\theta^2)(1-x)}}dx, \end{aligned}$$

and here the last integral is 1 (with the substitution $x \to 1/x$ it transforms to

$$\int_1^{\theta^{-2}}\frac{1}{\pi\sqrt{(x-1)(\theta^{-2}-x)}}dx = 1).$$

This proves (3.21). □

Proof of Corollary 6.3.2. We first show that $\lambda_1 < \lambda_2 \le 1$ implies

$$C_{\lambda_1} \subseteq C_{\lambda_2}. \tag{3.22}$$

With $\theta_1 := (1-\lambda_1)\setminus(1+\lambda_1) > \theta_2 := (1-\lambda_2)\setminus(1+\lambda_2)$ it follows from (3.7) that $J_1 := J(\theta_1) \subseteq J_2 := J(\theta_2)$. From (3.21) we therefore deduce that

$$h(\theta_1;z) \le h(\theta_2;z) \tag{3.23}$$

for $z \in J_2$ and for z in a neighborhood of $z = 0$. Since the difference of the two functions $h(\theta_1;\cdot)$ and $h(\theta_2;\cdot)$ is harmonic in $\overline{\mathbb{C}}\setminus(J_2\cup\{0\})$, it follows from the minimum principle for harmonic functions that (3.23) holds for all $z \in \mathbb{C}$. This implies (3.22).

Let $N \subseteq \mathbb{N}$ be an infinite subsequence such that

$$\frac{n_j}{m_j} \to \lambda_0 \quad \text{as} \quad j\to\infty,\ j \in N.$$

Then by (3.11) we have

$$\frac{1}{\lambda^*} \le \min\left(\frac{1}{\lambda_0},\lambda_0\right) \le \max\left(\frac{1}{\lambda_0},\lambda_0\right) \le \lambda^*,$$

and therefore $C_{1/\lambda^*} = C_{\lambda^*} \subseteq C_{\lambda_0}$ because of (3.22). From Theorem 6.3.1(a) we therefore know that

$$\lim_{j\to\infty, j\in N} [m_j/n_j](z) = f(z)$$

locally uniformly for $z \in C_{\lambda^*}$. Since the domain C_{λ^*} is independent of the selection of the subsequence N, the corollary is generally proved. □

Proof of Theorem 6.3.1 for the case $m \geq n$. (a) From Lemma 6.3.3 we know that

$$Q_{mn}(z) = \frac{1}{\gamma_n(\mu_{mn})} p_n(\mu_{mn}; z),$$

where μ_{mn} was defined in (3.17). Despite many similarities, we cannot apply Theorem 3.3.2 since the assumptions made there with respect to the varying measure μ_{mn} are not satisfied here. Therefore we give an independent proof.

Let $N \subseteq N(\lambda)$ be an infinite subsequence such that the two limits

$$(3.24) \qquad \frac{1}{n}\nu_{p_n(\mu_{mn};\cdot)} \xrightarrow{*} \nu_0, \qquad \frac{1}{n}\log \gamma_n(\mu_{mn}) \to c_0 \in \mathbb{R} \cup \{\infty\}$$

exist as $m + n \to \infty$, $(m, n) \in N$. It follows from Theorem 2.1.1 that $\operatorname{supp}(\nu_0) \subseteq S(\mu) \subseteq [0, 1]$. By the assumption (3.4) we have

$$\frac{n}{m} \to \lambda, \quad \frac{2n}{m+n} \to \frac{2\lambda}{1+\lambda} = 1 - \theta, \quad \frac{m-n}{m+n} \to \theta, \ \lambda \leq 1, \text{ and } \theta \geq 0.$$

From these limits together with the limits assumed in (3.24) and the principle of descent (see Appendix III) it follows that

$$(3.25) \qquad \limsup_{\substack{m+n\to\infty \\ (m,n)\in N}} \frac{1}{m+n} \log |p_n(\mu_{mn}; z)^2 z^{m-n}| \leq (1 - \theta)(c_0 - p(\nu_0; z)) + \theta \log |z|$$

locally uniformly in $\mathbb{C}$ and

$$(3.26) \qquad \lim_{\substack{m+n\to\infty \\ (m,n)\in N}} \frac{1}{m+n} \log |p_n(\mu_{mn}; z)^2 z^{m-n}| = (1 - \theta)(c_0 - p(\nu_0; z)) + \theta \log |z|$$

locally uniformly for $z \in \mathbb{C} \setminus (I(S(\mu)) \cup \{0\})$. In (3.26) we have used the fact that the orthonormal polynomials $p_n(\mu_{mn}; \cdot)$ have all their zeros in $I(S(\mu))$ (Theorem 2.1.1).

We first prove that

$$(3.27) \qquad c_0 - p(\nu_0; z) + \frac{\theta}{1-\theta} \log |z| \geq 0$$

for all $z \in \mathrm{supp}(\nu_0)$. The proof will be carried out indirectly. There are strong similarities to the proof of (1.1.7) and (1.1.10) in Theorem 1.1.4. However, since the situation is now much simpler, it will be sufficient to consider a two point variation of zeros of $p_n(\mu_{mn};\cdot)$.

We assume that (3.27) is false. Then there exists a $c_1 > 0$ such that the set

$$U = \{z \in (0,1) \mid c_0 - p(\nu_0; z) + \frac{\theta}{1-\theta} \log|z| < -c_1\}$$

has common points with $\mathrm{supp}(\nu_0)$. Since U is open in $\mathbf{R}$, there exist a $c_2 > 0$ and two intervals $I_1, I_2 \subseteq U$ of the form $I_2 = [x_0, y_0]$, $I_1 = [x_0+c_2, y_0-c_2]$ with $\nu_0(\mathrm{Int}(I_1)) > 0$. This implies for $(m,n) \in N$ and $m+n$ sufficiently large the existence of two zeros $x_{mn} < y_{mn}$ of $p_n(\mu_{mn};\cdot)$ in I_1.

Consider the monic polynomials

$$\tilde{Q}_{mn}(z) = \frac{1}{\gamma_n(\mu_{mn})} p_n(\mu_{mn}; z) \frac{(z - x_{mn} + c_2)(z - y_{mn} - c_2)}{(z - x_{mn})(z - y_{mn})},$$

$(m,n) \in N$, which differ from Q_{mn} only by the variation of two of its zeros. It is easy to verify that there exists a $c_3 < 1$ such that

$$\text{(3.28)} \qquad \left| \frac{(z - x_{mn} + c_2)(z - y_{mn} - c_2)}{(z - x_{mn})(z - y_{mn})} \right| \le c_3 \quad \text{for all } z \in S(\mu) \setminus I_2,$$

$(m,n) \in N$, and $m+n$ sufficiently large. With the principle of descent it follows, as in (3.25), that

$$\limsup_{m+n\to\infty,(m,n)\in N} |\tilde{Q}_{mn}(z)^2 z^{m-n}|^{1/2n} \le e^{-c_1-c_0}$$

uniformly for $z \in I_2$ since $I_2 \subseteq U$. This implies that for $(m,n) \in N$ and $m+n$ sufficiently large we have

$$\text{(3.29)} \qquad |\tilde{Q}_{mn}(z)^2 z^{m-n}| < \frac{1-c_3}{\mu(I_2)\gamma_n(\mu_{mn})^2}$$

for all $z \in I_2$. With (3.28) and (3.29) we derive an upper estimate for $\tilde{Q}_{mn}$ in the $L^2(\mu_{mn})$ norm. We have

$$\begin{aligned} \|\tilde{Q}_{mn}\|^2_{L^2(\mu_{mn})} &= \int_{S(\mu)\setminus I_2} + \int_{I_2} \tilde{Q}_{mn}(x)^2 x^{m-n} d\mu(x) \\ &< \frac{c_3}{\gamma_n(\mu_{mn})^2} \int p_n(\mu_{mn};\cdot)^2 d\mu_{mn} + \mu(I_2) \frac{1-c_3}{\mu(I_2)\gamma_n(\mu_{mn})^2} \\ &= \frac{1}{\gamma_n(\mu_{mn})}, \end{aligned}$$

which contradicts the minimality of the monic orthogonal polynomial Q_{mn} in $L^2(\mu_{mn})$. Therefore (3.27) is proved.

From (3.27) we deduce that

$$(3.30) \qquad (1-\theta)(c_0 - p(\nu_0; z)) + \theta \log|z| \geq h(\theta, z)$$

for all $z \in \mathbb{C}$. Indeed, if we use the representation of $h(\theta;\cdot)$ in (3.8) then we get for the difference

$$\begin{aligned} d(z) &= (1-\theta)(c_0 - p(\nu_0; z)) + \theta \log|z| - h(\theta, z) \\ &= \text{constant} + p(\nu_1 - (1-\theta)\nu_0; z), \end{aligned}$$

where ν_1 is the measure supported on $[\theta^2, 1]$ and given by the density function

$$\frac{1}{\pi}\sqrt{\frac{x-\theta^2}{x^2(1-x)}}.$$

Since $h(\theta; z) \leq 0$ for all $z \in [0,1]$ (see (3.21)), it follows from (3.27) that $d(z) \geq 0$ for all $z \in \operatorname{supp}(\nu_0)$. In particular, the potential $p(\nu_0; z)$ of ν_0 on $\operatorname{supp}(\nu_0)$ is bounded from above by a constant plus $(1-\theta)^{-1}$ times the potential of $p(\nu_1; z)$, and since the latter is bounded we can conclude that the potential $p(\nu_0; z)$ is bounded from above. Hence ν_0 has finite logarithmic energy. By the principle of domination then the inequality $d(z) \geq 0$ extends from $\operatorname{supp}(\nu_0)$ to the whole plane (note that ν_1 has total mass $1-\theta$) and this proves (3.30).

From (3.26) and (3.30) we deduce that

$$(3.31) \qquad \liminf_{m+n\to\infty,(m,n)\in N(\lambda)} \frac{1}{m+n} \log|p_n(\mu_{mn}; z)^2 z^{m-n}| \geq h(\theta; z)$$

locally uniformly for $z \in \overline{\mathbb{C}}\setminus[0,1]$ and for the special subsequence N instead of $N(\lambda)$. Since the right-hand side of (3.31) is independent of N, (3.31) is generally proved. With Lemma 6.3.4 then (3.9) follows not only for $z \in C_\lambda$, but also locally uniformly for $z \in \mathbb{C} \setminus [0,1]$.

(b) We first assume that $m-n$ is even for all $(m,n) \in N(\lambda)$. Then we have

$$\frac{1}{\gamma_n(\mu_{mn})} = \|Q_{mn}\|_{L^2(\mu_{mn})} = \|Q_{mn}(z) z^{\frac{m-n}{2}}\|_{L^2(\mu)}.$$

Since we have assumed that $\mu_\lambda = \mu|_J \in \mathbf{Reg}$, and since $\operatorname{supp}(\mu_\lambda) = J$ is regular with respect to the Dirichlet problem in $\overline{\mathbb{C}} \setminus J$, it follows from Corollary 5.3.2 that

$$(3.32) \qquad \limsup_{m+n\to\infty,(m,n)\in N(\lambda)} \left(\frac{|Q_{mn}(z)^2 z^{m-n}|}{\|Q_{mn}(z) z^{\frac{m-n}{2}}\|_{L^2(\mu)}} \right)^{1/(m+n)} \leq 1$$

for all $z \in J$.

If $m-n$ is odd, then for every compact set $K \subseteq \mathbb{C} \setminus \{0\}$ there exists a constant $k = k(K)$, which is independent of m, n, such that

$$\left(\frac{|Q_{mn}(z)^2 z^{m-n}|}{\|Q_{mn}\|^2_{L^2(\mu_{mn})}} \right)^{1/2} \leq k \frac{|Q_{mn}(z) z^{\frac{m-n+1}{2}}|}{\|Q_{mn}(z) z^{\frac{m-n+1}{2}}\|_{L^2(\mu)}} \quad \text{for } z \in K.$$

This inequality shows that (3.32) also holds true if the sequence $N(\lambda)$ contains elements (m, n) with $m-n$ odd.

Our next aim is to show that (3.32) implies equality in (3.30) for all $z \in \mathbb{C}$. From the limits in (3.24), the lower envelope theorem, and (3.32) it follows that

$$(1-\theta)(c_0 - p(\nu_0; z)) + \theta \log |z| \leq 0$$

for quasi every $z \in J$. Since $p(\nu_0, \cdot)$ is continuous in the fine topology, the inequality must hold for all $z \in J$ (see Appendix II). From (3.30) and $h(\theta; z) = 0$ for all $z \in J$, it then follows that the function d defined after (3.30) vanishes on J. Since J is the support of ν_1 and ν_1 has finite logarithmic energy, we can conclude from the principle of domination that $d(z) \leq 0$ for all $z \in \mathbb{C}$. Thus $d \equiv 0$ (cf. (3.30)), and we have shown equality in (3.30) for all $z \in \mathbb{C}$.

By (3.26) we deduce from this that

$$\text{(3.33)} \qquad \lim_{m+n\to\infty, (m,n)\in N(\lambda)} \frac{1}{m+n} \log |p_n(\mu_{mn}; z)^2 z^{m-n}| = h(\theta; z)$$

locally uniformly for $z \in \mathbb{C} \setminus [0, 1]$ and for the special subsequence N (instead of $N(\lambda)$) for which the limits in (3.24) exist. Since the right-hand side of (3.33) is independent of N, (3.33) is generally proved. With Lemma 6.3.3 and (3.33) assertion (i) in Theorem 6.3.1(b) then follows for $z \in C_\lambda$ and assertion (ii) follows for $z \in D_\lambda \setminus [0, 1]$.

Assertion (ii) is proved for all $z \in D_\lambda$ if we can show that no point of $[0, \theta^2)$ is the limit point of zeros of $p_n(\mu_{mn}; \cdot)$. This can be done by a variational argument similar to that used in the proof of the inequality (3.27), but now only one zero has to be varied. Let $N \subseteq \mathbb{N}(\lambda)$ be an infinite subsequence such that for every $(m, n) \in N$ there exists a zero $x_{mn} \in [0, \theta^2)$ of $p_n(\mu_{mn}; \cdot)$ so that $x_{mn} \to x_0 \in [0, \theta^2)$ as $m+n \to \infty$, $(m, n) \in N$. Then, by a variation similar to that in the proof of the inequality (3.27), it can be shown that this assumption leads to a contradiction. In that proof replace I_2 by $[0, x_0 + c_2]$ with some small $c_2 > 0$ and $\tilde{Q}_{mn}$ by

$$\frac{1}{\gamma_n(\mu_{mn})} p_n(\mu_{mn}; z) \frac{z - x_{mn} - c_2/2}{z - x_{mn}}.$$

Then the analogue of (3.29) holds true because we have equality in (3.30) and $h(\theta; z)$ is strictly negative on $[0, x_0 + c_2]$ (cf. the proof of (3.21)) and from these we get a contradiction to the L^2 minimality of orthogonal polynomials exactly as we did after (3.29). □

In the last part of the proof of Theorem 6.3.1 we have to consider ray sequences in the upper triangle of the Padé table; that is, we have to consider the case $n \geq m$. In this part of the proof we simultaneously use both the function f and its reciprocal $1/f$ and the Padé approximants $[m/n]$ to f together with approximants to $1/f$.

Lemma 6.3.5. (a) *If f is a Markov function of type* (3.5), *then*

$$\frac{1}{f(z)} = a_0 - a_1 z - g(z) \tag{3.34}$$

and

$$g(z) = \int \frac{d\nu(x)}{x - z}, \tag{3.35}$$

where ν is a positive measure with support $S(\nu) \subseteq [0, 1]$, that is, g is again a Markov function of type (3.5). *The constants a_0 and a_1 in* (3.34) *are given by*

$$a_1 = \frac{1}{\mu(\mathbf{R})} \quad \text{and} \quad a_0 = \frac{1}{\mu(\mathbf{R})^2} \int x d\mu(x). \tag{3.36}$$

The total variation of ν is

$$\nu(\mathbf{R}) = \frac{1}{\mu(\mathbf{R})^2} \int \left(x - \frac{1}{\mu(\mathbf{R})} \int y d\mu(y) \right)^2 d\mu(x). \tag{3.37}$$

(b) *For $n \geq m \geq 1$ let $[m/n]_f$ and $[n-1/m-1]_g$ be Padé approximants to f and g, respectively. Then we have the identities*

$$\frac{1}{[m/n]_f(z)} = a_0 - a_1 z - [n-1/m-1]_g(z) \tag{3.38}$$

and

$$f(z) - [m/n]_f(z) = f(z)[m/n](z)(g(z) - [n-1/m-1]_g(z)). \tag{3.39}$$

Proof. (a) In the development

$$f(z) = \int \frac{d\mu(x)}{x - z} = \frac{f_1}{z} + \frac{f_2}{z^2} + \cdots$$

the coefficients f_j are the moments

$$f_j = -\int x^{j-1} d\mu(z), \quad j = 1, 2, \ldots .$$

We have

$$\frac{1}{f(z)} = \frac{1}{f_1}z - \frac{f_2}{f_1^2} - \frac{f_3 - f_2^2/f_1}{f_1^2}\frac{1}{z} + O(z^{-2}) \quad \text{as} \quad |z| \to \infty.$$

A comparison of coefficients proves (3.36).

Since $f(z) \neq 0$ for all $z \notin [0,1]$, the function g is analytic in $\overline{\mathbb{C}} \setminus [0,1]$, and $g(\infty) = 0$. If we apply Cauchy's integral formula for g on contours close to $[0,1]$ and approach $[0,1]$ with these contours then we can get a representation like (3.35) with a possibly complex-valued measure. That the so-obtained ν is actually positive follows from the facts that $g(\overline{z}) = \overline{g(z)}$ and, as it can be easily seen by a convexity argument, that $\operatorname{Im}(g(z)) > 0$ for all z with $\operatorname{Im}(z) < 0$ and $\operatorname{Im}(g(z)) < 0$ for all z with $\operatorname{Im}(z) > 0$.

From the development of $1/f$ it further follows that

$$\begin{aligned}\nu(\mathbf{R}) = -\frac{f_3 - f_2^2/f_1}{f_1^2} &= \frac{1}{\mu(\mathbf{R})^2}\int x^2 d\mu(x) - \frac{1}{\mu(\mathbf{R})^3}\left(\int x d\mu(x)\right)^2 \\ &= \frac{1}{\mu(\mathbf{R})^2}\int \left(x - \frac{1}{\mu(\mathbf{R})}\int y d\mu(y)\right)^2 d\mu(x),\end{aligned}$$

which proves (3.37).

(b) Let p_{mn} and q_{mn} be the numerator and denominator polynomials of $[m/n]_f$. From (3.2) it follows that

$$p_{mn}(0) = 0 \tag{3.40}$$

and

$$-q_{mn}(0)\mu(\mathbf{R}) = p'_{mn}(0). \tag{3.41}$$

Multiplying (3.2) by $1/f$ we get

$$q_{mn}\left(\frac{1}{z}\right) - p_{mn}\left(\frac{1}{z}\right)\frac{1}{f(z)} = O(z^{-m-n}) \quad \text{as} \quad |z| \to \infty.$$

With the identity (3.34) we derive

(3.42)

$$\begin{aligned} &q_{mn}\left(\frac{1}{z}\right) - p_{mn}\left(\frac{1}{z}\right)\frac{1}{f(z)} \\ &= q_{mn}\left(\frac{1}{z}\right) - p_{mn}\left(\frac{1}{z}\right)(a_0 - a_1 z) + p_{mn}\left(\frac{1}{z}\right)g(z) \\ &= \frac{1}{z}\tilde{q}_{mn}\left(\frac{1}{z}\right) - \frac{1}{z}\tilde{p}_{mn}\left(\frac{1}{z}\right)g(z) = \frac{1}{z}O(z^{-m-n+1}) \quad \text{as} \quad |z| \to \infty,\end{aligned}$$

where we have set

$$\tag{3.43} \begin{aligned} \tilde{q}_{mn}\left(\frac{1}{z}\right) &= z\left(q_{mn}\left(\frac{1}{z}\right) - p_{mn}\left(\frac{1}{z}\right)(a_0 - a_1 z)\right), \\ \tilde{p}_{mn}\left(\frac{1}{z}\right) &= -z p_{mn}\left(\frac{1}{z}\right). \end{aligned}$$

It follows from (3.40) that $\tilde{p}_{mn} \in \Pi_{m-1}$, and from (3.36), (3.40), (3.41), and $n \geq m$ that $\tilde{q}_{mn} \in \Pi_{n-1}$. This together with the last row of (3.42) proves that

$$[n-1/m-1]_g(z) = \frac{\tilde{q}_{mn}(\frac{1}{z})}{\tilde{p}_{mn}(\frac{1}{z})}.$$

The identity (3.38) then follows from (3.43).

From (3.34) and (3.38) together we derive that

$$\tag{3.44} \frac{1}{[m/n]_f(z)} - \frac{1}{f(z)} = g(z) - [n-1/m-1]_g(z).$$

Multiplying this identity by $f(z)[m/n]_f(z)$ we get (3.39). □

Proof of Theorem 6.3.1 for the case of $n \geq m$. (a) If $n \geq m$ then the $[m/n]$ Padé approximant is contained in the upper triangle of the Padé table and correspondingly we have $\lambda \geq 1$ in (3.3). Simultaneously with the sequence of indexes $N(\lambda)$ we consider the sequence $N_1 = N_1(\lambda_1) := \{(n-1, m-1) \in \mathbb{N}^2 \mid (m,n) \in N(\lambda)\}$, and together with f we consider the function g defined in (3.34). The λ value associated with N is given by

$$\lambda_1 = \frac{1}{\lambda} \leq 1,$$

and correspondingly we have a θ value

$$\theta_1 = \frac{1-\lambda_1}{1+\lambda_1} = \frac{\lambda - 1}{\lambda + 1} = -\theta.$$

Hence, $|\theta_1| = |\theta|$, $h(\theta_1, z)$, $C_{1/\lambda} = C_\lambda$, and $D_{\lambda_1} = D_\lambda$.

The sequence

$$\tag{3.45} \{[n-1/m-1]_g \mid (n-1, m-1) \in N_1\}$$

belongs to the lower triangle of the Padé table, $\lambda_1 \leq 1$, and for such sequences (3.9) has already been proved. Therefore we know that the sequence converges to g locally uniformly in C_λ, and since $f(z) \neq 0$ for all $z \in C_\lambda$, we deduce from (3.44) that

$$\tfrac{1}{2}|f(z)| \leq |[m/n]_f(z)| \leq 2|f(z)|$$

for z on any compact set in C_λ and $(m,n) \in N(\lambda)$ and $m+n$ sufficiently large. With (3.39) we then have

(3.46)
$$\tfrac{1}{2}|f(z)|^2|g(z) - [n-1/m-1]_g(z)| \le |f(z) - [m/n]_f(z)| \le 2|f(z)|^2|g(z) - [n-1/m-1]_g(z)|.$$

Since (3.9) has already been proved for the sequence (3.45), (3.46) proves it for the sequence $\{[m/n]_f \mid (m,n) \in N(\lambda)\}$ with $n \ge m$.

(b) In order to carry out the same transfer of results from the lower to the upper triangle of the Padé table, we have to know whether the assumption $\mu|_J \in \mathbf{Reg}$ also carries over to the measure ν defined in (3.35). In order to verify a positive answer to this question we use the localization results from Chapter 5.

We consider the two diagonal sequences $\{[n/n]_f \mid n \in \mathbb{N}\}$ and $\{[n-1/n-1]_g \mid n \in \mathbb{N}\}$ of Padé approximants to f and g, respectively. From the identity (3.39) it follows that

$$\lim_{n\to\infty} \left(\frac{|f(z) - [n/n]_f(z)|}{|g(z) - [n-1/n-1]_g(z)|} \right)^{1/n} = \lim_{n\to\infty} |f(z)[n/n]_f(z)|^{1/n} = 1$$

locally uniformly for $z \in \overline{\mathbb{C}} \setminus [0,1]$ (recall that for the diagonal sequence $\{[n/n]_f\}$ the corresponding C_λ is the whole complement $\overline{\mathbb{C}} \setminus [0,1]$). With Lemma 6.3.3 this implies that

$$\lim_{n\to\infty} \left(\frac{|p_n(\mu;z)|}{|p_{n-1}(\nu;z)|} \right)^{1/n} = 1 \tag{3.47}$$

locally uniformly for $z \in \overline{\mathbb{C}} \setminus [0,1]$.

Since J is an interval, condition (5.3.2) is satisfied if we set K to be the disk with diameter J, and so Theorem 5.3.1 is applicable. In this theorem the connection between the asymptotic behavior of orthonormal polynomials $p_n(\mu;z)$ near J and the regularity of $\mu|_J$ has been investigated. From the assumption that $\mu|_J \in \mathbf{Reg}$ it follows by assertions (i) and (ii) of Theorem 5.3.1 that

$$\limsup_{n\to\infty} |p_n(\mu;z)|^{1/n} \le e^{g_{\overline{\mathbb{C}}\setminus J}(z;\infty)}$$

locally uniformly for $z \in \mathbb{C}$. From (3.47) then we can deduce

$$\limsup_{n\to\infty} |p_n(\nu;z)|^{1/n} \le e^{g_{\overline{\mathbb{C}}\setminus J}(z;\infty)} \tag{3.48}$$

locally uniformly for $z \in \mathbb{C} \setminus [0,1]$. The maximum principle easily yields then (3.48) for all $z \in \mathbb{C}$. Applying again Theorem 5.3.1, it follows from (3.48) that $\nu_\lambda = \nu|_J \in \mathbf{Reg}$. Therefore we know that the assumptions of part (b) are satisfied for the function g and the assertions (i) and (ii) are already proved for the sequence (3.45). From assertion (i) for the sequence (3.45) together with (3.46) and $f(z) \neq 0$ for all $z \in C_\lambda$ we can derive assertion (i) also for the sequence $\{[m/n]_f \mid (m,n) \in N(\lambda)\}$.

In the same way we deduce for $\lambda > 1$ assertion (ii). Since $a_0 - za_1$ is bounded in D_λ, we have by (3.10) and $\lambda_1 = 1/\lambda < 1$ that

$$\text{(3.49)} \qquad \lim_{\substack{m+n\to\infty \\ (n-1,m-1)\in N_1(\lambda_1)}} |[n-1/m-1]_g(z)|^{1/(m+n-2)}$$

$$= \lim_{\substack{m+n\to\infty \\ (n-1,m-1)\in N_1(\lambda_1)}} |[n-1/m-1]_g(z) - a_0 + a_1 z|^{1/(m+n)}$$

$$= \exp\left(-h\left(\frac{|1-\lambda_1|}{1+\lambda_1}; z\right)\right)$$

locally uniformly for $z \in D_\lambda$. With (3.38), $1 = \operatorname{sign}(\lambda_1)$, and $|1-\lambda_1|/(1+\lambda_1) = |1-\lambda|/(1+\lambda)$, the limit (3.10) then follows from (3.49) for the sequence $\{[m/n]_f \mid (m,n) \in N(\lambda)\}$. □

6.4 Weighted Polynomials in $L^p(\mu)$

Let $\Sigma \subseteq \mathbb{C}$ be a closed set, and let $w(z) = \exp(-Q(z))$, $Q\colon \Sigma \to (-\infty, \infty]$, be a continuous and not identically vanishing weight function on Σ with the property that $|z|w(z) \to 0$ as $z \to \infty$ and $z \in \Sigma$ for the case Σ is unbounded. In several recent investigations weighted polynomials of the form $w^n(z)P_n(z)$, $\deg(P_n) \leq n$, have played a significant role, and one of the most basic questions concerning them is where the norms of such polynomials "live."

To make this more precise we need the concept of the equilibrium distribution associated with w (for the subsequent discussion see also Appendix IX). It can be shown (see Appendix IX) that there is a unique probability measure ω_w with support in Σ, called the extremal or equilibrium measure associated with w, minimizing the weighted energy

$$I_w(\omega) = \int\int \left(\log\frac{1}{|z-t|} + Q(z) + Q(t)\right) d\omega(z)d\omega(t)$$

among all probability measures with support in Σ. Let $S_w = \operatorname{supp}(\omega_w) \subseteq \Sigma$ be the support of this extremal measure. It is known (see Appendix IX) that S_w is compact and Q is bounded on S_w.

To avoid technical difficulties, let us assume that Σ is of positive capacity at each of its point, that is, for every open set U the intersection $U \cap \Sigma$ is either empty or it is of positive capacity, and that Σ is regular with respect to the Dirichlet problem in $\overline{\mathbb{C}} \setminus \Sigma$. This is the case in every relevant application. Under these conditions the potential of ω_w is continuous and we have

$$(4.1)\qquad \begin{aligned} &p(\omega_w; z) \geq -Q(z) + F_w, \quad z \in \Sigma, \\ &\text{with equality for } z \in \mathcal{S}_w, \end{aligned}$$

where the constant F_w is determined by

$$F_w = I_w(\omega_w) - \int Q d\omega_w.$$

If $\deg(P_n) \leq n$ and

$$(4.2)\qquad w^n(z)|P_n(z)| \leq M \quad \text{for } z \in \mathcal{S}_w,$$

then (see (A.31) in Appendix IX)

$$(4.3)\qquad |P_n(z)| \leq M \exp(n(-p(\omega_w; z) + F_w)), \quad z \in \mathbb{C}.$$

Inequality (4.3) is the analogue of the Bernstein–Walsh lemma and (4.1)–(4.3) tell us that the supremum norm of every weighted polynomial $w^n P_n$ "lives" on $\mathcal{S}_w$, that is,

$$(4.4)\qquad \|w^n P_n\|_\Sigma = \|w^n P_n\|_{\mathcal{S}_w}.$$

It turns out that $\mathcal{S}_w$ is the smallest compact set with this property (see [SaTo2, Theorem II.2.2]).

Now what about the $L^p(\mu)$ norm if μ is a given measure with compact support in Σ? Of course, if μ is supported outside $\mathcal{S}_w$ then the $L^p(\mu)$ norm of weighted polynomials cannot live on $\mathcal{S}_w$, but if we assume that the restriction of μ to $\mathcal{S}_w$ is regular, then it does so (at least in the nth-root sense).

Theorem 6.4.1. *With the above notations and assumptions let us also suppose that the support of $\mu|_{\mathcal{S}_w}$ is $\mathcal{S}_w$, $\mu|_{\mathcal{S}_w} \in$ **Reg**, and that* $\mathrm{Int}(\mathrm{Pc}(\mathcal{S}_w)) = \emptyset$, *or at least that the support $\mathcal{S}_w$ is on the outer boundary of $\mathcal{S}_w$, that is, on the boundary of the unbounded component of $\mathbb{C} \setminus \mathcal{S}_w$. Then the $L^p(\mu)$ norms of weighted polynomials $w^n P_n$ essentially live on $\mathcal{S}_w$ in the sense that*

$$(4.5)\qquad \lim_{n\to\infty} \left(\sup_{\substack{\deg P_n \leq n \\ P_n \not\equiv 0}} \frac{\|w^n P_n\|_{L^p(\mu)}}{\|w^n P_n\|_{L^p(\mu|_{\mathcal{S}_w})}} \right)^{1/n} = 1.$$

Furthermore, if S is any compact subset of Σ and

$$(4.6)\qquad \lim_{n\to\infty} \left(\sup_{\substack{\deg P_n \leq n \\ P_n \not\equiv 0}} \frac{\|w^n P_n\|_{L^p(\mu)}}{\|w^n P_n\|_{L^p(\mu|_S)}} \right)^{1/n} = 1,$$

then $\mathcal{S}_w \subseteq S$.

We shall show after the proof that the assumptions of the theorem cannot be essentially weakened.

Remarks. (1) In many cases the continuity of w can be relaxed to upper semi-continuity but not always; for example, if w has a large jump at a mass point of μ outside $\mathcal{S}_w$ then (4.5) may be violated. The same remark applies if we want to relax the condition that Σ be of positive capacity at each of its point.

(2) Of course, the regularity of $\mu|_{\mathcal{S}_w}$ is not necessary for the conclusion (4.5); for example, every μ with $\text{supp}(\mu) \subseteq \mathcal{S}_w$ satisfies (4.5).

(3) After the proof of Theorem 6.4.1 we shall show by examples that the assumptions of the theorem cannot be essentially relaxed.

(4) The theorem holds for locally finite measures μ even if the total mass of μ is infinite, provided the quantity $\mu(\Delta_R(0))$ does not grow faster than a power of R as $R \to \infty$, and $|z|^\sigma w(z) \to 0$ as $z \to \infty$ for some $\sigma > 1$. The proof of this fact easily follows from (4.3) and the following proof.

Proof of Theorem 6.4.1. If $\mu|_{\mathcal{S}_w}$ is regular then (see Theorem 3.4.4, (i)$\Rightarrow$(vi), and Remark 3.4.5(1) made after it)

$$\lim_{n\to\infty} \left(\sup_{\substack{\deg P_n \le n \\ P_n \not\equiv 0}} \frac{\|w^n P_n\|_{\mathcal{S}_w}}{\|w^n P_n\|_{L^p(\mu|_{\mathcal{S}_w})}} \right)^{1/n} = 1.$$

Therefore (4.5) immediately follows from (4.4).

Now let S be a compact set not containing $\mathcal{S}_w$. Then S is disjoint from some $D_\delta(z_0)$ neighborhood of a point $z_0 \in \mathcal{S}_w$. Let $\Sigma_1 = \Sigma \setminus D_\delta(z_0)$ and $w_1 = w|_{\Sigma_1}$. We cannot have

$$p(\omega_{w_1}; z) - F_{w_1} \ge -Q(z)$$

for every $z \in \mathcal{S}_w \cap D_\delta(z_0)$ because then (4.3) applied first to w_1 rather than to w and then to w would yield for any polynomial P_n of degree $\le n$

$$\|w^n P_n\|_\Sigma = \|w^n P_n\|_{\mathcal{S}_{w_1}}. \tag{4.7}$$

But (4.7) is possible only for $\mathcal{S}_w \subseteq \mathcal{S}_{w_1}$ (see Theorem II.2.2 of [SaTo2]) and this cannot happen because $z_0 \in \mathcal{S}_w \setminus \Sigma_1 \subseteq \mathcal{S}_w \setminus \mathcal{S}_{w_1}$. Thus, from (4.1) applied to w_1 we can conclude the existence of $\varepsilon > 0$, $z_1 \in \mathcal{S}_w$, and $\rho > 0$ such that $D_\rho(z_1) \cap \mathcal{S}_{w_1} = \emptyset$ and for $z \in D_\rho(z_1)$ we have

$$p(\omega_{w_1}; z) - F_{w_1} \le -Q(z) - 2\varepsilon. \tag{4.8}$$

For each n let $Z^{(n)} = \{z_1^{(n)}, \ldots, z_n^{(n)}\}$ be a point system in Σ_1 maximizing

$$\prod_{\substack{1 \le i,k \le n \\ i \ne j}} |z_i^{(n)} - z_j^{(n)}| w_1(z_i^{(n)}) w_1(z_j^{(n)}).$$

We can choose $Z^{(n)}$ so that $Z^{(n)} \subseteq \mathcal{S}_{w_1}$ is satisfied (this follows from the extremality of $Z^{(n)}$ and of (4.3) applied to w_1; see, e.g., Theorem II.2.5 in [SaTo2]). But then the polynomials

$$\varphi_n(z) = \Psi_n(z)/\|w^n \Psi_n\|_{\Sigma_1},$$

$$\Psi_n(z) = \prod_{j=1}^{n} (z - z_j^{(n)})$$

satisfy

$$\|w^n \varphi_n\|_{\Sigma_1} \le 1 \tag{4.9}$$

and (see Corollary II.1.9 of [SaTo2])

$$\lim_{n\to\infty} |\varphi_n(z)|^{1/n} = \exp(-p(\omega_{w_1}; z) + F_{w_1})$$

locally uniformly outside $\mathcal{S}_{w_1}$. Together with (4.8) this shows that for large n

$$w^n(z)|\varphi_n(z)| > (1+\varepsilon)^n, \quad z \in D_\delta(z_1). \tag{4.10}$$

Since we have assumed that the support of $\mu|_{\mathcal{S}_w}$ is $\mathcal{S}_w$ and $z_1 \in \mathcal{S}_w$, we must have $\mu(D_\delta(z_1)) > 0$ and so we get from (4.9) and (4.10) that

$$\limsup_{n\to\infty} \left(\frac{\|w^n \varphi_n\|_{L^2(\mu)}}{\|w^n \varphi_n\|_{L^2(\mu|_S)}} \right)^{1/n} \ge 1 + \varepsilon,$$

proving that (4.6) is not true. This proves the second claim in the theorem. □

Now we show that the assumptions of Theorem 6.4.1 cannot be essentially relaxed.

Example 6.4.2. First we show that it is essential that the support of $\mu|_{\mathcal{S}_w}$ be on the outer boundary of $\mathcal{S}_w$. In fact, let $\Sigma = \{z \mid |z| \le 2\}$ and $w(z) = \exp(-||z| - \frac{1}{3}|)$. It can be shown (see, e.g., [SaTo2, Section III.6]) that in this case $\mathcal{S}_w$ is the ring $R = \{z \mid \frac{1}{3} \le |z| \le 1\}$. Now we set $\mu = \delta_0 + \mu_1 + \mu_2$ with μ_1 the Lebesgue measure on the unit circumference and $\mu_2 = \sum_{k=1}^{\infty} \eta_k \delta_{x_k}$ a discrete measure with support R, where the very fast decreasing coefficients η_k will be chosen later. It is obvious that the restriction of μ to $R = \mathcal{S}_w$ is regular with support equal to $\mathcal{S}_w$.

Now let

$$P_m(z) = \prod_{k=1}^{m} (z - x_k).$$

For $n \geq m$ it is a polynomial of degree at most n, and the $L^2(\mu)$ norm of the weighted polynomial $w^n P_k$ is at least as large as $e^{-n/3}|P_m(0)| = e^{-n/3}\prod_{k=1}^m |x_k|$, whereas its $L^2(\mu|_{S_w})$ norm is at most

$$e^{-2n/3}2^m\sqrt{\pi} + 2^m\left(\sum_{k=m+1}^{\infty}\eta_k\right)^{1/2},$$

so by appropriate choice of the sequence $\{\eta_k\}$ and the numbers $n = n(m)$ (4.5) will not be satisfied.

In this example the measure $\mu|_{S_w}$ was not supported on the outer boundary of S_w □

Example 6.4.3. Next we show that the assumption that the support of $\mu|_{S_w}$ is S_w is also indispensable. This is very easy: Let $\Sigma = [-1, 2]$, $w \equiv 1$ on $[-1, 1]$, and

$$w(x) = \frac{1}{x + \sqrt{x^2 - 1}}$$

if $x \in [1, 2]$. Since the logarithm of this w is the equilibrium potential of $[-1, 1]$ plus a constant $(-\log 2)$, it easily follows from (4.1) that ω_w must be equal to the equilibrium measure of $[-1, 1]$ (cf. the proof of Theorem 6.5.1) and so $S_w = [-1, 1]$.

Now let μ be the Lebesgue measure on the set $[0, 1/4] \cup [1, 2]$. It is obvious that $\mu|_{S_w}$ is regular, but with $P_n(z) = z^n$ the $L^2(\mu)$ norms of the weighted polynomials $w^n P_n$ do not live on $S_w = [-1, 1]$ in the sense of (4.5) because

$$\|w^n P_n\|_{L^2(\mu)} \geq \left(\tfrac{1}{2}\right)^n$$

(note that for $x \geq 1$ we have $w^n P_n(x) > \frac{1}{2}$), whereas

$$\|w^n P_n\|_{L^2(\mu|_{S_w})} \leq \left(\tfrac{1}{4}\right)^n.$$

□

Example 6.4.4. That $\mu|_{S_w} \in \mathbf{Reg}$ also cannot be dropped in Theorem 6.4.1 can be shown by an example very similar to the previous one. In fact, if we add to the measure μ of Example 6.4.3 a discrete measure $\sum_{k=1}^{\infty}\eta_k\delta_{x_k}$ with support equal to $[-1, 1]$, where the coefficients η_k decrease very fast, then arguments similar to that applied in Examples 6.4.2 and 6.4.3 show that for this measure (4.5) is not satisfied although all the assumptions of Theorem 6.4.1 except the regularity assumption $\mu|_{S_w} \in \mathbf{Reg}$ are true in this case. □

One could also show that in general the regularity of S_w with respect to solutions of Dirichlet problems is also a necessary condition but we do not elaborate on this.

6.5 Regularity and Weighted Chebyshev Constants

Let μ be a measure satisfying the condition of Theorem 3.2.3(vi); that is, let us assume that $S(\mu) = \partial\Omega$ and that $\partial\Omega$ is regular with respect to the Dirichlet problem in Ω. Let $\Sigma = S(\mu)$ and consider a positive continuous weight w on Σ as in the preceding section. The constant F_w from Section 6.4 gives the solution of the weighted Chebyshev problem: Find the limit of the nth root of the weighted Chebyshev numbers

$$t_n^w := \inf\{\|w^n P_n\|_\Sigma \mid P_n(z) = z^n + \cdots\},$$

namely, (see (A.30) in Appendix IX)

$$\lim_{n\to\infty} (t_n^w)^{1/n} = e^{-F_w}.$$

If we take into account Theorem 3.2.3, (i)$\Rightarrow$(vi), we can see that for the case $\mu \in \mathbf{Reg}$, in the nth-root sense the $L^\infty(\mu)$ Chebyshev numbers t_n^w are asymptotically the same as the $L^2(\mu)$ Chebyshev numbers

$$\inf\{\|w^n P_n\|_{L^2(\mu)} \mid P_n(z) = z^n + \cdots\}.$$

But these are the leading coefficients of the nth orthonormal polynomials with respect to the varying measures $d\mu_n = w^{2n} d\mu$; hence F_w is also closely related to leading coefficients of orthonormal polynomials with respect to varying measures. This gives the first half of the following theorem.

Theorem 6.5.1. *Let $S(\mu) = \partial\Omega$ and suppose that $\partial\Omega$ is regular with respect to the Dirichlet problem in Ω. If $g_n \geq 0$ are weight functions on $S(\mu)$ such that $g_n^{1/n}$ uniformly converges to w^2, where w is continuous and positive on $S(\mu)$, then for the case $\mu \in$ **Reg** for the varying measures $d\mu_n = g_n d\mu$ we have*

$$\lim_{n\to\infty} \gamma_n(\mu_n)^{1/n} = e^{F_w}, \tag{5.1}$$

where F_w is the constant introduced in (4.1) in the preceding section in connection with w on $\Sigma := S(\mu)$.
*If in addition w is given by $w(z) = \exp(\vartheta p(\sigma; z))$ where σ is a probability measure of compact support in Ω and $0 \leq \vartheta \leq 1$, then (5.1) is equivalent to $\mu \in$ **Reg**.*

Example 6.5.2. Let $W_\lambda(x) = \exp(-|x|^\lambda)$, $\lambda > 0$, $x \in \mathbb{R}$. Then for the corresponding orthonormal polynomials $p_n(W_\lambda^2; z) = \gamma_n(W_\lambda^2)z^n + \cdots$ we have

$$\lim_{n\to\infty} \gamma_n(W_\lambda^2)^{1/n} n^{1/\lambda} = 2^{2/\lambda} e^{1/\lambda} \Gamma(\lambda)^{1/\lambda} \Gamma(\lambda/2)^{-2/\lambda}. \tag{5.2}$$

In fact, the substitution $x \to n^{1/\lambda}t$ carries $W_\lambda^2(x)$ into $(w_\lambda(t))^{2n}$ with $w_\lambda(t) = \exp(-|t|^\lambda)$; hence with μ the Lebesgue measure on $\mathbf{R}$ (5.1) can be applied. The corresponding extremal measure ω_{w_λ} is called the Ullman distribution and that $\exp(F_{w_\lambda})$ equals the right-hand side of (5.2) was calculated in [MhSa2]. That we can do this despite the fact that μ has $\mathbf{R}$ as its support (and not a compact set) is due to the fact that W_λ decreases rapidly (say the preceding proof works in this case as well). □

Proof of Theorem 6.5.1. We only have to verify the very last assertion concerning the equivalence. This is done via the equivalence of (i) and (ii) in Theorem 3.3.3. In fact, let σ' be the balayage of σ onto $S(\mu)$. We claim that the extremal measure ω_w of Section 6.4 is given by

$$\omega_w = \vartheta\sigma' + (1-\vartheta)\omega_{S(\mu)}, \tag{5.3}$$

where, as usual, $\omega_{S(\mu)}$ denotes the equilibrium measure of $S(\mu)$. In fact, if the measure on the right-hand side of (5.3) is denoted by ν then we have

$$\begin{aligned} p(\nu; z) &= \vartheta\left(p(\sigma; z) + \int g_\Omega(x;\infty)d\sigma(x)\right) + (1-\vartheta)\log\frac{1}{\operatorname{cap}(S(\mu))} \\ &= \log w(z) + c \end{aligned}$$

for all $z \in S(\mu)$, where

$$c = \vartheta \int g_\Omega(x;\infty)d\sigma(x) + (1-\vartheta)\log\frac{1}{\operatorname{cap}(S(\mu))}.$$

Thus, by (4.1) we have

$$p(\omega_w; z) - F_w + c \geq p(\nu; z)$$

for all $z \in \operatorname{supp}(\nu)$, so this inequality extends to all $\mathbf{C}$ by the principle of domination. In quite similar manner we have by (4.1)

$$p(\omega_w; z) - F_w + c \leq p(\nu; z)$$

for all $z \in \operatorname{supp}(\omega_w)$, so this inequality extends again to all $\mathbf{C}$ by the principle of domination (it easily follows from the definition of ω_w that it has finite logarithmic energy, so the principle of domination is applicable). Hence

$$p(\omega_w; z) - F_w + c = p(\nu; z)$$

for all z and it follows that $\omega_w = \nu$ as we claimed.

We also have that

$$F_w = \vartheta \int g_\Omega(x;\infty)d\sigma(x) + (1-\vartheta)\log\frac{1}{\operatorname{cap}(S(\mu))},$$

so the equivalence of (5.1) and $\mu \in \mathbf{Reg}$ immediately follows from the equivalence of the statements (i) and (ii) in Theorem 3.3.3 (set $\gamma = 2\vartheta$).

□

As a side product of the preceding proof it also follows that $\mu \in \mathbf{Reg}$ is equivalent to the fact that in $\mathbf{C} \setminus \mathrm{Co}(S(\mu))$ the nth-root asymptotics for the orthonormal polynomials $p_n(\mu_n; \cdot)$ with respect to the varying measures μ_n are given by

$$\lim_{n\to\infty} |p_n(\mu_n; z)|^{1/n} = \exp(F_w - p(\omega_w; z))$$

(cf. Theorems 1.1.4 and 3.3.3). Furthermore, for the case $\mathrm{Int}(\mathrm{Pc}(S(\mu))) = 0$ the asymptotic distribution of the zeros of $p_n(\mu_n; \cdot)$ is ω_w.

6.6 Regularity and Best $L^2(\mu)$ Polynomial Approximation

In this section we shall relate the regularity of μ to best $L^2(\mu)$ polynomial approximation of analytic functions. To avoid unnecessary complications we shall always assume that $\partial\Omega$ is regular with respect to the solutions of the Dirichlet problems in Ω.

Let

$$E_n(f)_{L^p(\mu)} = \inf_{\deg P_n \le n} \|f - P_n\|_{L^p(\mu)}, \quad 1 \le p \le \infty,$$

be the best $L^p(\mu)$ approximation of f by polynomials of degree at most n. A well-known result of S. N. Bernstein says that $\{E_n(f)_{L^\infty(\mu)}\}_{k=1}^\infty$ tends geometrically to zero if and only if f is analytic on $S(\mu)$ (see [Wa]). More precisely, if

$$\alpha(f) = \inf\{\delta \mid f \text{ is analytic in the region } \{x \mid g_\Omega(z;\infty) < \log 1/\delta\}\}$$

and

$$\varepsilon(f)_{L^p(\mu)} = \limsup_{n\to\infty} (E_n(f)_{L^p(\mu)})^{1/n},$$

then Bernstein's formula

$$\alpha(f) = \varepsilon(f)_{L^\infty(\mu)} \tag{6.1}$$

holds for every f. Note that in this result the measure μ actually does not play any role; the statement only refers to the compact set $S(\mu) = \mathrm{supp}(\mu)$. Of course, the situation will radically change if we consider the case $p < \infty$. We shall show in what follows that the $L^p(\mu)$ analogue of the preceding result is true if and only if $\mu \in \mathbf{Reg}$.

Besides $\alpha(f)$ and $\varepsilon(f)_{L^p(\mu)}$ we also introduce

$$\varphi(f)_\mu = \limsup_{n\to\infty} |\hat{f}(n)|^{1/n},$$

where

$$\hat{f}(n) = \int f p_n(\mu; \cdot) d\mu$$

are the Fourier coefficients of f with respect to the orthonormal polynomials $\{p_n(\mu; \cdot)\}$.

Let us also agree to write $\|\cdot\|_{S(\mu)}$, $E_n(f)_{S(\mu)}$, and so forth instead of $\|\cdot\|_{L^\infty(\mu)}$, $E_n(f)_{L^\infty(\mu)}, \ldots$.

Consider the following statements.

A_μ : $\quad \alpha(f) < 1 \quad \Leftrightarrow \quad \varepsilon(f)_{L^2(\mu)} < 1 \quad$ *for* $\quad f \in L^2(\mu)$;

that is, geometric order of best $L^2(\mu)$ approximation of a function f is equivalent to the analyticity of f (i.e., f coincides μ–a.e. with a function that is analytic on $S(\mu)$).

A^*_μ : $\quad \alpha(f) = \varepsilon(f)_{L^2(\mu)} \quad$ *for every* $\quad f \in L^2(\mu)$.

F_μ : $\quad \alpha(f) < 1 \quad \Leftrightarrow \quad \varphi(f)_\mu < 1 \quad$ *for* $\quad f \in L^2(\mu)$;

that is, analyticity is equivalent to geometric decrease of Fourier coefficients.

F^*_μ : $\quad \alpha(f) = \varphi(f)_\mu \quad$ *for every* $\quad f \in L^2(\mu)$.

C_μ: *If* $\|f - S_n(f)\|_{L^2(\mu)}$, *where* $S_n(f)$ *denotes the nth partial sum of the Fourier series of* f *in* $\{p_k(\mu; \cdot)\}$, *converges geometrically to zero as* $n \to \infty$, *then* $S_n(f)$ *uniformly converges on* $S(\mu)$;

that is, geometric convergence of the Fourier series in $L^2(\mu)$ implies its uniform convergence.

Now each of these is equivalent to the regularity of μ.

Theorem 6.6.1. *Assume that* $\partial\Omega$ *is regular with respect to the Dirichlet problem in* Ω. *Then the following six statements are equivalent:* $\mu \in$ **Reg**, A_μ, A^*_μ, F_μ, F^*_μ, C_μ.

Remarks. (1) Instead of the $L^2(\mu)$ norm we could use the $L^p(\mu)$ norm in the definition of A_μ, A^*_μ so long as $1 \le p < \infty$, and Theorem 6.6.1 would still hold.

(2) In C_μ we could have requested pointwise convergence of the Fourier series on $S(\mu)$ without uniform convergence and Theorem 6.6.1 would still be valid. Furthermore, from C_μ it always follows that the Fourier series geometrically converges to the function in the supremum norm as well.

Proof of Theorem 6.6.1. It is a standard matter that $\varepsilon(f)_{L^2(\mu)} = \varphi(f)_\mu$ for every $f \in L^2(\mu)$; therefore the statements concerning F_μ and F^*_μ follow from those on A_μ and A^*_μ.

Suppose first that $\mu \in \mathbf{Reg}$. Then we have by Theorem 3.2.3

$$\limsup_{n\to\infty}(\|P_n\|_{S(\mu)}/\|P_n\|_{L^2(\mu)})^{1/n} = 1 \tag{6.2}$$

for every sequence of polynomials $\{P_n\}$, $\deg(P_n) \le n$. Now if $\{S_n(f)\}$ are the partial sums of the Fourier series of f (with respect to $\{p_n(\mu;\cdot)\}$) then

$$\limsup_{n\to\infty} \|S_{n+1}(f) - S_n(f)\|_{L^2(\mu)}^{1/n} \le \varepsilon(f)_{L^2(\mu)}.$$

Together with (6.2) this tells us that

$$\limsup_{n\to\infty} \|S_{n+1}(f) - S_n(f)\|_{S(\mu)}^{1/n} \le \varepsilon(f)_{L^2(\mu)}$$

and this easily implies $\varepsilon(f)_{L^\infty(\mu)} \le \varepsilon(f)_{L^2(\mu)}$. The converse inequality is evident, so $\mu \in \mathbf{Reg} \Rightarrow A_\mu^*$ follows from this and (6.1).

$A_\mu^* \Rightarrow A_\mu$ is trivial; therefore to verify the equivalence of the first five statements in the theorem ($\mu \in \mathbf{Reg}, A_\mu, A_\mu^*, F_\mu, F_\mu^*$) it only remains to show that $A_\mu \Rightarrow \mu \in \mathbf{Reg}$. Suppose $\mu \notin \mathbf{Reg}$. Then there exist an $\alpha < 1$ and a subsequence $\mathcal{N} \subseteq \mathbf{N}$ of the natural numbers such that for each $n \in \mathcal{N}$ there are polynomials P_n, $\deg(P_n) \le n$, satisfying

$$\|P_n\|_{L^2(\mu)} \le \alpha^n, \qquad E_{n-1}(P_n)_{S(\mu)} = 1. \tag{6.3}$$

In fact, we claim that

$$P_n = p_n(\mu;\cdot)/E_{n-1}(p_n(\mu;\cdot))_{S(\mu)}$$

will do the job. To see this we only have to remark that

$$E_{n-1}(p_n(\mu;\cdot))_{S(\mu)} = \gamma_n(\mu)\|T_n\|_{S(\mu)},$$

where $\{T_n\}$ are the Chebyshev polynomials associated with $S(\mu)$, and that

$$\lim_{n\to\infty} \|T_n\|_{S(\mu)}^{1/n} = \operatorname{cap}(S(\mu))$$

(see [Ts, Chapter III, Theorem III.26]), whereas $\mu \notin \mathbf{Reg}$ implies

$$\limsup_{n\to\infty}(\gamma_n(\mu))^{1/n} > \frac{1}{\operatorname{cap}(S(\mu))}.$$

Choose now a sequence $\{n_k\} \subseteq \mathcal{N}$ with the property $n_{k+1} > 4n_k$ and set

$$f = \sum \frac{1}{n_k} P_{n_k}.$$

We get from (6.3)

$$E_n(f)_{L^2(\mu)} \le \left\| f - \sum_{k=1}^{l} \frac{1}{n_k} P_{n_k} \right\|_{L^2(\mu)} \le \sum_{k=l+1}^{\infty} \|P_{n_k}\|_{L^2(\mu)} \tag{6.4}$$

$$\le C\alpha^{n_{l+1}} \le C\alpha^n$$

if l is chosen to satisfy $n_l < n \leq n_{l+1}$. At the same time it is not difficult to see that if n is fixed in (6.3) then there is a $\delta = \delta(n) > 0$ such that for every subset S of $S(\mu)$ with the property $\mu(S(\mu) \setminus S) < \delta$ we have $E_{n-1}(P_n)_S \geq \frac{3}{4}$. If we also consider the fact that for $n \in \mathcal{N}$ and

$$A_n = \{z \mid |P_n(z)| \geq 1\}$$

we have $\mu(A_n) \leq \alpha^{2n}$ (cf. (6.3)), it easily follows that for sufficiently fast increasing $\{n_k\} \subseteq \mathcal{N}$ with $S_l = S(\mu) \setminus \bigcup_{k=l+1}^{\infty} A_{n_k}$ we get

$$E_{n_l-1}(f)_{S(\mu)} \geq E_{n_l-1}\left(\frac{1}{n_l}P_{n_l}\right)_{S_l} - \sum_{k=l+1}^{\infty} \frac{1}{n_k} \geq \frac{3}{4}\frac{1}{n_l} - \sum_{k=l+1}^{\infty} \frac{1}{n_k} \geq \frac{1}{4}\frac{1}{n_l},$$

and hence A_μ does not hold by (6.1).

Until now we have verified the equivalence of $\mu \in \mathbf{Reg}$, A_μ, A_μ^*, F_μ, and F_μ^*. As for C_μ, it follows from $\mu \in \mathbf{Reg}$ because this is what we verified in the proof $\mu \in \mathbf{Reg} \Rightarrow A_\mu^*$. That $C_\mu \Rightarrow \mu \in \mathbf{Reg}$ also holds is shown by the function

$$f_1 = \sum_{k=1}^{\infty} P_{n_k} = \sum_{k=1}^{\infty} p_{n_k}(\mu;\cdot)/E_{n_k-1}(p_{n_k}(\mu;\cdot))_{S(\mu)}$$

(cf. (6.3)), where $\{n_k\} \subseteq \mathcal{N}$. In fact, for this we have exactly as in (6.4)

$$E_n(f_1)_{L^2(\mu)} \leq C\alpha^n,$$

and at the same time for the partial sums of the Fourier expansion of f_1 we have

$$\|S_{n_k}(f_1) - S_{n_k-1}(f_1)\|_{S(\mu)} \geq E_{n_k-1}(P_{n_k}) = 1,$$

that is, if $\mu \notin \mathbf{Reg}$ then C_μ does not hold either. □

Before closing this section we show that the regularity of $\partial\Omega$ in Theorem 6.6.1 is essential to make the conclusions of the theorem.

Example 6.6.2. We construct a measure μ and a function f defined on the support of μ such that f cannot be extended to an analytic function in a neighborhood of $S(\mu)$ but its best $L^2(\mu)$ polynomial approximation satisfies

$$\limsup_{n\to\infty} E_n(f)_{L^2(\mu)}^{1/n} < 1.$$

Of course, then $\partial\Omega$ cannot be regular with respect to the Dirichlet problem in Ω.

Consider the construction given in Example 3.5.3, and let

$$I(n) = (2^{-n}, 2^{-n} + \exp(-n^3)).$$

By the construction given in Example 3.5.3 there exist a $\gamma > 1$ and three sequences $\{n_m\}$,$\{M_m\}$,$\{P_{M_m}^{(m)}\}$ such that

$$\deg(P_{M_m}^{(m)}) \leq M_m \quad \text{and} \quad \|P_{M_m}^{(m)}\|_{L^2(\mu)} < 2$$

but

$$\|P_{M_m}^{(m)}\|_{L^\infty(I(n_{m+1}))} > \gamma^{2M_m}\left(2 + \max_{1\leq s<m} \|P_{M_s}^{(s)}\|_{L^\infty(I(n_{m+1}))}\right).$$

Here μ is the restriction of the linear Lebesgue measure to the set $S = \cup_m I(n_m)$. Without loss of generality we can assume $\{M_m\}$ to be increasing so fast that

$$\gamma^{-M_s}\|P_{M_s}^{(s)}\|_{L^\infty(I(n_{m+1}))} \leq 2^{-s}$$

is satisfied for all m and $s > m$. In fact, we only have to mention that by Nikolskii's inequality (see [Ti, 6.9.6]) the supremum norm of $|P_{M_s}^{(s)}|$ on $I(n_{m+1})$ is bounded by $O(M_s)$ times its L^2 norm (the O depends on n_{m+1}), and this latter one is assumed to be less than 2 (recall that in Example 3.5.3 the numbers $M_{m+1}, M_{m+2}, \ldots$ were chosen to be large enough after the selection of n_{m+1}).

Set

$$f = \sum_{m=1}^{\infty} \gamma^{-M_m} P_{M_m}^{(m)}.$$

We claim that this f and μ satisfy the preceding requirements. In fact, let $M_k \leq n < M_{k+1}$, $k = 1, 2, \ldots$. Then

$$E_n(f)_{L^2(\mu)} \leq \sum_{m=k+1}^{\infty} \gamma^{-M_m}\|P_{M_m}^{(m)}\|_{L^2(\mu)} \leq \frac{2\gamma}{\gamma-1}\gamma^{-M_{k+1}} \leq \frac{2\gamma}{\gamma-1}\gamma^{-n},$$

so the rate of $L^2(\mu)$ polynomial approximation to f is geometric. Since the series representing f uniformly converges on any interval of S, f is continuous on $S \setminus \{0\}$; hence any analytic function that coincides μ–almost everywhere with f must actually coincide with f on $S \setminus \{0\}$. On the other hand, for the supremum norm of f we get

$$\|f\|_{\sup, I(n_{m+1})} \geq \gamma^{M_m}\left(2 + \max_{1\leq s<m} \|P_{M_s}^{(s)}\|_{L^\infty(I(n_{m+1}))}\right)$$

$$- \sum_{s=1}^{m-1} \gamma^{-M_s}\|P_{M_s}^{(s)}\|_{L^\infty(I(n_{m+1}))} - \sum_{s=m+1}^{\infty} \gamma^{-M_s}\|P_{M_s}^{(s)}\|_{L^\infty(I(n_{m+1}))} \geq \gamma^{M_m}$$

if m is large enough. But here the right-hand side tends to infinity as $m \to \infty$, so f cannot agree on $S = S(\mu)$ with any function analytic in a neighborhood of S. □

6.7 Determining Sets

In connection with the Erdős–Turán Criterion J. Ullman [Ul1] and P. Erdős and G. Freud [ErFr] studied and solved the following problem: For what Borel sets $A \subseteq [-1,1]$ is it true that if w is a weight function on $[-1,1]$ that is positive on A and $d\nu(x) = w(x)dx$, then the orthogonal polynomials $\{p_n(\nu;\cdot)\}$ belonging to ν have arcsine zero distribution. Obviously A must have positive Lebesgue measure in every interval $[a,b] \subseteq [0,1]$, which implies $c_\nu > 0$ and $\mathrm{supp}(\nu) = [-1,1]$ for all ν in question. But for $c_\nu > 0$ the polynomials $\{p_n(\nu;\cdot)\}$ have arcsine distribution if and only if ν is regular (see Corollary 2.2.3), so the preceding problem can be reformulated as: For what sets $A \subseteq [-1,1]$ is it true that if ν is a measure on $[0,1]$ such that $\nu'(x) > 0$ for almost every $x \in A$ then ν is regular. Setting $d\mu(x) = \chi_A(x)dx$, where

$$\chi_A(x) = \begin{cases} 1 & \text{if } x \in A; \\ 0 & \text{otherwise,} \end{cases} \tag{7.1}$$

this problem is equivalent to characterizing for a measure μ the implication

$$\Gamma(\nu) \subseteq \Gamma(\mu) \Rightarrow \nu \in \mathbf{Reg}.$$

Recall that $\Gamma(\mu)$ denotes the set of carriers of μ.

Using our results from Chapter 1 we can easily characterize all such measures μ.

Theorem 6.7.1. *The following statements are equivalent for a measure* μ.

(1) *If* ν *is any measure with* $\mathrm{supp}(\nu) \subseteq S(\mu)$ *and* $\Gamma(\nu) \subseteq \Gamma(\mu)$, *then* $\nu \in$ **Reg**.

(2) $c_\mu = \mathrm{cap}(S(\mu))$.

(3) *For every* $\varepsilon > 0$ *there is a* $\delta > 0$ *such that* $\mu(E) < \delta$ *implies*

$$\mathrm{cap}(S(\mu) \setminus E) > \mathrm{cap}(S(\mu)) - \varepsilon.$$

Theorem 6.7.2. *Let* A *be a Borel subset of* S, *where* S *consists of a finite number of rectifiable Jordan curves such that* $\mathbb{C} \setminus S$ *is connected, and let* m *denote the arc measure on* S. *Then the following statements are equivalent.*

(i) *If* ν *is a measure such that* $\mathrm{supp}(\nu) \subseteq S$ *and*

$$\frac{d\nu(s)}{dm(s)} > 0$$

for $m-$*almost every* $s \in A$ *then* $\nu \in$ **Reg**.

(ii) *With the assumptions of* (i) *the orthogonal polynomials* $\{p_n(\nu;\cdot)\}$ *have asymptotic zero distribution* ω_S.

(iii) *If $B \subseteq A$ such that $m(B) = 0$ then*

$$\mathrm{cap}(A \setminus B) = \mathrm{cap}(S).$$

(iv) *For every $\varepsilon > 0$ there is a $\delta > 0$ such that $B \subseteq A$, $m(B) < \delta$ implies*

$$\mathrm{cap}(A \setminus B) \geq \mathrm{cap}(S) - \varepsilon.$$

Remark. For $S = [-1, 1]$ and m equal to Lebesgue measure, Theorem 6.7.2 reduces to the aforementioned results of Ullman and Erdős and Freud. In this special case Ullman called sets A with property (ii) determining sets whereas Erdős and Freud used the terminology of having minimal capacity $\frac{1}{2}$ for property (iii).

Proof of Theorem 6.7.1. (1)⇒(2) because of Theorem 1.1.9 (cf. also Theorem 3.2.1).

(2)⇒(1) because Corollary 1.1.7 (or Ullman's Criterion), for $c_\mu = \mathrm{cap}(S(\mu))$, and $\Gamma(\nu) \subseteq \Gamma(\mu)$ imply $c_\nu = \mathrm{cap}(S(\mu))$.

(3)⇒(2) is obvious, so it is only left to prove that (2) implies (3).

Assume that (3) is false, that is, for some $\varepsilon > 0$ there are sets $E_n \subseteq S(\mu)$ such that $\mu(E_n) < 2^{-n}$ but

$$\mathrm{cap}(S(\mu) \setminus E_n) \leq \mathrm{cap}(S(\mu)) - \varepsilon.$$

Then

$$E^* = \limsup_{n\to\infty} E_n = \bigcap_{m=1}^{\infty} \bigcup_{n=m}^{\infty} E_n$$

has zero μ–measure. Therefore $C := S(\mu) \setminus E^*$ is a carrier of μ. But

$$\begin{aligned} \mathrm{cap}(C) &= \mathrm{cap}\left(\bigcup_{m=1}^{\infty} \bigcap_{n=m}^{\infty} (S(\mu) \setminus E_n)\right) \\ &= \lim_{m\to\infty} \mathrm{cap}\left(\bigcap_{n=m}^{\infty} (S(\mu) \setminus E_n)\right) \leq \mathrm{cap}(S(\mu)) - \varepsilon, \end{aligned}$$

where, in the second equality, we have applied the monotonicity of the sequence of sets $\{\cap_{n=m}^{\infty}(S(\mu) \setminus E_n)\}_{m=1}^{\infty}$ (cf. [La, Chapter II, Lemma 2.10]). This means, however, $c_\mu < \mathrm{cap}(S(\mu))$ and (2)⇒(3) is also verified. □

Proof of Theorem 6.7.2. If we set

$$d\mu(s) = \chi_A(s) ds$$

as before (7.1), and apply Theorem 6.7.1 we get the equivalence of (i), (iii), and (iv) from the corresponding statements (1), (2), and (3) in Theorem

6.7.1, provided we can show that each of (i), (iii), and (iv) implies $S(\mu) = S$. In proving this we can argue as follows. If $S(\mu) \neq S$ then on some subarc I of S the set A has zero arc measure: $m(I \cap A) = 0$. It is easy to show that because $\mathbb{C} \setminus S$ is connected and $\text{cap}(I) > 0$, we have then

$$\text{cap}(S \setminus I) < \text{cap}(S). \tag{7.2}$$

This proves that each of (iii) and (iv) implies $S(\mu) = S$ by setting there $B = I$.

To prove that (i) also implies $S(\mu) = S$ we choose as the measure ν in (i) the restriction $m|_{S\setminus I}$ of m to $S \setminus I$. From (7.2) and Corollary 1.1.7 we can see

$$\liminf_{n\to\infty} \gamma_n(\nu)^{1/n} > \frac{1}{\text{cap}(S)}. \tag{7.3}$$

Now we modify ν by setting

$$\tilde{\nu} := \nu + \sum c_j \delta_{x_j},$$

where $\{x_j\}$ is a dense subset of I. Applying a standard argument, the positive coefficients $\{c_j\}$ can be chosen in such a way that

$$\limsup_{n\to\infty} \gamma_n(\tilde{\nu})^{1/n} > \frac{1}{\text{cap}(S)}$$

also holds. But $\text{supp}(\tilde{\nu}) = S$, so this $\tilde{\nu}$ is not regular although the assumption of (i) is satisfied for it. This proves the implication (i) $\Rightarrow S(\mu) = S$ and with it the equivalence of (i), (iii), and (iv) has been verified.

Finally, we consider (i) and (ii). In the sequel we shall show $c_\nu > 0$ for every measure ν satisfying the assumption in (i), hence (i)$\Rightarrow$(ii) follows from Theorem 2.2.1(b) together with the just established fact that $\text{supp}(\nu) = S$ for every ν in (i). If we apply Theorem 2.1.1 instead of Corollary 1.1.7 in the preceding proof establishing the implication (i) $\Rightarrow S(\mu) = S$, we can similarly conclude that (ii) $\Rightarrow S(\mu) = S$ and that if (ii) holds then A has positive arc measure on every arc of S. We shall show in a moment that then $c_\nu > 0$ for every measure ν satisfying the condition in (i), and this will complete the proof of (ii)$\Rightarrow$(i) because of Corollary 2.2.3.

In proving that $c_\nu > 0$ for every ν satisfying the assumption in (i), we remark first of all that if ν is a measure with $d\nu(s)/dm(s) > 0$ for m–almost every $s \in A$ and C is a carrier of ν then $m(C) > 0$ (recall that $m(A) > 0$ has already been established). Of course, then for at least one of the connected arcs, say, for S_0, in S the intersection $C_0 = C \cap S_0$ also has positive m measure (actually, C must have positive arc measure on every subarc $I \subseteq S$). Let $\gamma(t)$, $t \in [0,1]$, be a parametric representation of S_0, and let

$$T_0 = \{t \mid \gamma(t) \in C_0\}.$$

Since

$$0 < m(C_0) = \int_{T_0} |\gamma'(t)|dt, \tag{7.4}$$

we must have $\text{meas}(T_0) > 0$, and hence $\text{cap}(T_0) > 0$. Now pick a $t_0 \in T_0$, $\delta > 0$, and $\varepsilon > 0$ such that for $t \in T_0 \cap [t_0 - \delta, t_0 + \delta]$ we have, say,

$$\text{Re}(\gamma(t) - \gamma(t_0)) \geq \varepsilon(t - t_0), \tag{7.5}$$

and at the same time $\text{cap}(T_0 \cap [t_0 - \delta, t_0 + \delta]) > 0$. The existence of such t_0, δ, ε follows from (7.4) with possibly replacing Re or $t - t_0$ by Im or $t_0 - t$ in (7.4). Inequality (7.5) shows that $t \to \gamma(t)/\varepsilon$ is an expansive mapping on $[t_0 - \delta, t_0 + \delta]$, hence the image of T_0 under γ is of positive capacity (cf. [Ts, Chapter III, Theorem III.26]) and this proves $\text{cap}(C) > 0$ and together with it also $c_\nu > 0$ (see Lemma 1.2.2). □

Appendix

We give a brief review of those aspects of the theory of logarithmic potentials that are relevant to the main body of the monograph.

The readers can be recommended to turn for details to the texts by N. S. Landkof [La], M. Tsuji [Ts], L. L. Helms [He], W. K. Hayman and P. B. Kennedy [HaKe], W. K. Hayman [Ha], J. L. Doob [Do], C. J. de la Vallée-Poussin [dVP], E. Hille [Hi], C. Pommerenke [Po], or O. D. Kellogg [Ke]. Unfortunately, none of these could serve as our standard reference for either they consider logarithmic potentials a little tangentially by putting the emphasis on Riesz or Green potentials or their exposition is of a different character or not as deep as we would need.

We restrict ourselves to the basic definitions and facts with two exceptions: We treat the Green functions and equilibrium measures corresponding to general Borel sets with full details. The reason for doing this is that our whole theory rests on these notions and they are omitted from most textbooks in the generality needed for us.

In our terminology "measure" means nonnegative (=positive) finite Borel measure. We will always use the adjective "signed" if signed measures are allowed.

A.I Energy and Capacity

Let μ be a signed Borel measure with compact support on $\mathbb{C}$. Its *logarithmic energy* is defined by

$$I(\mu) = \int\int \log \frac{1}{|z-t|} d\mu(t) d\mu(z), \tag{A.1}$$

provided that the integral has meaning, the simplest case of which is when $I(|\mu|)$ is finite (here $|\mu|$ denotes the total variation measure of μ).

It can be shown ([La, Theorem 1.16]) that $I(\mu) \geq 0$ if either

(a) $\mu(\mathbb{C}) = 0$, or
(b) $\text{supp}(\mu)$ is contained in a circle of radius 1

holds. Furthermore, in both cases $I(\mu) > 0$ unless $\mu = 0$.

A measure μ is said to be of *finite logarithmic energy* if $I(\mu) < \infty$.

If $K \subseteq \mathbb{C}$ is a compact set, then we set

$$V_K = \inf\{I(\mu) \mid \text{supp}(\mu) \subseteq K, \ \mu \geq 0, \ \|\mu\| = 1\}, \tag{A.2}$$

that is, the infimum is taken for all probability Borel measures supported on K. The (logarithmic) *capacity* of K is defined as

$$\text{cap}(K) := e^{-V_K}. \tag{A.3}$$

To extend the notion of capacity to arbitrary Borel sets E we introduce the notion of inner and outer capacity:

$$\text{cap}_i(E) = \sup\{\text{cap}(K) \mid K \subseteq E, K \text{ compact}\}$$

and

$$\text{cap}_o(E) = \inf\{\text{cap}_i(U) \mid E \subseteq U, U \text{ open}\}.$$

Then every Borel set E is capacitable in the sense that $\text{cap}_i(E) = \text{cap}_o(E)$ ([La, Theorem 2.8]) and we can simply write $\text{cap}(E)$ for this common value.

A property is said to hold *quasi everywhere* (qu.e.) if it holds outside a set of zero capacity. A countable union of sets of (outer) capacity zero has again zero capacity ([La, Chapter II, No. 6, rem. (1)]).

It follows from the definitions that if μ has finite logarithmic energy and E is of zero capacity, then $\mu(E) = 0$. In particular, if a property holds quasi everywhere, then it holds μ–almost everywhere provided μ has finite logarithmic energy.

A.II Potentials, Fine Topology

The *logarithmic potential* of a measure μ is defined by

$$p(\mu; z) = \int \log \frac{1}{|z-t|} d\mu(t). \tag{A.4}$$

Logarithmic potentials are *superharmonic functions*, that is,

(α) they are lower semicontinuous, and
(β) if h is harmonic on the closure of a bounded domain D and $p(\mu; z) = h(z)$ for $z \in \partial D$, then $p(\mu; z) \geq h(z)$ inside D.

Conversely, *F. Riesz's representation theorem* ([Ts, Theorem II.24]) asserts that if H is superharmonic in a domain D, then there is a unique measure μ such that for every subdomain $D^* \subseteq D$ with $\overline{D^*} \subseteq D$ we have

$$H(z) = h(z) + \int_{D*} \log \frac{1}{|z-t|} d\mu(t), \tag{A.5}$$

where h is a harmonic function on D^*.

Since the minimum of two superharmonic functions is again superharmonic, it follows that the minimum of two logarithmic potentials is again a logarithmic potential.

In the definition of superharmonicity condition β can be replaced by

$$p(\mu; z) \geq \frac{1}{2\pi} \int_0^{2\pi} p(\mu; z + re^{it})dt, \quad \text{for all } z \in \mathbb{C},\ r > 0. \tag{A.6}$$

The *fine topology* on $\mathbb{C}$ is the coarsest topology on $\mathbb{C}$ for which all superharmonic functions are continuous. This is finer than ordinary planar topology. However, if U is a connected open set then the boundary of U in the fine and Euclidean topologies coincide. In fact, this follows from the fact that if x is not in the fine closure of a set E, then there is an arbitrary small $r > 0$ such that $\mathbb{C} \setminus E$ contains the circle $C_r(x)$ of radius r around x ([He, Theorem 10.14]).

The following four statements are equivalent for a Borel set E ([He, Theorem 10.2, Corollary 10.10], [La, Chapter III, Section 1]):

(a) $\text{cap}(E) = 0$,
(b) there exists a μ such that $p(\mu; z) = \infty$ for $z \in E$,
(c) E is a discrete set in the fine topology,
(d) E has no limit point in the fine topology.

A.III Principles

The following five principles for logarithmic potentials are frequently used. In each of them the measures are assumed to be nonnegative Borel measures on $\mathbb{C}$.

Principle of descent. *Let μ and $\{\mu_n\}_{n=1}^{\infty}$ have their support in a compact subset of $\mathbb{C}$ and $\mu_n \xrightarrow{*} \mu$ as $n \to \infty$ in the weak* topology. Then*

$$p(\mu; z) \leq \liminf_{n\to\infty} p(\mu_n; z)$$

locally uniformly on $\mathbb{C}$.

See [La, Theorem 1.3]. In our terminology *"locally uniformly"* means that for any sequence $\{z_n\}$ of points with $z_n \to z$ we have

$$p(\mu; z) \leq \liminf_{n\to\infty} p(\mu_n; z_n).$$

If $p(\mu; \cdot)$ happens to be continuous in a domain D, then the limit-inferior relation in the principle of descent holds uniformly on compact subsets of D (this automatically follows from the assumed continuity of $p(\mu; \cdot)$ and the fact that the preceding relation holds locally uniformly on D).

Let us also remark that weak* convergence $\mu_n \xrightarrow{*} \mu$ means

$$\int f d\mu_n \to \int f d\mu \qquad \text{as} \quad n \to \infty$$

for every f continuous on $\overline{\mathbb{C}}$.

Lower envelope theorem. *With the same assumptions as in the principle of descent we have*

$$p(\mu; z) = \liminf_{n\to\infty} p(\mu_n; z)$$

for quasi every $z \in \mathbb{C}$.

Compare [La, Theorem 3.8].

Principle of domination. *Suppose that* μ, ν *are positive measures of compact support,* $\|\nu\| \leq \|\mu\|$, *and* μ *is of finite logarithmic energy. If for some constant* c *the inequality*

$$p(\nu; z) + c \geq p(\mu; z)$$

holds $\mu-$*almost everywhere, then it holds everywhere in* $\mathbb{C}$.

Compare [La, Theorem 1.27].

First maximum principle. *If* μ *is of compact support and*

$$p(\mu; z) \leq M \tag{A.7}$$

for every $z \in \text{supp}(\mu)$, *then* (A.7) *holds for every* $z \in \mathbb{C}$.

See [Ts, Theorem III.1] or [La, Theorem 1.6], but this is also an immediate consequence of the principle of domination.

Second maximum principle. *Let* D *be a subdomain of* $\overline{\mathbb{C}}$ *such that* $cap(\partial D) > 0$, *and let* h *be a subharmonic function on* D *that is bounded from above. If for quasi every points* x *of* ∂D *we have*

$$\limsup_{z\to x, z\in D} h(z) \leq M,$$

then $h \leq M$ *in* D. *Furthermore, we have* $h < M$ *in* D *unless* h *is constant.*

See [HaKe, Theorem 5.16]. In particular, this yields a strong form of the maximum principle for harmonic functions.

Related to the maximum principles is the following inequality.

Harnack's inequality. *Let D be a subdomain of $\overline{\mathbb{C}}$, $z_0 \in D$ an arbitrary point, and $K \subseteq D$ a compact subset of D. Then there is a constant $M = M(z_0, K, D)$, such that for every* positive *harmonic function h on D the inequality*

$$h(z) \le Mh(z_0)$$

holds for every $z \in K$.

See the inequality [HaKe, Theorem 1.18], and apply a standard "chains of overlapping disks" argument.

Finally, we mention the following *uniqueness theorem* ([La, Theorem 1.12′]): If two potentials corresponding to measures with compact support coincide almost everywhere with respect to planar Lebesgue measure, then the two measures are the same.

A.IV Equilibrium Measures

If $K \subseteq \mathbb{C}$ is compact with positive capacity, then there is a unique measure μ for which the infimum in (A.1) is attained. We shall denote this extremal measure by ω_K and it is called the *equilibrium measure* of K. For its potential

$$p(\omega_K; z) = \int \log \frac{1}{|z-t|} d\omega_K(t)$$

we have

$$\text{(A.8)} \qquad \begin{aligned} &(\alpha) \qquad p(\omega_K; z) \le \log \frac{1}{\operatorname{cap}(K)} \quad \text{for every } z \in \mathbb{C}, \\ &(\beta) \qquad p(\omega_K; z) = \log \frac{1}{\operatorname{cap}(K)} \quad \text{for quasi every } z \in K \end{aligned}$$

(see [Ts, Theorem III.12]). We would like to extend the notion of equilibrium measure to an arbitrary bounded Borel set E. This was done by de la Vallée-Poussin in [dVP], but since the existence and uniqueness of equilibrium measures play a central role in our investigations and the aforementioned extension is beyond the scope of most standard texts, we have found it necessary to give complete details.

Theorem A.1. *If E is a bounded Borel subset of $\mathbb{C}$ of positive capacity, then there exists a unique probability measure ω_E with the following properties:*

$$
\text{(A.9)} \qquad
\begin{aligned}
&\text{(i)} && p(\omega_E; z) \le \log \frac{1}{\operatorname{cap}(E)} && \text{for } z \in \mathbb{C}, \\
&\text{(ii)} && p(\omega_E; z) = \log \frac{1}{\operatorname{cap}(E)} && \text{for quasi every } z \in E.
\end{aligned}
$$

Unlike the compact case when the set (more precisely, its outer boundary) carries the equilibrium measure, in the present case we only can assert that $\operatorname{supp}(\omega_E)$ is contained in the closure of E.

Proof. We first prove the result in the case that E is an F_σ set, and then extend it to the general case.

First assume that E is a bounded F_σ set with $\operatorname{cap}(E) > 0$. Let $E = \cup_{n=1}^{\infty} K_n$, where the K_n are compact, $K_n \subset K_{n+1}$, and let ω_n be the equilibrium measure for K_n, $n = 1, 2, 3, \ldots$. Because $\operatorname{cap}(E) > 0$, we may assume without loss of generality that $\operatorname{cap}(K_n) > 0$, $n = 1, 2, 3, \ldots$. By properties (A.8) of the equilibrium measures

$$
\begin{aligned}
p(\omega_n; z) &\le \log(1/\operatorname{cap}(K_n)) \quad \text{for all } z \in \mathbb{C}, \\
p(\omega_n; z) &= \log(1/\operatorname{cap}(K_n)) \quad \text{for all } z \in K_n \setminus Z_n,
\end{aligned}
$$

where $\operatorname{cap}(Z_n) = 0$, $n = 1, 2, 3, \ldots$. Let μ be a weak* limit of the sequence $\{\omega_n\}_{n=1}^{\infty}$. Replacing ω_n by a subsequence if necessary, we may assume that $\lim_{n\to\infty} \omega_n = \mu$. We will show subsequently that μ is an equilibrium measure for E.

From the principle of descent we infer

$$
\begin{aligned}
p(\mu; z) &\le \liminf_{n\to\infty} p(\omega_n; z) \\
&\le \liminf_{n\to\infty} \log(1/\operatorname{cap}(K_n)) \\
&= \log(1/\operatorname{cap}(E)),
\end{aligned}
$$

so condition (i) is satisfied. For $z \in E \setminus \{Z_0 \cup \bigcup_{n=1}^{\infty} Z_n\}$, where Z_0 is the exceptional set of capacity zero in the lower envelope theorem for the convergence $\omega_n \xrightarrow{*} \mu$, it follows from the lower envelope theorem that

$$
\begin{aligned}
p(\mu; z) &= \liminf_{n\to\infty} p(\omega_n; z) \\
&= \liminf_{n\to\infty} \log(1/\operatorname{cap}(K_n)) \\
&= \log(1/\operatorname{cap}(E)).
\end{aligned}
$$

Since $\cup_{n=0}^{\infty} Z_n$ is of zero capacity, μ also satisfies condition (ii), so μ is an equilibrium measure for the bounded F_σ set E.

To prove uniqueness, let E be a bounded F_σ set of positive capacity, and let μ_1 and μ_2 be equilibrium measures for E. It follows from the defining properties (A.9) (i) and (ii) that $\mu_3 = (\mu_1 + \mu_2)/2$ is also an equilibrium measure for E.

Let $A = \{z \mid p(\mu_3; z) = \log(1/\mathrm{cap}(E))\}$. Our next aim is to prove that $\mu_3(A) = 1$. If $\{K_n\}$ are as before, then, by Section A.II, the energy of the signed measures $\mu_3 - \omega_{K_n}$ is nonnegative. This implies that

$$\begin{aligned}\int p(\mu_3; \cdot)d\mu_3 &\geq \int p(\omega_n; \cdot)d\mu_3 + \int p(\mu_3; \cdot)d\omega_n - \int p(\omega_n; \cdot)d\omega_n \\ &= 2\int p(\mu_3; \cdot)d\omega_n - \int p(\omega_n; \cdot)d\omega_n \\ &= 2\log(1/\mathrm{cap}(E)) - \log(1/\mathrm{cap}(K_n)),\end{aligned}$$

where we have used the defining properties (A.8) and (A.9) of the equilibrium measures and the fact that any set of zero capacity has ω_{K_n} measure zero. Letting here $n \to \infty$ we can conclude

$$\int p(\mu_3; \cdot)d\mu_3 \geq \log(1/\mathrm{cap}(E)),$$

from which $\mu_3(A) = 1$ follows from the defining property (A.9)(i) of an equilibrium measure.

From

$$1 = \mu_3(A) = \tfrac{1}{2}[\mu_1(A) + \mu_2(A)]$$

it follows that $\mu_1(A) = \mu_2(A) = 1$; moreover if $p(\mu_3; z) = \log(1/\mathrm{cap}(E))$ for a z, then necessarily we must have

$$p(\mu_1; z) = p(\mu_2; z) = \log(1/\mathrm{cap}(E)).$$

Thus

$$I(\mu_1 - \mu_2) = \int_A (p(\mu_1; \cdot) - p(\mu_2; \cdot))d(\mu_1 - \mu_2) = 0,$$

and by Section A.I we may conclude that $\mu_1 = \mu_2$.

We continue the proof of Theorem A.1 by demonstrating the existence of an equilibrium measure for E, where E is any bounded Borel set of positive capacity. Let K_n be a compact subset of E satisfying $\mathrm{cap}(K_n) > \mathrm{cap}(E) - 1/n$ and $K_n \subset K_{n+1}$, $n = 1, 2, 3, \ldots$, and let $E_1 = \cup_{n=1}^{\infty} K_n$. We will show that ω_{E_1} is an equilibrium measure for E.

First, note that

$$p(\omega_{E_1}; z) \leq \log(1/\mathrm{cap}(E_1)) = \log(1/\mathrm{cap}(E))$$

for every $z \in \mathbb{C}$, so condition (i) is also satisfied. We argue by contradiction to show that condition (ii) is also satisfied. Assume on the contrary that $p(\omega_{E_1}; z) < \log(1/\mathrm{cap}(E))$ for all $z \in B$, where B is a subset of E with $\mathrm{cap}(B) > 0$. Let K be a compact subset of B with $\mathrm{cap}(K) > 0$, and let

$E_2 = E_1 \cup K$. It follows that ω_{E_2} is also an equilibrium measure for E_1, so $\omega_{E_2} = \omega_{E_1}$, since E_1 is an F_σ set. Thus we have

$$p(\omega_{E_1}; z) = p(\omega_{E_2}; z) = \log(1/\text{cap}(E))$$

for some $z \in B$, and this contradicts the choice of B. This contradiction shows that (A.9)(ii) is also satisfied and ω_{E_1} is an equilibrium measure for E.

Now let μ_1 and μ_2 be two equilibrium measures for the Borel set E. There exists an F_σ subset F of E such that $\text{cap}(F) = \text{cap}(E)$, and it readily follows that μ_1 and μ_2 are both equilibrium measures for F. The uniqueness of the equilibrium measure for a bounded F_σ set of positive capacity implies that $\mu_1 = \mu_2$. □

A.V Green Functions

Let U be a subdomain of $\overline{\mathbb{C}}$ and let $a \in U$. The *Green function* $g_U(z; a)$ corresponding to U with pole at $a \in U$ has the following properties.

(α) $g_U(z; a)$ is nonnegative, subharmonic in $\mathbb{C}\setminus\{a\}$, and harmonic in $U\setminus\{a\}$.
(β) $g_U(z; a) + \log|z - a|$ remains bounded as $z \to a$.
(γ) $g_U(z; a) = 0$ for quasi every $z \in \overline{\mathbb{C}} \setminus U$.

g_U exists if and only if ∂U is of positive capacity and it is always unique. Furthermore, g_U is a symmetric function in its arguments.

For $a = \infty$ assumption (β) takes the form that

$$g_U(z; \infty) - \log|z|$$

remains bounded near infinity, and it turns out that in this case

$$(\beta)' \qquad g_U(z; \infty) = \log|z| - \log(\text{cap}(\mathbb{C} \setminus U)) + o(1) \quad \text{as} \quad |z| \to \infty$$

holds.

If U is the complement of a compact set K of positive capacity, then it follows from (A.8) that $g_U(z; \infty)$ has the form

$$g_{\mathbb{C}\setminus K}(z; \infty) = -p(\omega_K, z) - \log(\text{cap}(K)),$$

where ω_K is the equilibrium measure of K.

We need to extend the notion of Green function with pole at ∞ for an arbitrary Borel set $B \subseteq \overline{\mathbb{C}}$ with bounded complement. In this case the defining properties of $g_B(z; \infty)$ are

(i) $g_B(z; \infty)$ is nonnegative, subharmonic in $\mathbb{C}$, and harmonic in $\text{Int}(B) \setminus \{\infty\}$.
(ii) $g_B(z; \infty) = \log|z| - \log(\text{cap}(\mathbb{C} \setminus B)) + o(1)$ as $|z| \to \infty$.
(iii) $g_B(z; \infty) = 0$ for quasi every $z \in \mathbb{C} \setminus B$.

Of course, if $\operatorname{cap}(\mathbb{C} \setminus B) = 0$, then we set $g_B(z;\infty) \equiv \infty$. Otherwise we have the following theorem.

Theorem A.2. *If B is a Borel set such that $\mathbb{C} \setminus B$ is bounded and of positive capacity, then the Green function g_B with properties* (i)–(iii) *exists and is uniquely determined.*

Proof. The existence easily follows from Theorem A.1. In fact, if $E = \mathbb{C} \setminus B$ and we set

$$g_B(z;\infty) := -p(\omega_E; z) - \log(\operatorname{cap}(E)) \tag{A.10}$$

then (i)–(iii) follows from the corresponding properties of ω_E (recall also that ω_E is supported on $\overline{E}$).

To prove uniqueness, suppose that g_B satisfies (i)–(iii). On applying Riesz's representation theorem (see Section A.II) we get a measure μ such that for every $R > 0$

$$-g_B(z;\infty) = h_R(z) + \int_{|t|<R} \log \frac{1}{|z-t|} d\mu(t), \quad |z| < R,$$

where h_R is a harmonic function in the disk $D_R(0) = \{z \mid |z| < R\}$. We also know that μ does not have mass in an open set U if g_B is harmonic in U (see [Ts, Theorem II. 25]). Thus, μ is restricted to $\overline{\mathbb{C} \setminus B}$, and it follows that if $E := \mathbb{C} \setminus B \subseteq D_{R_0}(0)$, then for $R_0 < R_1 < R_2$ and $|z| < R_1$ we have $h_{R_1}(z) = h_{R_2}(z)$. Hence, by letting $R \to \infty$ we get the representation

$$-g_B(z;\infty) = h(z) + \int \log \frac{1}{|z-t|} d\mu(t), \quad z \in \mathbb{C}, \tag{A.11}$$

with a harmonic function h and a nonnegative measure μ with compact support. (A.11) and (ii) imply that h is bounded from above on $\mathbb{C}$ if $\|\mu\| \le 1$, whereas it is bounded below if $\|\mu\| > 1$, and in either case h must be constant. Thus, (A.11) and (ii) yield that $\|\mu\| = 1$ and then that $h = \log(\operatorname{cap}(\mathbb{C} \setminus B))$.

We can summarize our findings as follows: If g_B satisfies (i)–(iii), then there is a probability measure μ with bounded support such that

$$g_B(z;\infty) = -p(\mu; z) - \log(\operatorname{cap}(E)).$$

Thus, (i) and (iii) show that μ is the equilibrium measure of $E = \mathbb{C} \setminus B$. Now the uniqueness of g_B follows from the uniqueness of equilibrium measures (see Theorem A.1). □

A.VI Dirichlet's Problem

Let U be a subdomain of $\overline{\mathbb{C}}$ and suppose that $\mathbb{C} \setminus U$ is of positive capacity. Suppose that f is a continuous function defined on the boundary ∂U of U.

Dirichlet's problem consists of finding a function u that is harmonic in U, is continuous on $\overline{U}$, and has $u|_{\partial U} = f$. We say that U is *regular with respect to the Dirichlet problem* if the above u exists for every f. Sometimes we also use the terminology that ∂U (or $\overline{\mathbb{C}} \setminus U$) is regular with respect to the Dirichlet problem in U.

The following statements are equivalent:

(a) U is regular with respect to the Dirichlet problem,
(b) any Green function $g_U(z; a)$ associated with U is continuous throughout $\mathbb{C} \setminus \{a\}$,
(c) $\mathbb{C} \setminus U$ is dense in itself in the fine topology.

See [He, Corollary 10.12].

The *generalized Dirichlet problem* on U can be formulated as follows. Let f be a real-valued function on ∂U. For simplicity assume that $\infty \notin \partial U$. Consider the so-called upper and lower classes

$$\mathcal{U}_f = \{u \mid u \text{ superharmonic in } U; \liminf_{y \to x} u(y) \geq f(x) \text{ for all } x \in \partial U,\ u \text{ bounded from below on } U\}$$

and

$$\mathcal{L}_f = \{u \mid u \text{ subharmonic in } U; \limsup_{y \to x} u(y) \leq f(x) \text{ for all } x \in \partial U,\ u \text{ bounded from above on } U\}.$$

We define the upper and lower solutions of the generalized Dirichlet problem with boundary values f as $\overline{H}_f = \inf\{u \mid u \in \mathcal{U}_f\}$ and $\underline{H}_f = \sup\{u \mid u \in \mathcal{L}_f\}$. If $\underline{H}_f = \overline{H}_f$, then this common function H_f is called the solution of the generalized Dirichlet problem with boundary function f. It can be shown (see [He, Theorem 8.13]) that this is always the case if f is a Borel measurable function that is bounded from above or below. Furthermore in the definitions of the classes $\mathcal{U}_f$ and $\mathcal{L}_f$ we could have requested the limit relations $\liminf_{y \to x} u(y) \geq f(x)$ and $\limsup_{y \to x} u(y) \leq f(x)$ only for quasi every $z \in \partial U$ without changing the lower or upper functions.

A point $x \in \partial U$ is called *regular* if for every continuous f on ∂U we have $\lim_{z \to x, z \in U} H_f(z) = f(z)$. Then for all bounded f that are continuous at x the same relation holds; furthermore the following statements are equivalent for an $x \in \partial U$:

(α) x is a regular point,
(β) x is a fine limit of $\mathbb{C} \setminus U$,
(γ) there exists a positive superharmonic function u in a neighborhood of x in U such that $\lim_{y \to x, y \in U} u(y) = 0$.

See [He, Theorem 10.11 and Corollary 10.12]. The last property is often referred to as the barrier property.

It is obvious that U is regular with respect to the Dirichlet problem if and only if every point of ∂U is regular. On the other hand, for every U quasi every point on the boundary is a regular boundary point (see [Ts, Theorem III.33]).

A.VII Balayage

Consider in $\mathbb{C}$ an open set G with compact boundary ∂G and let μ be a measure with $\text{supp}(\mu) \subseteq \overline{G}$. The problem of *balayage* (or "sweeping out") consists of finding a new measure μ' *supported on* ∂G such that $\|\mu\| = \|\mu'\|$ and

$$p(\mu; z) = p(\mu'; z) \quad \text{for quasi every } z \notin G. \tag{A.12}$$

For bounded G such a measure always exists ([La, Chapter IV, Section 2]), but for unbounded G we must replace (A.12) by

$$p(\mu; z) = p(\mu'; z) + c \quad \text{for quasi every } z \notin G, \tag{A.13}$$

where the constant c turns out to be equal to

$$-\int_\Omega g_\Omega(z; \infty)\, d\mu(y),$$

where Ω is the component of G that contains the point infinity ([La, (4.2.6)]). If we also require the measure μ' to vanish on the set of irregular points of ∂G, then μ' is unique ([La, Chapter IV, Theorem 4.6, Corollary 2]) (see the uniqueness theorem in [La, Chapter IV, Section 14, p. 245]). Furthermore, if G is connected and it is regular with respect to the Dirichlet problem, then in (A.12)–(A.13) we have equality for all $z \notin G$.

Besides (A.12)–(A.13) we also know that

$$p(\mu'; z) \le p(\mu; z), \tag{A.14}$$

$$p(\mu'; z) \le p(\mu; z) + \int_\Omega g_\Omega(y; \infty)\, d\mu(y), \tag{A.15}$$

respectively, hold for all $z \in G$.

The balayage measures μ' also have the property that if h is any function that is continuous on $\overline{G}$ and harmonic in G, then

$$\int_{\partial G} h\, d\mu' = \int_{\overline{G}} h\, d\mu. \tag{A.16}$$

With the concept of balayage it is easy to describe the Green function of a domain G of compact boundary and $\operatorname{cap}(\partial G) > 0$. In fact, if G is bounded, then

$$g_G(z;a) = \log \frac{1}{|z-a|} - p(\delta'_a; z), \tag{A.17}$$

where δ'_a denotes the balayage of the Dirac mass δ_a at a onto the boundary of G. On the other hand, if G is unbounded (and so $\infty \in G$), then

$$g_G(z;a) = \log \frac{1}{|z-a|} - p(\delta'_a; z) + g_G(a;\infty), \quad a \neq \infty, \tag{A.18}$$

and

$$g_G(z;\infty) = -p(z;\delta'_\infty) - \log(\operatorname{cap}(\partial G)). \tag{A.19}$$

Furthermore, the equilibrium measure of a compact set K is the same as the balayage of δ_∞ onto the outer boundary of K.

A.VIII Green Potential and Condenser Capacity

Let $U \subseteq \overline{\mathbb{C}}$ be a domain and μ a measure in $\overline{\mathbb{C}}$. The *Green potential* of μ in U is defined as

$$g(\mu, U; z) := \int g_U(z;x)d\mu(x).$$

If $\operatorname{cap}(\overline{\mathbb{C}} \setminus U) = 0$, then $g(\mu, U; \cdot) \equiv \infty$.

Let $K, S \subseteq \overline{\mathbb{C}}$ be two disjoint compact sets. The pair (K, S) is called a *condenser*. For probability measures μ_1, μ_2 with $\operatorname{supp}(\mu_1) \subseteq K$ and $\operatorname{supp}(\mu_2) \subseteq S$ and the signed measure $\sigma = \mu_1 - \mu_2$ consider the energy

$$I(\sigma) = \int\int \log \frac{1}{|z-t|} d\sigma(z)d\sigma(t)$$

defined in (A.1), and set

$$\begin{aligned} V_{KS} = \inf\{I(\sigma) \mid \sigma = \mu_1 - \mu_2,\ &\operatorname{supp}(\mu_1) \subseteq K, \operatorname{supp}(\mu_2) \subseteq S, \\ &\mu_1,\ \mu_2 \geq 0,\ \|\mu_1\| = \|\mu_2\| = 1\}. \end{aligned} \tag{A.20}$$

The *condenser capacity* is defined as

$$\operatorname{cap}(K,S) = \frac{1}{V_{KS}}. \tag{A.21}$$

The capacity is symmetric in K and S, and $\operatorname{cap}(K,S) = 0$ if and only if either $\operatorname{cap}(K) = 0$ or $\operatorname{cap}(S) = 0$.

As in the case of logarithmic capacity, there exists an equilibrium distribution, which now is a pair of two probability measures: If K and S are of positive capacity, then there uniquely exists a pair of probability measures

$(\omega_{KS}, \omega_{SK})$ with $\operatorname{supp}(\omega_{KS}) \subseteq K$, $\operatorname{supp}(\omega_{SK}) \subseteq S$, and two constants V_S and V_K such that

(i) $$V_K - V_S = V_{KS},$$

(ii) $$V_S \le p(\omega_{KS} - \omega_{SK}; z) \le V_K \quad \text{for all } z \in \overline{\mathbb{C}},$$

(iii) $$p(\omega_{KS} - \omega_{SK}; z) = \begin{cases} V_S & \text{qu.e. on } S; \\ V_K & \text{qu.e. on } K \end{cases}$$

(see [Ba, Theorem 1] and [SaTo2, Chapter VII]). The measures ω_{KS} and ω_{SK} are called *equilibrium distributions of the condenser* (K, S) *on* K *and* S, respectively. With $\sigma = \omega_{KS} - \omega_{SK}$ the minimum is assumed in (A.20).

The properties (i) and (iii) imply that ω_{KS} (ω_{SK}) is the balayage of ω_{SK} (ω_{KS}) onto K (S).

We call

$$p_{KS}(z) = p(\omega_{KS} - \omega_{SK}; z)$$

the *condenser potential.* If $U = \mathbb{C} \setminus S$ is connected and $\operatorname{cap}(S) > 0$, then we have the representation (cf. (A.17) and (A.18))

(A.22) $$p_{KS}(\cdot) = g(\omega_{KS}, U; \cdot)$$

if $\infty \notin U$, and

(A.23) $$p_{KS}(\cdot) = g(\omega_{KS}, U; \cdot) - g(\omega_{KS}, U; \infty)$$

otherwise.

From [Ba, Theorem 1] (cf. also [SaTo2, Chapter VII]) it follows that if for two probability measures μ_1 and μ_2 in $\overline{\mathbb{C}}$ of compact support and V_1, $V_2 \in \mathbb{R}$ we have

(A.24) $$p(\mu_2 - \mu_1; \cdot) \begin{cases} \le V_1 & \text{qu.e. on } S; \\ \ge V_2 & \text{qu.e. on } K, \end{cases}$$

then

(A.25) $$V_2 - V_1 \le V_{KS} = \frac{1}{\operatorname{cap}(K, S)}.$$

It is easy to see (e.g., by adding to μ_j discrete measures of the form const. $\cdot\, \delta_x$ with $|x| \to \infty$) that (A.24) implies (A.25) for all positive measures μ_1 and μ_2 of compact support in $\mathbb{C}$ with $\|\mu_1\| \le 1$ and $\|\mu_2\| \le 1$.

The condenser capacity is outer continuous: If for a $\delta > 0$ K_δ (respectively, S_δ) denote the sets of points the distance of which to K (respectively, S) is at most δ, then

$$\lim_{\delta \to 0} \operatorname{cap}(K_\delta, S_\delta) = \operatorname{cap}(K, S).$$

A.IX The Energy Problem in the Presence of an External Field

Let $\Sigma \subseteq \mathbb{C}$ be a closed set and let $w\colon \Sigma \to \mathbb{R}_+$ be a weight function on Σ. We call w ***admissible*** if the following three conditions are satisfied:

(i) w is upper semicontinuous,
(ii) $w > 0$ on a set of positive capacity, and
(iii) if Σ is unbounded, then $w(z)z \to 0$ as $z \to \infty$, $z \in \Sigma$.

We shall set $w(z) = \exp(-Q(z))$ and for admissible w consider the problem of minimizing the weighted energy

$$I_w(\omega) = \int\int \left(\log \frac{1}{|z-t|} + Q(z) + Q(t)\right) d\omega(z)d\omega(t)$$

for all probability measures ω with support in Σ. There is a unique probability measure ω_w with support in Σ, called the *extremal* or *equilibrium measure* associated with w, minimizing the weighted energy (see [SaTo2, Section I.1]). This ω_w has finite logarithmic energy. Let $\mathcal{S}_w = \operatorname{supp}(\omega_w) \subseteq \Sigma$ be the support of the extremal measure. Then (see [SaTo2, Sections I.1 and I.3]) $\mathcal{S}_w$ is compact and of positive capacity, and Q is bounded on $\mathcal{S}_w$.

There exists a constant F_w such that for the potential of ω_w we have ([SaTo2, Section I.1])

$$p(\omega_w; z) \geq -Q(z) + F_w \quad \text{for quasi every } z \in \Sigma \tag{A.26}$$

and

$$p(\omega_w; z) \leq -Q(z) + F_w \quad \text{for every } z \in \mathcal{S}_w. \tag{A.27}$$

The constant F_w is determined by

$$F_w = I_w(\omega_w) - \int Q d\omega_w.$$

Another representation for F_w is (see [SaTo2, Section III.2])

$$\begin{aligned} F_w &= \log \frac{1}{\operatorname{cap}(\mathcal{S}_w)} + \int Q d\omega_{\mathcal{S}_w} \\ &= \inf_{K \subseteq \Sigma} \left\{ \log \frac{1}{\operatorname{cap}(K)} + \int Q d\omega_K \right\}, \end{aligned} \tag{A.28}$$

where the infimum is taken over all compact subsets of Σ.

F_w is connected to the weighted Chebyshev problem with varying weights w^n as follows (for the following results see [SaTo2, Sections II.2 and II.3]). First of all, if $P_n(z) = z^n + \cdots$ is any monic polynomial of degree n then

$$\|w^n P_n\|_\Sigma \geq \|w^n P_n\|_{\mathcal{S}_w} \geq \exp(-F_w). \tag{A.29}$$

On the other hand, if

$$t_n^w = \inf\{\|w^n P_n\|_\Sigma \mid P_n(z) = z^n + \cdots\}$$

is the nth weighted Chebyshev number, then the limit, which is called the *Chebyshev constant* for the weight w,

$$\lim_{n\to\infty} (t_n^w)^{1/n} = t^w \tag{A.30}$$

exists and is equal to $\exp(-F_w)$.

If $\deg(P_n) \le n$ and

$$w^n(z)|P_n(z)| \le M \quad \text{for qu.e. } z \in S_w$$

then

$$|P_n(z)| \le M \exp(n(-p(\omega_w; z) + F_w)) \quad \text{for qu.e. } z \in \mathbb{C}. \tag{A.31}$$

Inequality (A.31) is the analogue of the Bernstein–Walsh lemma, and (A.26) and (A.31) tell us that the supremum norm of every weighted polynomial $w^n P_n$ "essentially lives" on S_w. It turns out that S_w is the smallest compact set with this property.

If we consider the weights w^λ with $\lambda > 0$, then in the range of the λs for which w^λ is admissible, the support sets S_{w^λ} form a decreasing family as λ increases (see [SaTo2, Section III.2]).

The classical case corresponds to Σ being compact and $w \equiv 1$ on Σ. Of course, then the energy problem just discussed coincides with the one in Appendix I, and in this case $\omega_w \equiv \omega_\Sigma$, $F_w = \log(1/\text{cap}(\Sigma))$, and $t^w \equiv t^1 = \text{cap}(\Sigma)$.

Notes and Bibliographical References

Preface

Apparently the first study of nth-root asymptotic behavior was done by Faber [Fa] in 1922 in connection with polynomial expansions. Of course, in the spirit of his time he used quite strong assumptions. In his paper the measure μ was defined on $[-1, 1]$ by a density function, which was assumed to be bounded away from zero. Faber's paper appeared about the same time as Szegő's first papers ([Sz1], [Sz2]) on power asymptotics. Both authors were well aware of their different approaches and gave mutual references.

Major contributions to the further development of the subject of nth-root asymptotic behavior have been given by Erdős and Turán [ErTu], Widom [Wi1], Ullman [Ul1], and Erdős and Freud [ErFr], and Korovkin [Ko1,2]. A good survey of recent developments of the whole theory (not only nth-root theory) of asymptotic behavior of orthogonal polynomials can be found in [Lu1] and [Ne2].

Chapter 1

Lemma 1.1.3 is due to H. Widom; see [Wi1, Lemma 4].

The upper asymptotic bound (1.1.6) of Theorem 1.1.4 was proved by Ullman in Theorem 1.5 of [Ul1] for weight measures with $S(\mu) = [-1, 1]$, and in Theorem 3 of [Ul3] for general weight measures μ in $\mathbb{C}$. Theorem 1.1.8 seems to be new, but, of course, it is a rather obvious generalization of the upper bound (1.1.6).

The lower asymptotic bound (1.1.7) in Theorem 2.4 has been proved by Widom [Wi1] for special weight measures μ in $\mathbb{C}$, which he calls admissible, that is, measures satisfying Widom's criterion (in its original formulation, as given in Corollary 4.1.7). For these measures the lower bound can be established by a comparison with monic Chebychev polynomials of the set $S(\mu)$. For weight measures μ on $\mathbb{R}$ the lower bound (1.1.7) has been stated without proof in [Ul2].

The lower estimate in (1.1.13) was proved in [ErFr] for the special case $S(\mu) = [-1, 1]$.

The sharpness of the estimates is shown in [UWZ] and [UlWy] for the special case $S(\mu) = [-1, 1]$. In its generality Theorem 1.1.9 is new. In connection with Theorem 1.1.9 and Corollary 1.1.7 we mention the following result of M. F. Wyneken [Wy1]: Let $\text{supp}(\mu) \subseteq \mathbb{R}$ and $[A, B]$ be a subinterval of $[1/\text{cap}(S(\mu)), 1/c_\mu]$. Then there is a measure ν carrier related to μ such that the accumulation points of the sequence $\{\gamma_n(\nu)^{1/n}\}$ are exactly $[A, B]$. It can also be shown that the set of accumulation points must be a subinterval of $[1/\text{cap}(S(\mu)), 1/c_\mu]$.

Lemma 1.3.1 is due to Ullman [Ul1].

Lemma 1.3.2 is new in the form as it is stated; however, it can be rather directly deduced from Widom's original proof for Lemma 1.1.3 via a compactness argument. We have chosen a new, potential-theoretic argument for the proof that is close in spirit to the rest of the book. We mention that Lemma 1.3.2 can be very helpful in proving the analogue of Lemma 1.1.3 for many other extremal polynomials, for example, for polynomials having minimal weighted $L^p(\mu)$ norm with any weight.

Most of the proofs in Chapter 1 and many from later chapters are based on the results of Section 1.2.

Chapter 2

Some of the results of Section 2.2 belong to the folklore of the theory of orthogonal polynomials although they have never been investigated in the details presented in Section 2.2. See, for example, the survey paper [Sa2].

Theorem 2.1.1 is a consequence of Lemma 1.3.2 and it is due to Widom [Wi1].

For a simple proof of the Turán problem see [NeTo]. In connection with its generalization given in Theorem 2.1.3 it is an interesting problem that if S is given then precisely what subsets of $\text{Co}(S)$ can be the weak* limit points of zeros of orthogonal polynomials corresponding to a measure with support equal to S?

Concerning Theorems 2.1.3 and 2.1.4 we raise the following problem: Can it happen that the support of the measure μ of orthogonality is the unit circle and the support of the asymptotic distribution of the orthogonal polynomials $p_n(\mu; \cdot)$, or of a subsequence of them, is a prescribed closed set in the unit disk? Or more generally, what kind of measures can be obtained as weak* limits of zeros in this situation?

Theorem 2.1.4 solves a problem of Ullman and Wyneken (problem (c) of [UlWy]) asking for a characterization of possible weak* limits of zeros. In connection with the case $c_\mu > 0$ Ullman and Wyneken showed in [UlWy] that if $C \subseteq [0,1]$ is any carrier of μ then there is a measure μ' with the same carriers as μ such that the limit distribution of the orthogonal polynomials corresponding to μ' is the equilibrium distribution of C. They asked (see problem (a) of [UlWy]) if in the case $c_\mu > 0$ the weak* limits of zeros must always be equilibrium distributions of some carriers of μ. Example 2.1.5 shows that the answer for this problem is "no."

Lemma 2.1.6 follows from the general theory of weighted potentials (see [SaTo2]) outlined in Appendix IX.

Theorem 2.2.1(a) was used in [Ul2] and [UWZ] to show that for the case

$$\liminf_{n\to\infty} \gamma_n(\mu)^{1/n} < \limsup_{n\to\infty} \gamma_n(\mu)^{1/n}$$

the zeros of the orthogonal polynomials $\{p_n(\mu;\cdot)\}$ have continuum many different weak* limits. In fact, this follows from Theorem 2.2.1(a) and from the fact that the limit points of the sequence $\{\gamma_n(\mu)^{1/n}\}$ are always an interval.

Example 2.2.7 answers the following problem of [UlWy] in the negative: If $\{\gamma_n(\mu)^{1/n}\}$ converges to $\lambda \in (1/\mathrm{cap}(S(\mu)), 1/c_\mu)$, does $\{\nu_{p_n(\mu;\cdot)}\}$ converge?

The construction of Section 2.3 uses some ideas of the theory of weighted potentials (cf. Appendix IX and also the notes on Section 6.5). The points $\{x_0^{(k,n)}, x_1^{(k,n)}, \dots, x_n^{(k,n)}\}$ from step IV are called Fekete points associated with the weight w_k (see [SaTo2]). Note also that in step V of the proof we essentially verify the formula (A.30) from Appendix IX.

Chapter 3

The results in Section 3.1 must be compared with those announced in [Ul2]. Ullman considers asympotics for the monic orthogonal polynomials $q_n(\mu;\cdot)$ for measures μ with support in $[-1,1]$. Let us consider the statements

(a) $\lim_{n\to\infty} |q_n(\mu;z)|^{1/n} = |z+\sqrt{z^2-1}|$ for $z \in \mathbb{C}\setminus[-1,1]$,
(b) $\limsup_{n\to\infty} |q_n(\mu;z)|^{1/n} = \frac{1}{2}$ for qu.e. $z \in [-1,1]$,
(c) the orthogonal polynomials $q_n(\mu;\cdot)$ have arcsine zero distribution,
(d) $\lim_{n\to\infty} \gamma_n(\mu)^{1/n} = 2$.

In this setting (a),(b), and (c) are equivalent and (d) implies each of them. Thus, the theorem in Section 3.6 is nothing else than the extension of Ullman's results to the case when the support can be a general set on the complex plane.

[ErFr] contains the equivalence of (i) and (ii) of Theorem 3.1.1 in the special case $S(\mu) = [-1,1]$. The same equivalence when the support is the unit circle was proved in [LSS].

The results of Sections 3.2–3.5 are new.

Orthogonal polynomials with respect to varying weights, such as the ones in statement (vi) of Theorem 3.2.1 or those in Section 3.3, are important in multipoint Padé approximation and in best rational approximation (cf. Sections 6.1–6.3). Such polynomials have been the subject of some recent investigations (see, e.g., [GoLo], [Lo]).

Asymptotics for the leading coefficients of the polynomials $p_n(\mu; z)$, $n \in \mathbb{N}$, orthonormal with respect to measures μ of the type defined in (3.5.1) and (3.5.2) have been studied in [VAMa] and [Lu2] with a higher degree of precision than obtained and needed in Section 3.5. For example, Lubinsky (and by a somewhat different argument Van Assche and Magnus) verified that by properly choosing the coefficients $\{a_n\}$ in (3.5.2), one can get a purely discrete measure μ for which the recurrence coefficients $\{\gamma_n(\mu)/\gamma_{n+1}(\mu)\}$ converge to $\frac{1}{2}$ as $n \to \infty$.

Chapter 4

The results of Sections 4.2–4.6 are new, as well as parts of Section 4.1 (the generalization of and the relation among the Erdős–Turán, Ullman, and Widom criteria).

Let us mention that Ullman's original criterion was independently found also by P. P. Korovkin [Ko1] in a slightly different form.

The results in Chapter 4 solve a problem of D. S. Lubinsky [Lu1] that asked for the extension of Ullman's criterion to "weights spread over several intervals or with even more complicated supports."

Lemma 4.4.1 is probably known but we could not find a reference to it.

We have proved in Section 4.6 that neither criterion Λ^* nor criterion Λ implies the other one. In connection with this we mention the following open problem: We have found no regular μ for which

$$\liminf_{r \to 0} \operatorname{cap}(\{z \mid \mu(\Delta_r(z)) \geq r^L\})$$

does not tend to $\operatorname{cap}(S(\mu))$ as $L \to \infty$. If there is no such μ then criterion Λ^* is almost necessary and sufficient for regularity.

Chapter 5

The results of Chapter 5 are new. Chronologically they were preceded by the results of [St3], which paper contains a different method for proving localization theorems in the special case when the support is on the real line (for yet another method that works in this special case see [To2]).

Related to the localization results of Chapter 5 is the following theorem on the local density of zeros (see [UlTo]): *Suppose that $S(\mu)$ has empty interior and connected complement, and let $K \subseteq S(\mu)$ be a compact subset* *of $S(\mu)$. If $\mu|_K$ is regular (with respect to nth-root asymptotic behavior), then for any weak* limit point ν of the zeros of $p_n(\mu;\cdot)$ we have*

$$\nu|_K \geq \omega_{S(\mu)}|_K.$$

In other words, if $K_1 \subseteq K$ is any compact set and U is any neighborhood of K_1, then asymptotically there are at least $n\omega_{S(\mu)}(K_1)$ zeros of $p_n(\mu;\cdot)$ in U.

As a corresponding upper bound for the density of the zeros on K we mention

$$\nu|_K \leq \omega_K,$$

which means that if $K_1 \subseteq K$ is any compact set and U is any neighborhood of K_1, then asymptotically there are at most $n\omega_K(U)$ zeros of $p_n(\mu;\cdot)$ in K_1.

Chapter 6

Section 6.1

The connection between rational interpolation and orthogonal polynomials is classical. This is not only the case for the more special situation of continued fractions, where it has long been known that the denominator polynomials satisfy an orthogonality relation. Jacobi [Ja] already knew the orthogonality (6.1.9) of denominators of interpolating rational functions, representing, however, the integral by a sum of residues. From the orthogonality he deduced different determinantal representations for denominators and numerators. Jacobi did not investigate the convergence problem, neither did he care about the possibility that in general not all interpolation conditions can be satisfied by rational interpolants. Apparently, the possibility of interpolation defects was for the first time realized, or at least mentioned, by Kronecker in [Kr].

The first convergence result relevant for the subject of Section 6.1 goes back to A. Markov [Ma], and was formulated for Chebyshev continued fractions. Their approximants are rational interpolants at the point infinity. In [GoLo] Gonchar and López proved part (a) of Theorem 6.1.6 and proved exact rates of convergence under the assumption that $S(\mu)$ consists of finitely many closed intervals and the measure μ has a derivative $\mu' > 0$ almost everywhere on $S(\mu)$. This last condition corresponds to the regularity criterion of Corollary 4.1.2.

Part (b) of Theorem 6.1.6 is new.

Section 6.2

A discussion of the uniqueness problem for best rational approximants with real or complex coefficients can be found in the paper [SaVa] by Saff and

Varga. A comparison of the minimal error of real versus complex rational functions can be found in [GuTr], [Le], [RuVa1,2], and [SaVa].

We have already mentioned earlier the conjecture that for the case of a Markov function f and a compact set V with symmetry (6.2.2) the best rational approximants $r_n^*(f, V; \cdot)$, $n \in \mathbb{N}$, are all unique. This conjecture is still open.

W. Barrett [Bar1,2] and T. Ganelius [Ga] have investigated the rate of convergence of best rational approximants to Markov functions with measures that belong to the Szegő class. Gonchar proves part (a) of Theorem 6.2.2 in [Go] for sets $V \subseteq \mathbb{R}$, and gives the exact rate of convergence under the assumption that $S(\mu)$ consists of finitely many closed intervals and the measure μ has a derivative $\mu' > 0$ almost everywhere on $S(\mu)$. In its general form Theorem 6.2.2 is new.

Section 6.3

The convergence and divergence regions for nondiagonal sequences of Padé approximants have been investigated in [St1] for more general measures μ than in section 6.3 (see also [St2]). A very elementary (non-potential-theoretic) proof of part (a) of Theorem 6.3.1 for ray sequences in the lower triangle of the Padé table is contained in [GrM]. In [Sa1] the connection between nondiagonal Padé approximants and incomplete polynomials has been studied.

Section 6.4

The theory of weighted potentials has a long history. The existence of the equilibrium measure in the presence of a continuous outer weight function was already proved by Gauss in the nineteenth century (see [He]). The recent interest in weighted potentials is due to the works of H. N. Mhaskar, E. B. Saff [MhSa1,2,3], E. A. Rahmanov [Ra4], D. S. Lubinsky [LuSa1,2], and A. A. Gonchar [GoRa1]. Weighted potentials have played a prominent role in the solution of Freud's conjecture by Lubinsky, Mhaskar, and Saff [LMS] and in the solution of the 1/9-th conjecture by Gonchar and Rahmanov [GoRa2]. A detailed study of weighted potentials can be found in [SaTo2].

In connection with Theorem 6.4.1 we mention that it is possible and in many situations it is desirable to consider stronger forms of (6.4.5). In [MhSa3] it was stated that if Σ is a finite subinterval of $\mathbb{R}$ and w is positive and continuous on Σ then for every $\eta > 0$ there are constants c_1 and c_2 and a compact set $\Delta \subseteq \Sigma$ of Lebesgue measure $< \eta$ such that

$$\|w^n P_n\|_{L^p(\Sigma)} \le (1 + c_1 e^{-c_2 n}) \|w^n P_n\|_{L^p(S_w \cup \Delta)}. \tag{N.1}$$

However, the Σ and w used in Example 6.4.3 show together with the classical Chebyshev polynomials

$$P_n(x) = \frac{1}{2}\left\{(x + \sqrt{x^2 - 1})^n + (x - \sqrt{x^2 - 1})^n\right\},$$

that such a general statement is impossible to hold (cf. [MhSa4]), hence the nth root is natural in Theorem 6.4.1 to get a general statement. Of course, using special properties of the weight w we can often sharpen Theorem 6.4.1 in the sense of (N.1). This is the case, for example, for weight functions $w(x) = \exp(-|x|^\lambda)$, $\lambda > 1$, $x \in \mathbb{R}$; see [LMS], [LuSa1], or [Lu3]. See also [SaTo2] for general conditions ensuring (N.1).

Section 6.5

The expression $c^w := \exp(-I_w(\omega_w))$, where $I_w(\omega_w)$ is the minimal weighted energy of Section 6.4, is called the weighted capacity associated with the weight w. Another classical notion of potential theory is the transfinite diameter of a set, the weighted analogue of which is the limit

$$\delta^w := \lim_{n\to\infty} \delta_n^w,$$

where

$$\delta_n^w := \sup \left(\prod_{1\le i<j\le n} |z_i - z_j| w(z_i) w(z_j) \right)^{2/(n(n-1))}$$

with the supremum taken for all n-tuples $\{x_1, \ldots, x_n\} \subseteq \Sigma$. In the classical case ($w \equiv 1$, Σ compact) the three quantities capacity ($\mathrm{cap}(\Sigma)$), Chebyshev constant ($t^w :=$ limit of $\{(t_n^w)^{1/n}\}$), and transfinite diameter (δ^w) coincide (see [Ts, Theorem III.26]). In the weighted case there is only a slight modification (see [SaTo2, Theorems II.1.3 and II.3.1])

$$c^w = \delta^w = t^w \exp\left(-\int Q \, d\omega_w\right).$$

Far better asymptotics than (6.5.2) are available these days in the literature, due mainly to D. S. Lubinsky and E. B. Saff [LuSa1], for the leading coefficients for orthogonal polynomials with respect to so-called Freud weights $\exp(-|x|^\lambda)$. For asymptotics on the orthogonal polynomials themselves with respect to Freud weights see [Ra4] and [Lu3].

Section 6.6

Some of the equivalences in Theorem 6.6.1 were proved in [SaTo1] for $\mathrm{supp}(\mu) \subseteq [-1, 1]$, where also infinite measures were treated. Similar questions for the case when $\mathrm{supp}(\mu)$ is the unit circle were considered in [LSS].

Section 6.7

If one considers orthogonal polynomials with respect to a weight function and not a measure then much depends on the set where the weight function is positive. Such an attitude has led the authors of [Ul1] and [ErFr] to consider weight functions that are positive on a given set $E \subseteq [-1,1]$ and to investigate properties of orthogonal polynomials that depend exclusively on E. The results of Section 6.7 extend earlier investigations in this direction.

Appendix

In the proof of Theorem A.1 we follow the presentation of [Wy2].

In connection with Appendix VIII we mention that it is usual to put additional assumptions on V and S in the definition of a condenser. For example, in [Ba] it is assumed that there is an open set U separating V and S in the sense that U is the only connected component of $\mathbb{C}\backslash\{V \cup U\}$ that has a common boundary with both V and S. However, the results that we use in Section 6.2 in connection with the condenser capacity are valid without any further assumption on V and S, $V \cap S = \emptyset$ (see, e.g., [SaTo2, Chapter VII]).

Bibliography

[AlVi] M. P. Alfaro and L. Vigil. Solution of a problem of P. Turán on zeros of orthonormal polynomials on the unit circle. *J. Approx. Theory* **53**(1988): 195–7.

[An1] A. Ancona. Demonstration d'une conjecture sur la capacité et l'effilement. *C. R. Acad. Sci. Paris* **297**(1983): 393–5.

[An2] A. Ancona. Sur une conjecture concernant la capacité et l'effilement. In *Colloque du Théorie du Potential*, Lecture Notes in Mathematics, Vol. 1096, eds. G. Mokobodzki and D. Pinchon. Springer-Verlag, New York, 1984.

[Ba] T. Bagby. The modulus of a plane condenser. *J. Math. Mech.* **17**(1967): 315–29.

[Bak] G. A. Baker, Jr. *Essentials of Padé Approximants.* Academic Press, New York, 1975.

[BaGM] G. A. Baker, Jr. and P. R. Graves-Morris. *Padé Approximants*, Parts I and II. Encycl. Math. Vols. 13 and 14. Cambridge University Press, Cambridge, 1981.

[Bar1] W. Barrett. On best rational approximations to functions defined as integrals. *J. Inst. Maths. Applics.* **13**(1974): 107–16.

[Bar2] W. Barrett. On the convergence of sequences of rational approximations to analytic functions of a certain class. *J. Inst. Maths. Applics.* **7** (1971): 308–23.

[Br] D. Braess. *Nonlinear Approximation Theory.* Springer Series in Computational Mathematics, Vol. 7. Springer-Verlag, Berlin, 1986.

[Ca] L. Carleson. Mergelyan's theorem on uniform polynomial approximation. *Math. Scand.* **15**(1964): 167–75.

[Ch] T. S. Chihara. *An Introduction to Orthogonal Polynomials.* Gordon and Breach, New York, 1978.

[dVP] C. J. de la Vallée-Poussin. *Le Potentiel Logarithmique.* Gauthier–Villars, Paris, 1949.

[Do] J. L. Doob. *Classical Potential Theory and Its Probabilistic Counterpart.* Grundlehren der Mathematischen Wissenschaften, Vol. 262. Springer-Verlag, New York, 1983.

[ErFr] P. Erdős and G. Freud. On orthogonal polynomials with regularly distributed zeros. *Proc. London Math. Soc.* **29**(1974): 521–37.

[ErTu] P. Erdős and P. Turán. On interpolation III. *Ann. Math.* **41**(1940): 510–53.

[Fa] G. Faber. Ueber nach Polynomen fortschreitende Reihen. *Sitzungsberichte der Bayerischen Akademie der Wissenschaften*, 1922, pp. 157–78.

[Fr] G. Freud. *Orthogonal Polynomials.* Pergamon Press, Oxford, 1971.

[Ga] T. Gandelius. Orthogonal polynomials and rational approximation of holomorphic functions. In *Studies in Pure Mathematics* (To the Memory of Paul Turán), ed. P. Erdős, pp. 237–43. Birkhäuser Verlag, Basel, 1978.

[Ge] Y. L. Geronimus. *Polynomials Orthogonal on a Circle and on an Interval.* Pergamon Press, Oxford, 1960.

[Go] A. A. Gonchar. On the speed of rational approximation of some analytic functions. *Mat. Sb.* **105**,147(1978); English transl.: *Math. USSR Sb.* **34**(1978): 131–45.

[GoLo] A. A. Gonchar and G. López. On Markov's theorem for multipoint Padé approximants. *Mat. Sb.* **105**,147(1978); English transl.: *Math. USSR Sb.* **34**(1978): 449–59.

[GoRa1] A. A. Gonchar and E. A. Rahmanov. Equilibrium measure and distribution of zeros of extremal polynomials. *Mat. Sb.* **125**,167(1984): 117–27.

[GoRa2] A. A. Gonchar and E. A. Rahmanov. Equilibrium distributions and degree of rational approximation of analytic functions. *Mat. Sb.* **134**,176 (1987). English transl.: *Math. USSR Sbornik* **62**(1989): 305–48.

[GrM] P. R. Graves-Morris. The convergence of ray sequences of Padé approximants of Stieltjes functions. *J. Comp. and Appl. Math.* **7**(1981): 191–201.

[GuTr] M. H. Gutknecht and L. N. Trefethen. Real vs. complex rational chebychev approximation on complex domains. In *ISNM 67, Numerical Methods of Approximation Theory*, Vol. 7, eds. L. Kollatz and H. Werner, pp. 87–97. Birkhäuser Verlag, Basel, 1983.

[Ha] W. K. Hayman. *Subharmonic Functions*, Vol. 2. London Math. Soc. Monographs, Vol. 20. Academic Press, London, 1989.

[HaKe] W. K. Hayman and P. B. Kennedy. *Subharmonic Functions*, Vol. 1. London Math. Soc. Monographs, Vol. 9. Academic Press, London, 1976.

[He] L. L. Helms. *Introduction to Potential Theory.* Wiley-Interscience, New York, 1969.

[Hi] E. Hille. *Analytic Function Theory*, Vol. II. Ginn and Company, Boston, 1962.

[Ja] C. G. I. Jacobi. Ueber die Darstellung einer Reihe gegebener Werte durch eine rationale gebrochene rationale Function. *J. reine u. angew. Math.* **30**(1846): 127–56.

[Ke] O. D. Kellogg. *Foundations of Potential Theory.* Grundlehren der Mathematischen Wissenschaften, Vol. 31. Springer-Verlag, Berlin, 1967 (reprint).

[Ko1] P. P. Korovkin. Asymptotic representation of polynomials minimizing an integral. In *Studies of Modern Problems of Constructive Theory of Functions*, ed. Riekstins, pp. 273–6. Fizmatgiz, Moscow, 1961 (Russian).

[Ko2] P. P. Korovkin. The asymptotic representation of polynomials orthogonal over a region. *Dokl. Akad. Nauk USSR(NS)* **58**(1947): 1883–5 (Russian).

[Kr] L. Kronecker. Zur Theorie der Elimination einer Variablen aus zwei algebraischen Gleichungen. *Monatsber. Kőnigl. Preuss. Akad. Wiss. Berlin*, 1881, pp. 535–600.

[La] N. S. Landkof. *Foundations of Modern Potential Theory.* Grundlehren der Mathematischen Wissenschaften, Vol. 190. Springer-Verlag, New York, 1972.

[Le] A. Levin. On the degree of complex rational approximation to real functions. *Const. Approx.* **2**(1986): 213–19.

[LSS] X. Li, E. B. Saff, and Z. Sha. Behavior of best L_p polynomial approximants on the unit interval and on the unit circle. *J. Approx. Theory* **63**(1990): 170–90.

[Lo] G. López. Asymptotics of polynomials orthogonal with respect to varying measures. *Constructive Approximation* **5**(1989): 199–219.

[Lu1] D. S. Lubinsky. A survey of general orthogonal polynomials for weights on finite and infinite intervals. *Acta Appl. Math.* **10**(1987): 237–96.

[Lu2] D. S. Lubinsky. Jump distributions on $[-1,1]$ whose orthogonal polynomials have leading coefficients with given asymptotic behavior. *Proc. Amer. Math. Soc.* **104**(1988): 516–24.

[Lu3] D. S. Lubinsky. *Strong Asymptotics for Extremal Errors and Polynomials Associated with Erdős-type Weights.* Pitman Research Notes, Vol. 202. Longman House, Harlow, UK, 1988.

[LMS] D. S. Lubinsky, H. N. Mhaskar, and E. B. Saff. A proof of Freud's conjecture for exponential weights. *Constructive Approximation* **4**(1988): 65–83.

[LuSa1] D. S. Lubinsky and E. B. Saff. *Strong Asymptotics for Extremal Polynomials Associated with Weights on* $\mathbb{R}$. Lecture Notes in Mathematics, Vol. 1305. Springer-Verlag, New York, 1988.

[LuSa2] D. S. Lubinsky and E. B. Saff. Uniform and mean approximation by certain weighted polynomials, with applications. *Constructive Approximation* **4**(1988): 21–64.

[Ma] A. Markov. Deux demonstrations de la convergence de certaines fractions continues. *Act. Math.* **19**(1895): 93–104.

[MhSa1] H. N. Mhaskar and E. B. Saff. Where does the sup norm of a weighted polynomial live? *Constructive Approx.* **1**(1985): 71–91.

[MhSa2] H. N. Mhaskar and E. B. Saff. Extremal problems for polynomials with exponential weights. *Trans. Amer. Math. Soc.* **285**(1984): 203–34.

[MhSa3] H. N. Mhaskar and E. B. Saff. Where does the L_p norm of a weighted polynomial live? *Trans. Amer. Math. Soc.* **303**(1987): 109–24.

[MhSa4] H. N. Mhaskar and E. B. Saff. Errata to "Where does the L_p norm of a weighted polynomial live?" *Trans. Amer. Math. Soc.* **308**(1988): 431.

[Mu] N. I. Muskhelishvili. *Singular Integral Equations.* P. Noordhoff N. V., Groningen, 1953.

[Ne1] P. Nevai. *Orthogonal Polynomials.* Memoirs Amer. Math. Soc., Vol. 213. Amer. Math. Soc., Providence, RI, 1979.

[Ne2] P. Nevai. Géza Freud, orthogonal polynomials and Christoffel functions. A case study. *J. Approx. Theory* **48**(1986): 3–167.

[NeTo] P. Nevai and V. Totik. Orthogonal polynomials and their zeros. *Acta Sci. Math.* (Szeged) **53**(1989): 99–104.

[Pe] O. Perron. *Die Lehre von den Kettenbrüchen*. Chelsea Publ. Comp., New York, 1956 (reprint).

[Pl] W. Plešnak. On the distribution of zeros of the polynomials of best L^2-approximation to holomorphic functions. *Z. Nauk. Univ. Jagiellońskiego* **22**(1981): 29–35.

[Po] C. Pommerenke. *Univalent Functions*. Vandenhoek and Ruprecht, Göttingen, 1975.

[Ra1] E. A. Rahmanov. On the asymptotics of the ratio of orthogonal polynomials. *Mat. Sb.* **103**,145(1977): 237–52; English transl.: *Math. USSR Sb.* **32**(1977): 199–213.

[Ra2] E. A. Rahmanov. Convergence of diagonal Padé approximants. *Mat. Sb.* **104**,146(1977): 271–91; English transl.: *Math. USSR Sb.* **33**(1977): 243–60.

[Ra3] E. A. Rahmanov. On the asymptotics of the ratio of orthogonal polynomials II. *Mat. Sb.* **118**,160(1982): 104–17; English transl.: *Math. USSR Sb.* **465**(1983): 105–17.

[Ra4] E. A. Rahmanov. On asymptotic properties of polynomials orthogonal on the real axis. *Mat. Sb.* **119**,161(1982): 163–203; English transl.: *Math. USSR Sb.* **47**(1984): 155–93.

[RuVa1] A. Ruttan and R. S. Varga. Real versus complex rational Chebyshev approximation on an interval $\gamma_{m,m+2} \leq \frac{1}{3}$. *Rocky Mount. J. Math.* **19**(1989): 375–81.

[RuVa2] A. Ruttan and R. S. Varga. A unified theory for real versus complex rational Chebyshev approximation on an interval. *Trans. Amer. Math. Soc.* **312**(1989): 681–97.

[Sa1] E. B. Saff. Incomplete and orthogonal polynomials. In *Approximation Theory IV*, eds. C. K. Chui et al., pp. 219–55. Academic Press, New York, 1983.

[Sa2] E. B. Saff. Orthogonal polynomials from a complex perspective. In *Orthogonal Polynomials and Its Applications*, NATO Advanced Study Conference, Columbus, USA, 1989, ed. P. Nevai. NATO ASI Series C, Vol. 294, pp. 363–93. Kluwer Academic Publishers, Dordrecht, 1989.

[SaTo1] E. B. Saff and V. Totik. Weighted polynomial approximation of analytic functions. *Proc. London Math. Soc.* **37**(1988): 455–63.

[SaTo2] E. B. Saff and V. Totik. *Logarithmic Potential with External Fields*. Springer-Verlag (in press).

[SaVa] E. B. Saff and R. S. Varga. Nonuniqueness of best complex rational approximations to real functions on real intervals. *J. Approx. Theory* **23**(1978): 78–85.

[St1] H. Stahl. *Beiträge zum Problem der Konvergenz von Padé Approximierenden*. Dissertation TU–Berlin, 1976.

[St2] H. Stahl. Nondiagonal Padé approximants to Markov functions. In *Approximation Theory V*, eds. C. K. Chui et al., pp. 51–574. Academic Press, New York, 1986.

[St3] H. Stahl. On regular nth root behaviour of orthonormal polynomials *J. Approx. Theory* **66**(1991): 125–61.

[StTo] H. Stahl and V. Totik. Nth root asymptotic behavior of orthonormal polynomials. In *Orthogonal Polynomials and its Applications*, NATO Advanced Study Conference, Columbus, USA, 1989, ed. P. Nevai. NATO

ASI Series C, Vol. 294, pp. 395–417. Kluwer Academic Publishers, Dordrecht, 1989.

[Sz1] G. Szegő. Ueber orthogonale Polynome, die zu einer gegebenen Kurve der komplexen Ebene gehören. *Math. Zeitschr.* **9**(1921): 218–70 (see also Gábor Szegő, *Collected Papers I*, ed. R. Askey, Birkhäuser Verlag, Basel, 1982, pp. 315–70).

[Sz2] G. Szegő. Ueber die Entwicklung einer analytischen Funktion nach Polynomen eines Orthogonalsystems. *Math. Ann.* **82**(1921): 188–212 (see also Gábor Szegő, *Collected Papers I*, ed. R. Askey, Birkhäuser Verlag, Basel, 1982, pp. 372–97).

[Sz3] G. Szegő. *Orthogonal Polynomials.* Colloquium Publications, Vol. 23. Amer. Math. Soc., Providence, RI, 1975.

[Ti] A. F. Timan. *Theory of Approximation of Functions of a Real Variable.* International Series of Monographs in Pure and Applied Mathematics, Vol. 34. Pergamon Press, Oxford, 1963.

[To1] V. Totik. Representation of functionals via summability methods. *Acta Sci. Math.* (Szeged) **48**(1985): 483–98.

[To2] V. Totik. Regular behavior of orthogonal polynomials and its localization. *J. Approx. Theory* **66**(1991): 162–69.

[Ts] M. Tsuji. *Potential Theory in Modern Function Theory.* Maruzen, Tokyo, 1959.

[Tu] P. Turán. On some open problems in approximation theory. *J. Approx. Theory* **29**(1980): 23–85.

[Ul1] J. L. Ullman. On the regular behaviour of orthogonal polynomials. *Proc. London Math. Soc.* **24**(1972): 119–48.

[Ul2] J. L. Ullman. A survey of exterior asymptotics for orthogonal polynomials associated with a finite interval and a study of the case of general weight measures. In *Approximation Theory and Spline Functions*, eds. S. P. Singh et al., pp. 467–78. D. Reidel, Dordrecht, 1984.

[Ul3] J. L. Ullman. Orthogonal polynomials for general measures I. In *Rational Approximation and Interpolation*, eds. P. R. Graves-Morris et al., pp. 524–8. Lecture Notes in Mathematics, Vol. 1105. Springer-Verlag, New York, 1984.

[Ul4] J. L. Ullman. Orthogonal polynomials for general measures II. In *Polynomes Orthogonaux et Applications*, Proceedings, Bar-le-Duc, 1984, eds. C. Brezinski et al., pp. 247–54. Lecture Notes in Mathematics, Vol. 1171. Springer-Verlag, New York, 1986.

[UlTo] J. L. Ullman and V. Totik. Local asymptotic distribution of zeros of orthogonal polynomials. Unpublished manuscript.

[UlWy] J. L. Ullman and M. F. Wyneken. Weak limits of zeros of orthogonal polynomials. *Constructive Approx.* **2**(1986): 339–47.

[UWZ] J. L. Ullman, M. F. Wyneken, and L. Ziegler. Norm oscillatory weight measures. *J. Approx. Theory* **46**(1986): 204–12.

[VA] W. Van Assche. *Asymptotics for Orthogonal Polynomials.* Lecture Notes in Mathematics, Vol. 1265. Springer-Verlag, New York, 1987.

[VAMa] W. Van Assche and A. P. Magnus. Sieved orthogonal polynomials and discrete measures with jumps dense in an interval. *Proc. Amer. Math. Soc.* **106**(1989): 163–73.

[Wa] J. L. Walsh. *Interpolation and Approximation by Rational Functions in the Complex Domain.* Colloquium Publications, Vol. 20. Amer. Math. Soc., Providence, RI, 1960.

[Wi1] H. Widom. Polynomials associated with measures in the complex plane. *J. Math. Mech.* **16**(1967): 997–1013.

[Wi2] H. Widom. Extremal polynomials associated with a system of curves in the complex plane. *Advances in Math.* **2**(1969): 127–232.

[Wy1] M. F. Wyneken. Norm asymptotics of orthogonal polynomials for general measures. *Constructive Approx.* **4**(1988): 123–32.

[Wy2] M. F. Wyneken. The potential theory of Borel sets and some applications to the study of orthogonal polynomials (in press).

[Wy3] M. F. Wyneken. On norm and zero asymptotics of orthogonal polynomials for general measures I. *Rocky Mount. J. Math.* **19**(1988): 405–13.

Index